INCOMMENSURABILITY AND RELATED MATTERS

BOSTON STUDIES IN THE PHILOSOPHY OF SCIENCE

VOLUME 216

INCOMMENSURABILITY AND RELATED MATTERS

Edited by

PAUL HOYNINGEN-HUENE

University of Hannover, Germany

and

HOWARD SANKEY

University of Melbourne, Australia

KLUWER ACADEMIC PUBLISHERS

DORDRECHT / BOSTON / LONDON

A C.I.P. Catalogue record for this book is available from the Library of Congress.

ISBN 0-7923-6989-0

Published by Kluwer Academic Publishers,
P.O. Box 17, 3300 AA Dordrecht, The Netherlands.

Sold and distributed in North, Central and South America
by Kluwer Academic Publishers,
101 Philip Drive, Norwell, MA 02061, U.S.A.

In all other countries, sold and distributed
by Kluwer Academic Publishers,
P.O. Box 322, 3300 AH Dordrecht, The Netherlands.

Printed on acid-free paper

Printed in the Netherlands.

TABLE OF CONTENTS

HOWARD SANKEY / PAUL HOYNINGEN-HUENE

INTRODUCTION

1. THE INCOMMENSURABILITY THESIS

The aim of this book is to assess the merits and current fortunes of one of the most controversial theses to emerge in the philosophy of science during the latter half of the twentieth century. This is the thesis of the incommensurability of scientific theories. The controversy about incommensurability dates to the year 1962, the year in which the thesis of incommensurability was first explicitly proposed by its two chief advocates, Paul Feyerabend and Thomas Kuhn.

It is convenient to treat the year 1962 as the year in which the incommensurability thesis first emerged because that is when the thesis was first asserted in print by Feyerabend and Kuhn. Feyerabend originally claimed that some successive theories may be incommensurable in his paper "Explanation, Reduction and Empiricism" (1962).[1] The claim was made in the course of his critique of the reductionist account of the relations between scientific theories proposed by logical empiricism. Kuhn ascribed a central role to incommensurability in his theory of the development of science as a sequence of revolutionary transitions between scientific paradigms, which he presented in his classic work *The Structure of Scientific Revolutions* (1962).[2]

It is, however, something of an oversimplification to take 1962 as the year in which the incommensurability thesis first emerged. For, in proposing the idea of incommensurability, Kuhn and Feyerabend were drawing on earlier developments in the philosophy and history of science, as well as in philosophy at large. In many respects, the incommensurability thesis is a product of the philosophical climate of the late 1950's and early 1960's. This was a time that saw the rise of the professional discipline of the history of science, the influence of Gestalt psychology on the philosophy of perception, the decline of the logical positivism of the Vienna Circle, the widespread influence of the later Wittgenstein and Quine's attack on the analytic/synthetic distinction.

Apart from being a product of its time, the incommensurability thesis is also one of the characteristic claims of a new movement in the philosophy of science that began to emerge in the late 1950's and early 1960's. Together with the thesis of the theory-dependence of observation, the rejection of a fixed scientific method, and insistence on the importance of the history of science to the philosophy of science, the incommensurability thesis is one of the leading claims of what came to be known as the *post-positivist* or *historical* philosophy of science. In addition to Kuhn and Feyerabend, other initial participants in this movement also included such figures as Norwood Russell Hanson, Michael Polanyi and Stephen Toulmin.[3]

P. Hoyningen-Huene and H. Sankey (eds.). Incommensurability and Related Matters, vii–xxxiv.

2. WHAT IS INCOMMENSURABILITY?

Before turning to the thesis of incommensurability, a word of caution is in order with regard to the *concept* of incommensurability itself. Productive discussion of the incommensurability thesis may at times be impeded by lack of consistent use or clear meaning of the term 'incommensurability'. The term has a standard use in mathematics, where it implies the absence of a common unit of measurement. To say that two magnitudes are incommensurable is to say that there is no common unit of measurement, whole units of which may be used to measure both magnitudes. But application of the mathematical concept to the case of alternative scientific theories is an extension of the concept that leaves considerable scope for alternative interpretations.

Discussion of the incommensurability of scientific theories rarely proceeds in accordance with the mathematical concept of incommensurability. Instead, discussion of incommensurability tends to be framed in terms of a range of concepts and considerations of a broadly semantic and epistemological nature. The discussion is frequently couched, for example, in terms of such factors as the incomparability of the content of scientific theories, variation in the meaning of scientific terms, translation failure between the vocabulary of theories, or absence of common standards of theory appraisal.

This raises the question of the relationship between the concept of incommensurability in the strict sense of lack of a common measure, and the various other claims which have framed the discussion of the incommensurability thesis. Is the incommensurability of scientific theories some single, unified relation between theories, of which the various associated factors constitute mere aspects or component parts? Or is it instead the case that there are a number of different things, such as the incomparability of the content of theories, or lack of shared evaluative standards, which are each a source of incommensurability in their own right?

To answer this question one way or the other is already to take a side in the dispute. The question of how to apply the concept of incommensurability in the present context is itself one of the questions at stake. Some parties to the dispute take incommensurability to be a relation that may obtain in its own right between theories, of which such things as meaning variance and lack of shared evaluative standards are mere aspects or constitutive parts. In contrast, other parties to the dispute treat the claim of incommensurability as consisting entirely in one or another of the various claims associated with talk of incommensurability, such as the claim that the content of alternative theories is unable to be compared due to meaning variance of the terms employed by the theories. Given such potential variation in use, it is important to bear in mind that not all parties to the dispute may understand the concept of incommensurability in the same way.[4]

3. ARE THERE DIFFERENT FORMS OF INCOMMENSURABILITY?

Let us now turn to the *thesis* of incommensurability. If one takes an overview of the critical literature concerned with the incommensurability thesis, it can hardly escape notice that this literature contains a variety of separate discussions that are conducted in quite different terms. Some authors write about the topic of meaning variance and content comparison. Some write about conceptual change and the intelligibility of alternative conceptual schemes. Others write about scientific realism and the continuity of reference of theoretical terms. And still others are concerned with the rationality of scientific theory choice, and the availability of objective standards of theory evaluation.

The need to address a variety of issues under the heading of incommensurability owes much to the original discussion by Kuhn and Feyerabend. In his original discussion of the topic in "Explanation, Reduction and Empiricism", Feyerabend took incommensurability to consist in absence of logical relations due to semantic variance of the terms used by theories, resulting in the inability to directly compare the content of theories (1981d, pp. 62–69, 92–93). By contrast, in *The Structure of Scientific Revolutions*, Kuhn treated incommensurability as a multi-dimensional relationship between paradigms, which involves methodological, semantic and perceptual components (Kuhn, 1970a, pp. 148–150). According to Kuhn, paradigms employ diverse standards of theory appraisal, and address different sets of scientific problems. The vocabulary employed by scientists changes meaning in the transition between paradigms. Scientists in rival paradigms perceive the world differently. Perhaps they even inhabit different worlds.

With so many themes already present in Kuhn's and Feyerabend's original discussion, it is no wonder that a host of issues emerged when other philosophers turned to the topic. To impose order on the discussion, we will introduce a distinction between two versions of the incommensurability thesis. The first version, which we will call the *semantic incommensurability thesis*, is the thesis that alternative scientific theories may be incommensurable due to semantic variance of the terms employed by theories. The second version, which we will call the *methodological incommensurability thesis*, is the thesis that alternative scientific theories may be incommensurable due to absence of common standards of theory appraisal. We will now sketch the main developments that have taken place in connection with each of these two versions of the incommensurability thesis.

4. SEMANTIC INCOMMENSURABILITY

The thesis of semantic incommensurability derives from the claim of Kuhn and Feyerabend that the meaning of the terms employed by theories varies with theoretical context. Both authors reject the empiricist idea of an independently meaningful, theory-neutral observation language. Instead, they claim that the meaning of the terms employed by scientific theories depends on the theoretical context in which the vocabulary is employed. Given the contextual nature of meaning, the meaning of scientific terms is subject to variation with the theory in which they occur.[5]

The thesis of meaning variance gives rise to the thesis of semantic incommensurability in the following way. Because the meaning of the terms employed by scientific theories varies with theoretical context, the vocabulary of such theories may fail to share common meaning. But if theories are unable to be expressed by means of a common vocabulary, the content of such theories cannot be directly compared. For in the absence of a shared, semantically neutral vocabulary, it is impossible for statements about the world asserted by one theory to either assert or deny the same thing as any statement made by the other theory.[6] Theories which are unable in this way either to agree or disagree with respect to any claim about the world are incommensurable in the sense that their content is unable to be directly compared due to semantic variance.

For simplicity, we have formulated the thesis of semantic incommensurability in terms of radical meaning variance which applies to the entirety of the terms employed by a theory. However, a more limited version of the thesis may also be formulated in terms of partial meaning variance restricted to a limited portion of the vocabulary employed by theories. The radical version of the meaning variance thesis tends to be associated with Feyerabend, whereas the partial meaning variance thesis tends to be associated with Kuhn.[7]

Response to the semantic incommensurability thesis divides into two main lines of criticism. On the one hand, advocates of what we will call the *referential response* argue that there are relations of co-reference between the terms of meaning variant theories which suffice for content comparison. On the other hand, advocates of what we will call the *translational response* argue that the idea of an untranslatable language, to which the meaning variance thesis gives rise, is an idea of which no coherent sense can be made. We will first discuss the referential response to semantic incommensurability.

The referential response was presented by Israel Scheffler in his 1967 book, *Science and Subjectivity*. Employing a Fregean distinction between sense and reference, Scheffler pointed out that, even if the sense of a scientific term varies with theory, it does not follow that the reference of the term also varies with theory. Terms may co-refer but differ in sense. Hence terms may retain stable reference despite variation in sense. But if terms employed by theories preserve reference through variation of sense, it remains possible to compare the theories with respect to content. For statements which theories make about the world may agree or disagree with respect to common states of affairs, provided only that their constituent terms refer to the same things, despite variation of sense.[8]

But Scheffler's point that reference need not co-vary with sense was not enough to settle the issue. On the one hand, there are a number of historical cases in which different theories seem to employ the same terms to refer to different things. On the other hand, the existence of radical conceptual change in science suggests that there has been widespread discontinuity of reference in the history of science, since radical conceptual change seems to imply variation of reference. Thus, while common reference may well suffice for the comparability of content, it remains to be shown how continuity of reference may be sustained in the transition between meaning variant theories.[9]

Rather than settle the issue, therefore, Scheffler's appeal to reference serves merely to shift the focus to the issue of reference. For it raises the question of how terms which occur in different theories may preserve reference through variation in the conceptual content which theories associate with them. More specifically, it raises the question of the determination of the reference of the terms that are employed by scientific theories.

Where Scheffler formulated the referential response in terms of a Fregean theory of reference, the subsequent emergence of the causal theory of reference seemed to offer a promising resolution of the issue. It was suggested by such authors as Saul Kripke and Hilary Putnam that reference is determined in a direct manner by means of causal relations between speaker and object, rather than by the descriptive content which speakers associate with the terms they employ (cf. Kripke, 1980; Putnam, 1975b;

1975c). On such a view, reference is fixed at the initial introduction of a term, and the reference of later use of the term traces back by a historical chain to its original use. Because reference is determined independently of descriptive content, the reference of terms employed by scientific theories may be preserved despite variation in the concepts associated with such terms. But if reference remains stable through variation of conceptual content, no problem of theory comparison arises, since reference is preserved even though terms may be associated with divergent conceptual content in the context of alternative theories.

So simple a resolution of the issue was, of course, too good to be true. There are several difficulties facing the causal theory which prevent employing it in unmodified form to sustain the referential response. First, it is implausible to suppose that reference is permanently fixed at the initial introduction of a term, since this excludes in principle the possibility of change of reference in the history of science (cf. Fine, 1975). Second, if the reference of natural kind terms is fixed by entirely non-descriptive means – e.g., by ostension of samples of a kind – then it is impossible to secure unambiguous reference to a specific natural kind as opposed to the numerous other kinds instantiated by the sample set. This is the so-called *qua* problem, since the problem is how to pick out an object *qua* member of a given kind (cf. Papineau, 1979; Devitt and Sterelny, 1999; Sterelny, 1983). Third, if the reference of theoretical terms is determined by specification of a causal relation between observed phenomena and the entities responsible for the phenomena, then it would be impossible for theoretical terms ever to fail to refer. Yet failure of reference would appear to be a routine occurrence in the history of science (cf. Enç, 1976; Kroon, 1985; Nola, 1980).

Such problems suggest that the causal theory of reference must be modified to allow variation in reference subsequent to initial term-introduction, as well as to include a role for descriptive content in the determination of reference. This has led to the development of various modified versions of the causal theory, such as the causal descriptive theory of reference. According to the causal descriptive theory of reference, causal relations between speaker and object must be supplemented by at least minimal descriptive apparatus to more fully determine reference.[10] Such descriptive apparatus, which may include specification of natural kind and the causal role of theoretical entities, is required in order to resolve the *qua* problem and to allow for the reference failure of theoretical terms. But such modified versions of the causal theory fail to deliver results as unequivocally contrary to the incommensurability thesis as was initially promised by the causal theory of reference in its original form. For incorporation of a role for description in reference determination, as well as allowance of reference change subsequent to initial term-introduction, yields considerable scope for reference to vary with theory change.[11]

Let us now turn to the translational response to the semantic incommensurability thesis. This response is directed against an implication of the meaning variance thesis that there may be translation failure between the vocabulary employed by incommensurable theories. The thesis of radical meaning variance suggests that the terms employed to express a theory might be unable to be translated by means of any of the terms employed by an alternative theory with which it is incommensurable. Taken to the extreme, meaning variance suggests that there might even be entire languages that fail to be intertranslatable.

The idea of an untranslatable language has seemed paradoxical to many philosophers. For if one cannot translate a foreign language at all, then what evidence can there be that what fails to be translated is in fact a language? Failure to translate is indeterminate between being evidence that a language is untranslatable and that it is not a language at all. If one provides, as evidence of untranslatability, examples of expressions of the untranslatable language, then the very act of providing such examples undermines the claim of untranslatability, since to present the examples *within* one's own language presupposes translation. Still worse, even to profess to understand what is said in an untranslatable language seems to imply the *tran*slatability of the language, since understanding a foreign language seems to presuppose translation into a language that one understands.

Such thoughts as these lie at the heart of Donald Davidson's famous article, "On the Very Idea of a Conceptual Scheme" (1984). The thrust of Davidson's critique, however, is not restricted to semantic incommensurability. He was "after larger game". According to Davidson, to make sense of the idea of a language independent of translation requires a distinction between a conceptual scheme and the content that is organized by a conceptual scheme. But, Davidson argues, no coherent sense can be made of the idea of a conceptual scheme. So no sense may be attached to the idea of an untranslatable language.

Some of the deep issues raised by Davidson may be avoided by taking into account two points made by Kuhn and Feyerabend with respect to translation. First, in later work Kuhn developed a local version of incommensurability, which restricts translation failure to narrow clusters of interdefined terms from rival theories (Kuhn, 1983a). Restricting untranslatability to such local clusters of terms, or even to the special vocabulary of theories, removes the need to make coherent sense of either a totally untranslatable language or the scheme/content dualism. Second, both Kuhn and Feyerabend sought to distinguish between translating a language and understanding it (Kuhn, 1983a; Feyerabend, 1987). While one might fail to *translate* from a foreign language into one's own, it need not follow that one must fail to *understand* the other language. The combination of these two points yields a refined version of semantic incommensurability, on which translation failure is restricted to specialized vocabularies within a language, which are capable of being understood by rival theorists. Such a refined version seems an unlikely target for Davidson's attack.

Our distinction between referential and translational responses to semantic incommensurability should not be taken to suggest that the two responses address completely unrelated aspects of the incommensurability thesis. While the question of the comparability of theories due to co-reference is distinct from that of the coherence of the idea of an untranslatable language, the answer given to one question may have consequences for the answer to the other. For the question of whether the terms of one theory refer to the same things as the terms of another is not unrelated to the question of whether the terms of one theory express the same meanings as those of another theory.

Indeed, one of the Editors of this volume, Howard Sankey, has developed an approach to semantic incommensurability which addresses both the issues of reference and translation. In his book, *The Incommensurability Thesis* (1994), Sankey adopts a modified version of the causal theory of reference, which allows reference change subsequent to initial term-introduction and grants a role to description in the determination of reference. Sankey argues that there may be translation failure between theories due to difference in means of reference determination, and he defends restricted translation failure against Davidson's translational response to incommensurability. Sankey's approach combines acceptance of the referential response with rejection of the translational response, since he argues that semantically variant theories may be compared by means of overlapping reference despite the inability to translate between such theories. Such an approach illustrates that while the referential and translational responses address different topics, it is possible to address the issues raised by both responses within a unified approach to semantic incommensurability.[12]

5. METHODOLOGICAL INCOMMENSURABILITY

We will now discuss the thesis of *methodological incommensurability*. Our primary focus will be on the development of Kuhn's views on this topic. For, while Feyerabend defended a range of well-known theses about the nature and limits of scientific method (e.g., 1975), he did not do so under the heading of incommensurability. As previously noted, unlike Kuhn, who originally took incommensurability to have methodological dimensions, Feyerabend restricted incommensurability to semantic relations between theories.

While the term 'incommensurable' is usually understood in the sense of semantic incommensurability, some authors do employ the term in a methodological sense. It is more common, however, for issues relating to methodological incommensurability to be dealt with under the rubric of the rationality of scientific theory choice and relativism due to variation in methodological standards. Still, the point of departure for discussion of relativism and rational theory choice is often Kuhn's claim that standards of theory appraisal vary with paradigm, which was treated by Kuhn in *The Structure of Scientific Revolutions* as one of the constitutive aspects of the incommensurability of paradigms. Feyerabend's own critique of a fixed scientific method, which he did not present under the rubric of incommensurability, marks another key reference point in this aspect of the discussion.

According to the thesis of methodological incommensurability, there are no shared, objective methodological standards of scientific theory appraisal. Standards of theory appraisal vary from one theory or paradigm to another. There are no external or neutral standards which may be employed in the comparative evaluation of competing theories. As a result, alternative scientific theories may be incommensurable due to absence of common methodological standards capable of adjudicating the choice between them.

The idea that scientific theories may be incommensurable in a methodological sense arises out of the rejection of the traditional view that there is a uniform, invariant scientific method, employed throughout science, which is the distinguishing feature of science. In apparent opposition to this traditional view, Kuhn claimed in *The Structure of Scientific Revolutions* that standards of theory appraisal depend on and vary with the currently dominant scientific paradigm. Paradigms "are the source of the methods, problem-field, and standards of solution accepted by a mature scientific community" (1970a, p. 103); "when paradigms change, there are usually significant shifts in the criteria determining the legitimacy of problems and of proposed solutions" (1970a, p. 109). But such criteria and standards do not govern choice between paradigms, which "cannot be determined merely by the evaluative procedures characteristic of normal science" (1970a, p. 94). Nor are there independent standards of theory choice, since, "as in political revolutions, so in paradigm choice – there is no standard higher than the assent of the relevant community" (1970a, p. 94). Kuhn's denial of paradigm-independent standards created the impression that the rationality of scientific theory acceptance is relative to prior choice of scientific paradigm, and the choice of paradigm is incapable of being governed by shared objective standards of theory appraisal.

Kuhn's emphasis on the fundamental commitment of a scientific community to a dominant paradigm drew attention to a conservative dimension of science. By contrast, Feyerabend highlighted the opposing elements of change, pluralism and rivalry with his principle of the proliferation of theories.[13] Like Kuhn, however, Feyerabend also held that the methodological rules and standards of science are subject to variation rather than remaining constant throughout the history of science. Moreover, Feyerabend argued that all proposed rules of scientific method have been justifiably violated at some stage in the history of science (cf. 1975, pp. 23–24). In his view, therefore, there is no single, invariant or binding scientific method that is applicable to all sciences throughout all periods in the history of science. His point, though, was not that there is no method, or that science obeys no rules, but that "all methodologies, even the most obvious ones, have their limits" (1975, p. 32; cf. 1978, p. 32). This, rather than an incitement to epistemic anarchy, is the content of his notorious slogan, "anything goes" (1975, p. 28).

On the picture of scientific theory choice that emerges in the work of Kuhn and Feyerabend, there is no fixed set of objective scientific standards to which appeal may be made to adjudicate the dispute between conflicting scientific theories. The choice between alternative scientific theories is not a decision that may be made on the basis of common methodological standards accepted by all parties to the dispute. Thus, the methodological views of Kuhn and Feyerabend appear to lead to a thorough-going epistemological relativism, on which scientists may rationally accept conflicting theories on the basis of alternative sets of methodological standards. Moreover, in the absence of higher-order standards which may adjudicate between theories, the choice between competing theories would appear to rest on ineliminable subjective or irrational factors, rather than on objective methodological considerations. Such generally relativistic features of Kuhn's and Feyerabend's treatment of method and rational theory choice have been found objectionable by numerous authors, who have subjected their views to sustained and searching critique.[14]

While Feyerabend did little to dispel the impression of relativism, Kuhn sought to distance himself from the relativistic implications of his original position. In later work, Kuhn explained that he had not meant to deny that the choice between alternative theories may be a rational process governed by methodological standards. Rather, he had meant only to insist that:

> There is no neutral algorithm for theory-choice, no systematic decision procedure which, properly applied, must lead each individual in the group to the same decision. (1970a, p. 200)

Lacking such an algorithm, rational theory choice involves ineliminable elements of judgement and deliberation. Nor had Kuhn meant to deny the existence of fixed standards of theory appraisal. Indeed, he lists a number of such standards: e.g., accuracy, consistency, simplicity, scope and fruitfulness (1977c, p. 322). Kuhn's point, rather, was that such standards "function not as rules, which determine choice, but as values, which influence it" (1977c, p. 331). Moreover, he says, scientists "may legitimately differ about their application to concrete cases", and "when deployed together, they repeatedly prove to conflict with one another" (1977c, p. 322). The result is that, despite adherence to a common set of standards, there may be rational disagreement between scientists who embrace opposing theories because they interpret or weight the same standards differently.

Kuhn's conception of rational disagreement governed by non-algorithmic standards offers a promising account of rational scientific theory choice. However, it leaves open a question which Kuhn was never able to resolve satisfactorily. This is the *meta*-methodological question of the normative ground of standards of theory appraisal. Kuhn originally seemed to ground epistemic normativity in social consensus (cf. 1970a, p. 94). However, at a later stage he sought to naturalize such normativity by grounding it in successful scientific practice (1970b, p. 237). Still later he offered a conceptual grounding for scientific norms whose rationality he took to be analytically insured by the very concept of science (1983b).[15]

The issue of the metamethodological justification of epistemic norms has in recent years come into sharper focus as a result of critical discussion of the views of Larry Laudan. Laudan distinguishes between intuitionist, conventionalist and naturalist metamethodological stances, and argues that meeting the relativist challenge requires a naturalist metamethodology that grounds normative methodology in empirical facts about means to epistemic ends. Whatever the fate of Laudan's own normative naturalist alternative, resolution of the issues surrounding methodological incommensurability will require development of an adequate metamethodological theory of the warrant of methodological norms.[16]

6. INCOMMENSURABILITY AND REALISM

In the two preceding sections we have discussed a number of major issues which arise in relation to the semantic and methodological versions of the incommensurability thesis. In light of the problems which arise for rational theory choice, the main import of both versions of the thesis may seem to be principally epistemological in nature. For if it is impossible to compare theories either with respect to content or by means of common standards, then it is unclear how a decision between such theories may be made on an objective, rational basis. But the controversy about incommensurability is not confined to epistemic issues relating to rational theory choice. It includes issues of a broadly metaphysical character as well. The semantic incommensurability thesis, in particular, leads to a number of contentious issues about the relation between theory and reality, which are of particular relevance to the debate between scientific realist and anti-realist approaches to the philosophy of science.

In our earlier discussion of the referential response to semantic incommensurability we took note of the possibility of discontinuity of reference in the transition between successive theories. Radical conceptual change may result in wholesale discontinuity of reference, so that no term of a later theory refers to any entity referred to by an earlier theory. Such wholesale discontinuity of reference conflicts with a scientific realist philosophy of science, since the realist holds that successive theories in the same domain typically provide alternative descriptions of the same entities and that progress in science consists in an increase in truths known about a common set of entities. But if later theories refer to none of the same entities as earlier theories, then the realist account of scientific progress as increase of truth about a common set of entities is untenable.

As we saw in the earlier discussion, the question of discontinuity of reference turns on the issue of how reference is determined. The degree to which reference may vary with theory depends on how sensitive reference is to variation in the descriptive content associated with terms. But the question of whether successive theories may sustain reference to a common domain of entities is not independent of issues of a metaphysical nature. For it is not simply a matter of how reference is determined, but of the ontological status of the entities to which reference is made. The realist holds that the entities to which the terms of a theory refer exist independently of the theory, and that the world investigated by natural science is an objective reality which exists independently of human thought. But such assumptions may not be shared by those anti-realist philosophers for whom the objects of reference and the world investigated by science depend in some way on human thought. Some anti-realist philosophers hold that the world and the objects it contains are constituted, either in whole or in part, by our theories, concepts or language. Such philosophers may deny that the terms of conceptually variant theories refer to the same objects, since such theories constitute their own domains of reference. In short, the question of whether later theories refer to the same entities as earlier theories raises metaphysical questions of a kind that tend to divide realism from anti-realism in the philosophy of science.

The ontological status of the entities referred to by theories is of particular relevance to the incommensurability thesis, since both Feyerabend and Kuhn tended toward an anti-realist metaphysics with at least a trace of idealism. While Feyerabend defended scientific realism against an instrumentalist view of theories, occasional remarks suggest that he may in fact have held a fundamentally anti-realist attitude to the relation between theory and reality.[17] The case of Kuhn is especially revealing, since a critical attitude to realism and a neo-Kantian tendency were persistent themes of his work.[18] We will illustrate the relevance of anti-realism to incommensurability with reference to Kuhn's image of a 'world-change'.

In *The Structure of Scientific Revolutions*, Kuhn employed the image of a change of world to describe the transformation brought about by change of paradigm. Consideration of past science, Kuhn remarked, might tempt a historian of science to "exclaim that when paradigms change, the world itself changes with them", since "it is rather as if the professional community had been suddenly transported to another planet" (1970a, p. 111). In a similar vein, Kuhn spoke of new entities coming into being in paradigm shift, as well as of scientists in different paradigms inhabiting different perceptual worlds. For example, Kuhn wrote, "pendulums were brought into existence by something very like a paradigm-induced gestalt switch" (1970a, p. 120), and "Lavoisier ... saw oxygen where Priestley had seen dephlogisticated air" (1970a, p. 118). Kuhn at one point described the idea that scientists "practice their trade in different worlds" as "the most fundamental aspect of the incommensurability of competing paradigms" (1970a, p. 150). Such remarks as these suggest that a somewhat idealist conception of the relation between theory and reality may underlie Kuhn's thinking about incommensurability.[19]

While some authors treat Kuhn's world-change image as a mere metaphor, others take it more seriously.[20] For example, one of the Editors of the present volume, Paul Hoyningen-Huene, interprets the image in a neo-Kantian fashion.[21] In his book, *Reconstructing Scientific Revolutions*, Hoyningen-Huene argues that Kuhn's metaphysical stance is in fact a dynamic Kantian position, which is based on a distinction between an unknowable "world-in-itself" and a "phenomenal world" that is jointly constituted out of input from "the world-in-itself" and the conceptual contribution of the human subject. Where Kuhn differs from Kant is in allowing the human conceptual contribution to vary with change of theory. Because of such conceptual variation, the world that changes in transition between theories is the phenomenal world of scientists, rather than the "world-in-itself", which is unaffected by such change. On this interpretation, incommensurable theories may be compared in a variety of ways, including by means of reference (1993, pp. 218–222). But the latter is not to be understood in realist fashion as reference to a shared domain of mind-independent objects. For, while incommensurable theories may refer to some of the same things, the things to which they refer are not mind-independent objects. Rather, they are objects in the overlapping phenomenal worlds of alternative theories, which are jointly constituted out of conceptual input from the human subject and external input from the "world-in-itself".[22]

An alternative interpretation of the world-change image has been offered by Ian Hacking, who understands it as an expression of nominalism. He sees Kuhn as a 'revolutionary nominalist', for whom the world contains no underlying natural kinds, the only kinds being those that result from imposition of systems of classification on the world (1983, p. 109). The individual entities that make up the world do not change. What changes in transition between theories is the system of kinds into which theories classify individual objects (1993, p. 306).[23] Hacking's nominalist interpretation of Kuhn allows for incommensurability in the sense of untranslatability between the kind terms of theories (1993, pp. 294–295). But his interpretation of Kuhn seems consistent with the referential response. For it allows that the objects diversely classified by theories do not themselves undergo change in the transition between theories, which permits theories to be compared by means of common reference to the same entities.

Such alternative interpretations of Kuhn raise interesting issues of an exegetical nature. But these are not our present concern. Rather, the alternative interpretations of Kuhn illustrate the point that the incommensurability thesis is not restricted to epistemic issues, but raises issues of a broadly metaphysical nature as well. It is not just the case that the semantic incommensurability thesis leads to problems of a metaphysical nature with respect to the scientific realist account of the nature of scientific progress. In addition, the main advocates of the incommensurability thesis appear to have held metaphysical viewpoints which are distinctly at odds with basic assumptions of scientific realist philosophy of science.

Such potential for metaphysical disagreement leads to an important point about the debate about incommensurability. Given the apparent anti-realist tendencies of Kuhn and Feyerabend, it is not unreasonable to suppose that they may base some aspects of the incommensurability thesis on anti-realist assumptions. That they may have done so suggests that the metaphysical stance toward the objects of reference and the nature of the world investigated by science is itself one of the points at issue in the dispute between alternative approaches to the incommensurability thesis. This has at least two important consequences. On the one hand, realists and anti-realists may beg the question against each other if they assume that the objects to which theories refer are either ontologically dependent or independent of human cognition.[24] On the other hand, given the role that may be played by underlying metaphysical assumptions, the dispute over incommensurability is not simply a narrow dispute between rival approaches within the philosophy of science. Rather, the dispute about incommensurability reflects a deep dispute between realist and anti-realist metaphysical perspectives.

7. OVERVIEW OF THE VOLUME

In the discussion so far, we have sought to introduce the subject matter of this volume by offering a general overview of the incommensurability thesis, as well as of the main themes which have emerged in the critical literature on the topic. We will now turn our attention to the more immediate task of introducing the contents of the volume by summarizing the papers contained herein. In our overview of the debate about incommensurability we have made a selective choice of topics. Hence, in our discussion of the contents of this volume, we will take the opportunity to draw attention to a number of other aspects of the debate on which we have not so far commented. In some cases, we will also provide additional background to the papers by commenting on their place within the context of the debate. The volume is divided into five sections, which reflect the central themes of the papers. We will adhere to this organization in the following overview of the volume.

7.1. Incommensurability, Meaning and Reference

The first section contains three papers which focus on broadly semantic aspects of the incommensurability thesis. The first two papers are by Richard Boyd and Martin Carrier, and the third is a joint paper written by Fred Kroon and Robert Nola. Within the broad area of focus, each of the papers has a more specific focus on a particular aspect of semantic incommensurability. Richard Boyd's paper addresses a form of semantic incommensurability which stems from variation in conceptual meaning. Martin Carrier discusses the proposal by the later Kuhn that there is local translation failure due to variation in taxonomic structure between scientific theories. In their paper, Fred Kroon and Robert Nola pursue a question which arises within the theory of reference of how the reference of theoretical terms is determined.

The title of the paper by Richard Boyd is “Reference, (In)commensurability and Meanings: Some (Perhaps) Unanticipated Complexities”. In his paper, Boyd develops a novel treatment of conceptual aspects of incommensurability. He distinguishes between a version of the semantic incommensurability thesis which asserts referential discontinuity between theories and a version which involves variation in what he calls ‘conceptual meaning’. Boyd considers that the standard referential response based on a causal theory of reference satisfactorily disposes of the extreme thesis of referential discontinuity between theories. But that is not the end of the matter. In Boyd’s view, there remains a significant and insufficiently appreciated source of conceptual incommensurability which is unaffected by the standard referential response. In particular, Boyd argues that the standard response overlooks the phenomenon of ‘malignant meaning’. According to Boyd, the conceptual meanings of key scientific terms employed in certain scientific disciplines are malignant in the sense that they incorporate incoherent concepts and fallacious or otherwise unreliable inferential practices. In some cases, such malignant meanings may include elements of social ideology external to the scientific discipline in question, which sustain continued use of the meanings within the discipline.

Malignant meaning poses an obstacle to communication between participants in alternative traditions within a discipline. As such, it raises problems of comparative theory appraisal of a kind similar to those which were earlier described as methodological incommensurability. Boyd notes that malignant meaning may prevent participants in a research tradition from understanding criticism directed at basic assumptions of their tradition. They may also be unable to engage productively in critical discussion with representatives of opposing viewpoints. Given such limits on communication, advocates of competing theories may fail to agree on mutually acceptable methods or standards by means of which to adjudicate between their opposing theories. Boyd illustrates the phenomenon of malignant meaning and the basis of the resulting incommensurability by means of cases drawn from the current literature on evolutionary psychology. He concludes by arguing that the lack of a basis for rational dialogue due to malignant meaning means that in certain circumstances the only possible critique of a scientific tradition is a political or social critique which derives from outside the tradition. Because of this, an adequate understanding of the epistemological nature of the incommensurability resulting from malignant meaning requires development of a political epistemology which reveals the epistemic nature of political critique.

With the paper by Martin Carrier, the focus shifts to the thesis of taxonomic incommensurability which Kuhn developed in his later work. As we have not so far touched on this issue, we will take this opportunity to comment briefly on the topic. In a series of papers published over the last decade and a half of his life, Kuhn developed a refined version of the semantic incommensurability thesis. According to Kuhn’s mature thought on the subject, revolutionary scientific change involves change in the taxonomic schemes which theories employ to classify the entities in their domains of application (Kuhn, 1987). Change in taxonomic scheme leads to semantic change affecting the sense and reference of central clusters of terms used by theories. Because the terms affected by taxonomic change are typically related in holistic fashion, change in taxonomic scheme gives rise to change in the meaning of a number of interrelated terms. In Kuhn's view, such semantic change is restricted to localized clusters of terms which refer specifically to the taxonomic categories affected by the change of structure (1983a, pp. 670–671). As a result of such holistic change of meaning, key terms from one theory may fail to be translatable into an interrelated cluster of terms in another theory. In Kuhn's later work, such localized translation failure between clusters of terms becomes the key feature of incommensurability (1983a; 1991, pp. 4–5).

Carrier's paper is entitled "Changing Laws and Shifting Concepts: On the Nature and Impact of Incommensurability". In the paper, Carrier presents an analysis of the relation of untranslatability found in Kuhn's later account of incommensurability. Carrier agrees with Kuhn that there is translation failure between theories, but argues against Kuhn that untranslatable statements may be compared by empirical means. According to Carrier, Kuhn's account of untranslatability is based on a contextual theory of meaning, on which the meaning of a scientific term depends on its use in the context of a theory, which is governed by the laws of nature in which the terms occur. Carrier employs the theoretical context account of meaning to show that certain basic concepts of phlogistic chemistry are incapable of translation into the language of the oxygen theory. While expressions from both theories may be applied in the same circumstances (e.g., 'phlogiston escape' and 'oxygen bonding'), it is not possible to translate them in a way that preserves both conditions of application and the inferential relations between statements which employ the expressions. But where Kuhn and others have supposed that the untranslatable content of theories may not be compared, Carrier argues that incommensurable theories may be capable of empirical comparison even in those areas where translation fails. Such comparison requires only that the observational consequences of competing theories be subjected to empirical test with respect to a common domain of phenomena. It does not require that the theories describe the phenomena using shared vocabulary. Hence, it is possible to conduct comparative empirical tests of incommensurable theories, even if the theories report the phenomena in untranslatable terms.

The title of the paper by Fred Kroon and Robert Nola is "Ramsification, Reference Fixing and Incommensurability". Their paper addresses the issue of the determination of the reference of theoretical terms. After reviewing the assumptions about reference that underlie the incommensurability thesis, the authors consider causal and causal descriptive approaches to the reference of theoretical terms. They see limitations in both the latter approaches. Hence, they turn their attention to an account of the reference of theoretical terms due to David Lewis, which adapts Ramsey sentences to the reference of such terms. On the Lewis-Ramsey account, the reference of a theoretical term 't' is fixed by means of the definite description $t = (!x)\ [T(x, O_1, O_2, \ldots O_m)]$, where 't' denotes whatever uniquely satisfies the open sentence $[T(x, O_1, O_2 \ldots O_m)]$, and denotes nothing if the open sentence is not uniquely satisfied.

The question arises of how much of a theory to include in the reference-fixing description. Kroon and Nola consider a modification of the Lewis-Ramsey account proposed by David Papineau, which involves a threefold distinction between those parts of a theory which do (T_y, y = yes), those which do not (T_n, n = no), and those which might contribute to the definition of a term (T_p, p = perhaps) (Papineau, 1996, p. 11). On such an analysis, a term may be imprecisely defined in the sense that there is a part of a theory, T_p, which does not play a definite role in fixing reference. Such imprecision need not render reference indeterminate, since T_y may fix a unique reference independently of T_p. However, reference is indeterminate if T_y fails to fix a unique reference independently of T_p. Such indeterminacy gives rise to two cases of relevance to incommensurability: first, T_y does not fix a unique reference, but does so when conjoined with part of T_p; second, the conjunction of T_y and T_p fails to refer, but removal of part of T_p secures a unique reference. According to Papineau, such indeterminacy cannot be resolved by appeal to further semantic or empirical facts. Rather, it is a matter of microsociological factors such as the conservative or radical attitudes toward established theory of the scientists who employ the term (1996, p. 19).

After considering an objection against Papineau due to Stephen Stich, Kroon and Nola turn to their own positive account of reference determination, which emphasizes the epistemic purpose of the practice of reference. Where advocates of the causal theory of reference treat perception as the principal causal relation constitutive of reference, Kroon and Nola note that perception is a cognitive relation which enables language-users to track items about which they wish to acquire knowledge. More generally, the purpose of reference is not simply to enable speakers to talk about objects or kinds, but to enable them to acquire and exchange information about the objects and kinds they talk about. In light of the epistemic purpose of reference, the authors hold that it is not urgent to decide on the causal or descriptive nature of reference. Whether causal or descriptive, they claim that reference is governed by two conditions: the Existence Condition, that speakers believe that the referent of a term exists; and the Fact-Finding Condition, that speakers introducing a term be able to acquire further information about its referent. Given these conditions, a referential practice is one on which the use of a term to refer to an item or set of items is epistemically warranted. Kroon and Nola argue that such epistemic factors must be introduced into the Lewis-Ramsey account to meet Stich's objection and motivate Papineau's threefold distinction. On the modified account which they propose, the reference of an imprecisely defined term may remain constant as knowledge of the referent evolves with time. Imprecise definition may result in indeterminacy of reference, which is resolved by later theoretical developments. Thus, rather than referential discontinuity between theories, Kroon and Nola allow that there may be semantic indeterminacy that is removed in the advance of science.

7.2. Realism and Incommensurability

Concern with scientific realism lies behind much discussion of semantic incommensurability, particularly in relation to the reference of theoretical terms. In the second section of the book, such concern with realism comes to the fore. The section contains two papers which address the relation between realism and incommensurability. The first paper is by Harold Brown, who argues that a realist account of scientific change requires that incommensurability occur as a result of conceptual change, but that the extent of conceptual variation possible is constrained by nature itself. In the second paper, Michael Devitt argues against an anti-realist version of the incommensurability thesis by arguing that realism is better grounded than the anti-realism on which that version of the thesis rests. It might, at first blush, seem that these two papers are to a certain extent in tension with each other, since one argues that realism requires incommensurability, while the other argues from a realist point of view against incommensurability. The difference is explained by the fact that Brown understands incommensurability in terms of untranslatability between conceptual systems, whereas Devitt thinks of incommensurability as the incomparability of alternative theories.

Harold Brown's paper is entitled "Incommensurability and Reality". Brown begins by explaining how incommensurable systems of scientific concepts may arise, illustrating this by means of a number of cases in which new concepts were proposed and then eliminated in the course of scientific change. Rather than frame the issue in semantic terms as a problem about change of meaning or reference, Brown approaches it in cognitive terms as a problem of conceptual change. In his view, concepts are mental representations, which are not to be identified with linguistic entities such as meanings. In investigating the world, scientists must develop appropriate concepts to represent the items under investigation. Scientific concept formation involves fallible hypotheses about the nature of the items that the concepts represent, which are subject to revision in the light of empirical investigation and theoretical advance. Brown's emphasis on the revisable, hypothetical character of concept formation reflects a naturalistic approach to conceptual change, on which the process of conceptual change is closely implicated in the process of theoretical change itself.

Some anti-realists hold that because our concepts are created by us this means that we may have no epistemic access to objects outside the mind. By contrast, Brown defends the realist view that science successfully pursues the aim of developing systems of concepts that correctly describe the items in domains under scientific investigation. The development of correct systems of concepts requires the introduction of new concepts and the elimination of old ones. Given the need for such conceptual change, Brown argues that pursuit of the realist aim of correct concepts entails that incommensurable conceptual systems may arise during the development of science. For the concepts employed by some later theories differ significantly from concepts employed by the earlier theories that they displace. But, while conceptual change gives rise to incommensurability, Brown argues that there are limits on the extent of incommensurability that are imposed by the world itself. The range of permissible conceptual systems is severely constrained by the precision and variety of evidence that must be accommodated by theories which embody alternative conceptual systems. Since the evidence derives from interactions, typically mediated by instruments, with parts of the natural world described by a conceptual system, conceptual variation is constrained by items in the domains to which the theories are applied. Because interaction with the external world provides evidential constraints on the range of permissible conceptual systems, evidence for a theory is evidence that its concepts accurately reflect the items they describe.

The title of Michael Devitt's paper is "Incommensurability and the Priority of Metaphysics". Devitt argues against a semantic version of the incommensurability thesis, which he takes to be associated with constructivist forms of anti-realism in the philosophy of science. Devitt understands the relevant version of the semantic incommensurability thesis as the thesis that alternative scientific theories may be unable to be compared with respect to agreement or disagreement due to differences in the meaning and reference of the terms employed by the theories. He characterizes the doctrine of Realism as the doctrine that the physical items of commonsense and science have objective, mind-independent existence. He contrasts such Realism with the anti-realist doctrine of Constructivism, which he characterizes as the Kantian doctrine that the only world to which we have epistemic access is a phenomenal world constructed by the imposition of concepts on an unknowable noumenal world of "things-in-themselves". Devitt argues that the doctrine of Constructivism leads to incommensurability, while Realism supports the comparability of theories. For while the Constructivist takes the terms of alternative theories to refer to distinct sets of objects in different phenomenal worlds, the Realist takes the terms of such theories to refer to a common domain of objects in a shared, mind-independent world. Since Constructivism leads to incommensurability while Realism does not, Devitt seeks to dispose of the incommensurability thesis by arguing in favour of Realism and against Constructivism.

In arguing for Realism, Devitt combines the commonsense of Moore with the naturalism of Quine. Realism is so intimately bound up with our commonsense experience that it may only be given up in the face of an overwhelming case against it. Anti-realist arguments for Constructivism typically proceed from speculations about knowledge and meaning to the conclusion that the only world which we may know and talk about is a world that is constructed by us. But, says Devitt, we should attach far less credence to speculations in epistemology and semantics than we do to the commonsense in which Realism is grounded. From a naturalistic perspective, moreover, such speculations are not to be treated as *a priori* claims, but as empirical claims like any other. But, treated as such, they enjoy considerably less empirical support than does Realism. Given the strength of the case for Realism that derives from commonsense and naturalism, and the comparative lack of support for the anti-realist speculations about knowledge and meaning, Devitt concludes that the incommensurability thesis is to be rejected.

7.3. Incommensurability, Rationality and Relativism

In the third section, the focus shifts from issues of meaning and realism raised by the semantic incommensurability thesis to issues of rationality and relativism associated with the methodological incommensurability thesis. The section contains two papers, one by Gerald Doppelt, the other by Dudley Shapere. In his paper, Doppelt defends a moderate relativism based on variation of standards of theory appraisal, which reflects Kuhn's claim that change of paradigm is not limited to change in the substantive content of theories, but extends to normative aspects of science as well. Shapere argues that the problems of rationality and relativism raised by the incommensurability thesis are in large part resolved by the piecemeal conception of scientific reasoning which he proposes.

Doppelt's paper, "Incommensurability and the Normative Foundations of Scientific Knowledge", seeks to articulate and defend a moderate epistemological relativist stance suggested by Kuhn's treatment of methodological change in science.[25] Doppelt opposes the interpretation of Kuhn as an extreme relativist for whom paradigm choice is an irrational decision between radically incommensurable theoretical alternatives. Instead, Doppelt argues, scientific theory choice is a rational process in which choice of theory is justified by appeal to standards of theory appraisal to which a group of scientists is normatively committed. Standards of theory appraisal are subject to variation in the history of science as well as between scientific communities. The result is that proponents of competing theories may appeal to alternative standards of appraisal in justifying their choice of theory. The question thus arises of whether there may be epistemic grounds for the choice between standards of theory appraisal. Doppelt allows that in some cases there may be epistemic grounds for choice between standards. But he denies that all such choices may be justified on epistemic grounds. Rather, the choice of standards of theory appraisal is ultimately a pragmatic choice driven by interest. Hence, the adoption of a given set of standards by an individual scientist or group of scientists is a decision that is to be explained on the basis of historical or sociological factors.

Doppelt develops his relativist position in dialectical fashion by considering a series of objections found in the literature on the topic. A distinction between neutral evaluative standards that are external to theory and theory-dependent internal standards plays a pivotal role in the early stages of his discussion. Against the objection that moderate relativism is self-refuting because it employs external standards to argue in *a priori* vein that there are no such standards, Doppelt replies that the position is based on empirical evidence of methodological variation in the history of science and does not imply that in principle there may be no external standards. He then addresses the point that the existence of rational debate between advocates of rival theories seems to entail the existence of shared external standards, since rational debate requires such standards. In reply, he notes that parties to a rational debate may recognize what their opponents count as reasons without accepting the same standards; moreover, their standards may even overlap since they may accept some but not all of the same standards. He next considers the possibility that there may indeed be external standards, arguing that any such standards would either be open to variant interpretation by rival theorists or too abstract to be of any real utility in theory choice.

From objections which draw on the idea of an external standard, Doppelt turns to an alternative anti-relativist strategy which attempts to show that some standards are better justified than others. He distinguishes between gradualist and normative naturalist versions of this strategy. According to gradualism, scientific change is piecemeal and continuous. Rational change of standards is insured by maintaining consistency between the epistemic aims and theories endorsed by a relevant group of scientists. But, Doppelt notes, the fact that standards may be justified in this way does not show them to be better justified than other standards. For there may be rival groups of scientists who justify opposing standards in relation to alternative sets of aims and theories. According to normative naturalism, methodological standards are subject to empirical evaluation with respect to their effectiveness in leading to the aims of scientific inquiry. But the aims of science are either variable and multiple or constant and universal. On the one hand, if the aims of science vary, the problem arises of how to adjudicate between the alternative aims pursued by scientists who employ different standards. On the other hand, if the aims of science are constant, the problem is to reconcile such constancy with the apparent variation of aims in the history of science. Moreover, if truth is proposed as the constant aim of science, then the aim is unable to serve as independent arbiter between conflicting standards, since there is no means of access to the truth that is independent of such standards.

In "Reasons, Radical Change, and Incommensurability in Science", Dudley Shapere presents a piecemeal model of scientific reason akin to the gradualism that Doppelt at one point criticizes. According to Shapere, the form of reason characteristic of science has been developed by scientists in the process of undertaking science itself. Scientific inquiry proceeds in a piecemeal fashion. In the process, a great variety of background beliefs become established. These then serve as reasons which scientists adopt as the basis for research or for the acceptance of a theory or experimental result. Shapere contrasts the piecemeal model with 'classical' and 'post-classical' philosophy of science. Classical empiricist philosophers of science sought explicit metascientific rules of method, which were universal and exempt from change with the advance of science. Post-classical philosophers of science, of whom Kuhn is the prime example, understand science as a developmental, historical process, and emphasize the important role played by prior belief in the practice of science. But, in Shapere's view, neither approach accounts adequately for the role of background belief in scientific inquiry. On the one hand, the classical approach fails to explain the rational basis of well-established background belief. On the other hand, Kuhn's version of the post-classical approach places undue emphasis on the unity of background belief, while failing to recognize either the diversity of such belief or the importance of that diversity.

In addition to the classical and post-classical approaches, another of Shapere's targets is language-dominated philosophy of science. He seeks to extricate the problem of incommensurability from the embrace of the philosophy of language. In his view, philosophical understanding of science is not advanced by viewing science through the lens of language. In effect, he regards the problem of incommensurability as an artifact of linguistic analysis, which arises from treating the issue of meaning as prior to that of reason. Rather than approach reason via the language of science, Shapere suggests that consideration of how experimental test of theories is undertaken in modern science reveals scientific reason at work. He maintains that the problem of incommensurability is dissolved once the nature of the reasoning employed in modern science is properly understood.

To illustrate, Shapere considers the case of the solar neutrino experiment, in which neutrinos from the sun combine with chlorine atoms to form a radioactive isotope of argon in an underground tank. The experiment was designed to test the theory that the measured energy of solar neutrinos was due to nuclear reactions in the interior of the sun. In both the conception and design of the experiment, scientists drew on a variety of background beliefs about nuclear reactions, the behavior of chemical elements, and factors which might interfere with the experiment. The background beliefs provide the reasons which justified scientists in proceeding in the manner they did in the design and execution of the solar neutrino experiment. The process which produces such background beliefs involves the piecemeal study of distinct scientific domains in isolation from each other, as well as an attempt to unify the results of such piecemeal study in a coherent manner. The success of this piecemeal approach justifies scientists in employing background beliefs as the basis for reasoning with regard to theory appraisal and experimental design and practice. In particular, Shapere notes that background beliefs may be used to construct chains of reasoning which justify the alteration and revision of the use of scientific terms that give rise to change of scientific concepts.

Thus, rather than the rationality of science being undermined by semantic change, scientific reasoning is itself what provides the rational justification for semantic change in science. In this way, the problem of incommensurability gives way to an analysis of scientific reason not couched in the terms of linguistic philosophy.

7.4. Incommensurability, Multiculturalism and Science Education

Amongst philosophers of science, discussion of the incommensurability thesis usually tends to focus on topics such as those of semantic variance, realism and progress, or rational theory choice, which we have canvassed so far. Despite its general significance, the broader cultural implications of the thesis have been little explored in the literature of the philosophy of science. The two papers in the fourth section explicitly address such cultural implications in relation to science education. Harvey Siegel pursues the implications of incommensurability for the issue of multiculturalism in science education, arguing that the culturally located nature of science in no way diminishes its claim to universality. Writing in response to Siegel, Hugh Lacey argues that the alternative values endorsed by different cultures may promote alternative strategies of scientific research, which give rise in turn to the possibility of legitimate multicultural science.

There is a tendency on the part of some writers about science education to suggest that the traditional beliefs of indigenous peoples about the natural world should occupy the same status as Western science within the science curriculum. These writers adopt the standpoint of multiculturalism. On the basis of the multiculturalist standpoint, they reject the claims of universality that are made on behalf of science. In his paper, "Incommensurability, Rationality and Relativism: In Science, Culture and Science Education", Harvey Siegel subjects the multiculturalist standpoint to critical scrutiny. Against the advocates of multiculturalism in science education, Siegel argues that science teachers should teach students the best available accounts of the natural world. If the best available accounts of the natural world are those produced by modern Western science, then those are the accounts that should be taught. Students should also be taught the capacity for independent, critical thought, as well the value of such thought, which are crucial features of science. Only by teaching the best available accounts of the natural world, and by imparting a capacity and value for critical thought, does one treat students in a dignified and respectful manner.

Siegel proposes that multiculturalism is to be understood in moral terms. It asserts that all cultures and their members are to be treated with dignity and respect. Such treatment is taken to require inclusion of indigenous views of nature in the science curriculum. In support of the latter, advocates of multicultural science education argue for the legitimacy of the belief systems of indigenous peoples, and deny that science has greater legitimacy than such belief systems. In their view, all belief systems are local; none are universal. On Siegel's analysis, the claims of the multiculturalists rest on a kind of incommensurability thesis. For they take it to be impossible to establish the greater legitimacy of Western science, since it is impossible to conduct a rational comparison of science with indigenous belief systems.

Siegel takes the multiculturalists to task on two main counts. On the one hand, Siegel argues that it is possible for a view to be both local and universal. For while a scientific claim may be produced in a particular local context, it may express a universal truth that holds irrespective of cultural context. On the other hand, Siegel argues that the standard criteria of scientific theory appraisal (e.g., predictive power, testability, simplicity, etc.), take precedence over the evaluative criteria of traditional cultures in determining the content of the science curriculum, since the standard criteria are the appropriate criteria for evaluating accounts of the natural world. The standard criteria of theory appraisal are not, however, to be adopted in a dogmatic spirit. The belief that the standard criteria are the appropriate ones for evaluating views of the natural world is a fallible belief. Moreover, the criteria should themselves be taught to students in a critical manner designed to cultivate the capacity for independent thought and judgment. Siegel does not claim that there may be no other criteria (e.g., social values), apart from the standard evaluative criteria, on the basis of which scientific claims may be judged. But he does deny that such criteria are relevant to the evaluation of the scientific status of such claims.

The scientific relevance of criteria other than the standard evaluative criteria is the main issue addressed by Hugh Lacey in his paper, "Incommensurability and 'Multicultural Science'". Lacey seeks to establish the legitimacy of at least some forms of multicultural science. To do so, he introduces a distinction between the standard evaluative criteria – which, following Kuhn, he treats as 'cognitive values' – and what he calls "strategies". By use of the term 'strategy', Lacey wishes to capture part of what Kuhn originally meant by 'paradigm', viz., that scientific research rests on an underlying conception of the "legitimate methods, problems and standards of solution" in an area (Kuhn, 1970a, p. 48). A strategy provides constraints which govern choice of theory, as well as selection of data. It also plays a crucial role in determining the phenomena which constitute the domain of investigation. Lacey illustrates the idea of a strategy with a distinction between materialist strategies, which focus on the material possibilities of things, Aristotelian strategies, which relate phenomena to their place in the cosmos, and agroecological strategies, which emphasize aspects of agroecosystems that cannot be reduced to the material possibilities of things (e.g., seeds).

Materialist strategies focus on the material nature and possibilities of things in abstraction from their role in human life, or any moral or social value that may be placed on them. According to Lacey, materialist strategies are the dominant strategy in modern science. However, the choice of strategy is not one that may proceed entirely in isolation from human social relations and cultural values. Rather, the choice of such a strategy must inevitably raise questions about matters of significance to human life, on which people of different cultures or ways of life may place a different value. This is precisely the point at issue between Lacey and Siegel. For, where Siegel dismisses criteria other than the standard evaluative criteria as of no relevance to science, Lacey argues that social values have intrinsic relevance to science because of their bearing on choice of scientific research strategy. Since pursuit of an alternative scientific strategy which reflects alternative cultural values might, in principle, give rise to empirically well-grounded results, there is no reason to dismiss the results of such a strategy as non-scientific. This opens up the possibility of a truly multicultural science, one which embraces a variety of alternative research strategies which are informed by the value systems of a multiplicity of cultures.

7.5. Incommensurability, Cognition and Conceptual Change

Research on the nature of cognition provides a source of inspiration and focus of interest for a number of different areas of philosophy. Cognitive science exerts a powerful influence on current work in epistemology, philosophy of language, and, especially, the philosophy of mind. It is also the source of an increasingly important influence on research on semantic incommensurability and conceptual aspects of scientific change. The two papers in the fifth and final section of this volume take up and develop these themes. Peter Barker applies the dynamic frame model of Lawrence Barsalou to the conceptual shifts that occurred in the course of the Copernican revolution. Nancy Nersessian presents her cognitive-historical approach to conceptual change, which she illustrates with reference to the historical development of the concept of an electromagnetic field.

Peter Barker's paper is entitled "Incommensurability and Conceptual Change During the Copernican Revolution". Barker places Kuhn's mature taxonomic conception of incommensurability within the context of Barsalou's dynamic frame account of concepts. Following Barsalou, Barker takes frames to represent conceptual systems as patterns of nodes. Frames have a hierarchical structure, which link superordinate concepts (e.g., 'celestial object') to subordinate concepts (e.g., 'moon') by means of the attributes of concepts (e.g., orbit shape), and the values of attributes (e.g., elliptical). Barker notes that the dynamic frame account captures the family resemblance character of scientific concepts, which Kuhn emphasized in his account of concept acquisition through exposure to similarity and dissimilarity classes. However, the frame account is a more general account than Kuhn's, since it is not restricted to analysis of taxonomic structure, nor are relations within a frame restricted to taxonomic relations of kind-membership. Within the frame account, incommensurability arises where there is a mismatch between the nodes of subordinate concepts in alternative frames for the same superordinate concept. As illustration, Barker applies the frame model to the conceptual transformation that took place during the Copernican revolution, drawing particular attention to Kepler's introduction of the new concept of an orbit. On the one hand, Ptolemaic and Copernican astronomy were minimally incommensurable, since both employed the same pattern of nodes, with only slight conceptual difference due to variation in the value assigned to a single attribute. On the other hand, Kepler's astronomy was strongly incommensurable with both Ptolemaic and Copernican astronomy, since introduction of the concept of an orbit required changes in the pattern of nodes, with resultant changes in the attributes and values of concepts.

The title of Nancy Nersessian's paper is "Concept Formation and Incommensurability". Nersessian's approach is based on an empirical study of the cognitive practices of scientists, which she describes as a cognitive-historical approach. The approach applies the analytic tools of cognitive science to historical case studies of conceptual change. In her paper, she applies the cognitive-historical approach to the development from Faraday to Einstein of the concept of an electromagnetic field. In order to analyze the conceptual changes which were involved, Nersessian employs the framelike device of a meaning schema, which specifies four components of the concept of an electromagnetic field, viz., the ontological status, function, mathematical structure and causal power of such a field. The meaning schema analysis permits Nersessian to illustrate the problem-solving nature of conceptual change in relation to the problem situations in which the concept of an electromagnetic field was embedded at various stages of development. It also enables her to reveal the chains of reasoning which link the stages in the conceptual development together into a continuous process. Nersessian argues that such chains of reasoning play a crucial role in the generation of new concepts. In addition, she identifies a special kind of analogical reasoning employed by Maxwell, which she calls 'generic abstraction'. On the account of conceptual change that she proposes, the process of conceptual innovation is a continuous process that proceeds on the basis of a connected chain of reasoning which is typically analogical in character. Thus, in Nersessian's view, the problem of incommensurability may in large part be credited to a mistaken analysis of the nature of the reasoning which underlies conceptual change.

7.6. Select Bibliography

A brief word is in order on the select bibliography which concludes the volume. The papers collected in this volume are based on some of the invited contributions to the conference *Incommensurability (and related matters)*, which was held in Hannover, Germany 13–16 June, 1999. The bibliography was developed as a resource for the conference. It was posted on the conference website to enable participants to familiarize themselves with the literature on the topic. In the months leading up to the conference, we received numerous proposals of entries from various interested parties. Accordingly, the bibliography began to take on a status beyond that of simply being a guide to the literature. Since the conference, the bibliography has been further developed in an attempt to improve coverage. While it may not be possible to create a truly comprehensive bibliography for the literature on a subject matter such as incommensurability, we do believe that the bibliography will serve as a useful and important resource in its own right. Hence the bibliography is included here in the hopes that, as with the papers contained in this volume, it will assist and promote further research and discussion of the topic of incommensurability.

8. Acknowledgements

As just mentioned, the papers contained herein derive from work presented at the conference *Incommensurability (and related matters)*, held in Hannover in June 1999. The conference was organized under the auspices of the Center for the Philosophy and Ethics of Science of the University of Hannover. It was attended by more than 120 enthusiastic participants from 20 countries around the world. For their enthusiastic participation, friendship and stimulating conversation, we wish to thank all of those who took part in the conference. We wish to express our gratitude to the *Deutsche Forschungsgemeinschaft* for financial support of the conference, as well as to the *Deutscher Akademischer Austauschdienst* for support of Howard Sankey's study visit to the Hannover Center. We are also indebted to our fellow members of the conference committee, Ronald N. Giere, Marcel Weber and Eric Oberheim, for their assistance in the organization of the conference. Special thanks are due to Eric Oberheim and Daniel Sirtes who prepared the index, bibliography and camera ready manuscript of the volume.

Philosophers who work in different intellectual traditions, and who base their work on opposing philosophical orientations, all too often find it difficult to engage in productive dialogue with one another. The two editors of this volume are themselves advocates of opposed philosophical viewpoints. Nevertheless, we hope that this volume will stand as testament to the potential for fruitful dialogue which may emerge from the meeting of such divergent orientations. In our own case, we have found the process of engaging in discussion between such alternative viewpoints to be a stimulating and rewarding experience. It is our hope that readers of the papers in this volume will benefit from our efforts to promote such discussion.

University of Melbourne
University of Hannover

NOTES

[1] For convenience, all subsequent reference to this paper will be to the reprinted version in Feyerabend's collected papers (Feyerabend, 1981d).

[2] All subsequent reference to this work will be to the standard second edition of Kuhn's *The Structure of Scientific Revolutions* (1970a), which includes the *Postscript –1969*.

[3] It was once common to draw a sharp contrast between the historical or post-positivist philosophy of science and the logical positivist and empiricist tradition that preceded it. However, recent research on the history of 20th century philosophy of science has led to a re-evaluation of the relationship between positivist and post-positivist philosophy of science. Some studies suggest that logical positivism had more in common than previously supposed with Kant than with British empiricism (e.g., Friedman, 1993; Parrini, 1998). Other studies show that the logical empiricist double-language model contained the seeds of post-positivist claims of meaning variance (English, 1978). Still others make much of Carnap's sympathetic reception of Kuhn's *The Structure of Scientific Revolutions* in his capacity as Editor of the *International Encyclopedia of Unified Science* (Reisch, 1991). Indeed, striking parallels have been drawn between the views of Carnap and Kuhn in seeking to explain this reception (Earman, 1993; Irzik and Grünberg, 1995). An important point that emerges from these studies is that a number of theses and tendencies associated with the incommensurability thesis may also be found in the earlier work of authors writing in the positivist tradition.

[4] Indeed, the issue of the appropriate analysis of the concept of incommensurability is one on which the Editors have themselves previously adopted differing views. Hoyningen-Huene treats incommensurability as a unified relation of which associated factors such as meaning and standard variance are aspects (Hoyningen-Huene, 1990, p. 488). Sankey denies unified analysis in the case of Kuhn (Sankey, 1993, pp. 760–765), and treats semantic incommensurability as incomparability of content (Sankey, 1997, p. 428).

[5] For comments indicative of their initial treatment of meaning variance and the absence of an independent observation language, see Kuhn (1970a, pp. 101–102, 125–129, 149, 200–204; 1970b, pp. 266–271) and Feyerabend (1965, pp. 168–170, 180; 1981c; 1981d, pp. 44–45, 64–68, 76ff)

[6] The point that sentences which lack common meaning are unable to contradict one another is conceded by Feyerabend (1981f, p. 115) in response to a comment by Dudley Shapere. For remarks by Kuhn on the limitations on direct comparison of content imposed by lack of a common language and inability to translate from one theory into another, see Kuhn (1970b, p. 266; 1976, p. 191; 1979, p. 416).

[7] For radical meaning variance, see Feyerabend (1981d, pp. 68, 93; 1981e, p. 97; 1981f, pp. 114–115). For partial meaning variance, see Kuhn (1970b, p. 267; 1983a, pp. 670–671).

[8] For present purposes, the requirement that terms refer to the same things in order for the claims of semantically variant theories to be comparable for content may be understood in a broad sense. Extensional overlap, rather than identity of extension, is all that is required for comparability of content. As shown by

Martin (1971; 1972), statements from alternative theories may conflict provided that they contain predicates whose extensions are related by either containment or intersection. A similar point may be made on the basis of Field's notion of partial denotation (1973).

[9] Examples that have been given of terms whose reference may have changed include 'atom', 'electron', 'mass', 'planet', 'compound' and 'gene'. For the claim that reference changes with theory, see Kuhn (1970a, p. 102; 1970b, p. 269) and Feyerabend (1981e, p. 98).

[10] The use of the expression 'causal descriptivism' is subject to some variation between authors. The expression is employed here in the sense in which it is employed in Sankey (1994, pp. 61–67). This use is similar to that of the expression 'descriptive causal' which is used by Devitt and Sterelny to refer to theories which are a hybrid of causal and descriptive theories of reference (1999, pp. 96–101). Thus, the present use contrasts with that of David Lewis (1984, p. 226), for whom a causal descriptive theory of reference is one on which reference is fixed by description of causal relations, rather than by causal relations themselves.

[11] For detailed discussion of the required modifications of the causal theory of reference, and the implications with respect to incommensurability, see Sankey (1994, chapter 2).

[12] For a critical review of Sankey (1994) by the other Editor of this volume and two co-authors, see Hoyningen-Huene, Oberheim and Andersen (1996).

[13] The principle of proliferation recommends that scientists "[i]nvent, and elaborate theories which are inconsistent with the accepted point of view, even if the latter should happen to be highly confirmed and generally accepted" (1981f, p. 105). On the contrast between Kuhn's monistic and Feyerabend's pluralist vision of science, see Feyerabend (1981g, p. 139).

[14] For a sample of the criticism raised against Kuhn's and Feyerabend's views of the nature of scientific method and the rationality of theory choice, see Lakatos (1970, p. 178), Laudan (1996, chapter 5), Siegel (1987, pp. 51–54) and Shapere (1984, pp. 46–47). For a sympathetic interpretation of Kuhn as a "moderate relativist", see Doppelt (1982), and for the suggestion that Kuhn's and Feyerabend's treatment of theory choice may in fact contain the germs of a new way of thinking about rationality, see Bernstein (1983).

[15] For more detailed treatment of metamethodological aspects of Kuhn's theory of method, see Nola and Sankey (2000b, pp. 26–30). For an attempt to develop Kuhn's approach, see Hoyningen-Huene (1992).

[16] For Laudan's normative naturalist alternative to intuitionist and conventionalist metamethodologies, see particularly Laudan (1996, chapter 7). For critical discussion of Laudan's views, see Doppelt (1990), Worrall (1988) and Siegel (1990). For further comment on Laudan's normative naturalism in relation to the general topic of methodological incommensurability, see Sankey (1996; 2000).

[17] See, for example, Feyerabend (1978, p. 70) for the claim that the world changes with theory and rejection of the assumption that "the objective world remains unaffected by our epistemic activities". See also Feyerabend's favourable comment on Hoyningen-Huene's neo-Kantian interpretation of Kuhn (Feyerabend, 1989, p. 405, fn.26). For further discussion, see Devitt (1991, chapter 9) and Sankey (1994, chapter 6).

[18] For remarks indicative of Kuhn's critical attitude toward realism, see, for example, Kuhn (1970a, p. 206; 1979, p. 418; 1991, p. 6; 1993, pp. 330–331). Kuhn's Kantian tendency is evident in, for example, his (1979, pp. 418–419; 1991, p. 12; 1993, p. 331).

[19] One author to have commented on Kuhn's idealist tendency is Scheffler (1967, p. 19). It is important to note that the form of idealism found in Kuhn is not an extreme idealism on which the world is a mental construct produced entirely by the mind. There are numerous respects in which Kuhn's position allows, and even requires, the existence of a mind-independent reality (cf. Brown, 1983; Devitt, 1991, p. 156). The form of idealism of relevance to Kuhn is of the Kantian 'constructivist' variety. For discussion of the issues surrounding Kuhn's idealism, see Hoyningen-Huene (1989; 1993, pp. 268–269).

[20] The world-change image is treated as a metaphor that does not commit Kuhn to a substantive metaphysical position in Sankey (1994, pp. 152–153).

[21] That Kuhn might be interpreted in a neo-Kantian fashion had of course been noted previously; see, e.g., Buchdahl (1969, p. 511, fn. 1) and Devitt (1991, chapter 9).

[22] For a critical review of Hoyningen-Huene (1993) by the other Editor of the present volume, see Sankey (1995). For discussion of the possibility of co-reference between theories on Hoyningen-Huene's neo-Kantian interpretation of Kuhn, see Sankey (1997, pp. 440–441).

[23] Kuhn expresses sympathy for the position described by Hacking, but rejects the proposed nominalist interpretation. Among other things, it is not simply the kinds to which objects belong, but the objects themselves, that vary with classificatory scheme (Kuhn, 1993, pp. 315–316).

[24] The possibility that such question-begging may occur in realist and anti-realist discussions of incommensurability has led to the suggestion that such discussion may be affected by a meta-level incommensurability (Hoyningen-Huene, Oberheim and Andersen, 1996; Oberheim and Hoyningen-Huene, 1997). For critical discussion, see Sankey (1997), as well as the paper in this volume by Michael Devitt.
[25] Doppelt originally proposed his moderate relativist treatment of Kuhn in his (1982), in which he interprets methodological incommensurability as deriving from differences in the problem-solving agendas of competing paradigms, combined with differential weighting of the significance of shared problems.

REFERENCES

Bernstein, R. (1983). *Beyond Objectivism and Relativism.* Philadelphia: University of Pennsylvania Press.

Brown, H. (1983). "Incommensurability." *Inquiry* **26**: 3–29.

Buchdahl, G. (1969). *Metaphysics and the Philosophy of Science. The Classical Origins: Descartes to Kant.* Cambridge, Mass.: MIT Press.

Davidson, D. (1984). "On the Very Idea of a Conceptual Scheme." In *Inquiries into Truth and Interpretation*, pp. 183–198, Oxford: Oxford University Press.

Devitt, M. (1991). *Realism and Truth.* 2nd edition. Blackwell: Oxford.

Devitt, M. and K. Sterelny. (1999). *Language and Reality.* 2nd edition. Oxford: Blackwell.

Doppelt, G. (1982). "Kuhn's Epistemological Relativism: An Interpretation and Defence." In J. Meiland and M. Krausz, eds., *Relativism: Cognitive and Moral.* Notre Dame: University of Notre Dame Press.

Doppelt, G. (1990). "The Naturalist Conception of Methodological Standards." *Philosophy of Science* **57**: 1–19.

Earman, J. (1993). "Carnap, Kuhn and the Philosophy of Scientific Method." In P. Horwich, ed., *World Changes: Thomas Kuhn and the Nature of Science*, pp. 9–36, Cambridge Mass.: M.I.T. Press.

Enç, B. (1976). "Reference of Theoretical Terms." *Nous* **10**: 261–282.

English, J. (1978). "Partial Interpretation and Meaning Variance." *Journal of Philosophy* **75**: 57–76.

Feyerabend, P. (1962). "Explanation, Reduction and Empiricism." In H. Feigl and G. Maxwell, eds., *Minnesota Studies in the Philosophy of Science, Volume 3, Scientific Explanation, Space and Time*, pp. 28–97, Minneapolis: University of Minnesota Press.

Feyerabend, P. (1975). *Against Method.* London: New Left Books.

Feyerabend, P. (1978). *Science in a Free Society.* London: New Left Books.

Feyerabend, P. (1981a). *Realism, Rationalism and Scientific Method: Philosophical Papers, Volume 1.* Cambridge: Cambridge University Press.

Feyerabend, P. (1981b). *Problems of Empiricism: Philosophical Papers, Volume 2.* Cambridge: Cambridge University Press.

Feyerabend, P. (1981c). "An Attempt at a Realistic Interpretation of Experience." In Feyerabend, 1981a, pp. 17–36.

Feyerabend, P. (1981d). "Explanation, Reduction and Empiricism." In Feyerabend, 1981a, pp. 44–96. (Reprint of Feyerabend 1962.)

Feyerabend, P. (1981e). "On the 'Meaning' of Scientific Terms." In Feyerabend, 1981a, pp. 97–103.

Feyerabend, P. (1981f). "Reply to Criticism: comments on Smart, Sellars and Putnam." In Feyerabend 1981a, pp. 104–131.

Feyerabend, P. (1981g). "Consolations for the Specialist." In Feyerabend, 1981b, pp. 131–161.

Feyerabend, P. (1987). "Putnam on Incommensurability." *British Journal for the Philosophy of Science* **38**: 75–81.

Feyerabend, P. (1989). "Realism and the Historicity of Knowledge." *Journal of Philosophy* **86**: 393–406.

Field, H. (1973). "Theory Change and the Indeterminacy of Reference." *Journal of Philosophy* **70**: 462–481.

Fine, A. (1975). "How to Compare Theories: Reference and Change." *Nous* **9**: 51–65.

Friedman, M. (1993). "Remarks on the History of Science and the History of Philosophy." P. Horwich, ed., *World Changes: Thomas Kuhn and the Nature of Science*, pp. 37–54, Cambridge Mass.: M.I.T. Press.

Hacking, I. (1983). *Representing and Intervening.* Cambridge: Cambridge University Press.

Hacking, I. (1993). "Working in a New World: The Taxonomic Solution." In P. Horwich, ed., *World Changes: Thomas Kuhn and the Nature of Science*, pp. 275–310, Cambridge Mass.: M.I.T. Press.

Horwich, P. (1993). *World Changes: Thomas Kuhn and the Nature of Science.* Cambridge, Mass.: MIT Press.

Hoyningen-Huene, P. (1989). "Idealist Elements in Thomas Kuhn's Philosophy of Science." *History of Philosophy Quarterly* **6**: 393–401.

Hoyningen-Huene, P. (1990). "Kuhn's Conception of Incommensurability." *Studies in History and Philosophy of Science* **21**: 481–491.

Hoyningen-Huene, P. (1992). "The Interrelations Between the Philosophy, History and Sociology of Science in Thomas Kuhn's Theory of Scientific Development." *British Journal for the Philosophy of Science* **43**: 487–501.

Hoyningen-Huene, P. (1993). *Reconstructing Scientific Revolutions: Thomas S. Kuhn's Philosophy of Science*, trans. A. Levine. Chicago: University of Chicago Press.

Hoyningen-Huene, P., E. Oberheim and H. Andersen (1996). "On Incommensurability." *Studies in History and Philosophy of Science* **27**: 121–141.

Irzik, G. and Grünberg, T. (1995). "Carnap and Kuhn: Arch Enemies or Close Allies?" *British Journal for the Philosophy of Science* **46**: 285-309.

Kripke, S. (1980). *Naming and Necessity*. Oxford: Blackwell.

Kroon, F. (1985). "Theoretical Terms and the Causal View of Reference." *Australasian Journal of Philosophy* **63**: 143–166.

Kuhn, T. (1962). *The Structure of Scientific Revolutions*. Chicago: University of Chicago Press.

Kuhn, T. (1970a). *The Structure of Scientific Revolutions*. 2nd edition. Chicago: University of Chicago Press.

Kuhn, T. (1970b). "Reflections on my Critics." In I. Lakatos and A. Musgrave, eds., *Criticism and the Growth of Knowledge: Proceedings of the International Colloquium the Philosophy of Science, London, 1965*, pp. 231–278.

Kuhn, T. (1976). "Theory-Change as Structure Change: Comments on the Sneed Formalism." *Erkenntnis* **10**: 179–199.

Kuhn, T. (1977a). *The Essential Tension*. Chicago: University of Chicago Press.

Kuhn, T. (1977b). "Second Thoughts on Paradigms." In Kuhn, 1977a, pp. 293–319.

Kuhn, T. (1977c). "Objectivity, Value Judgment, and Theory Choice." In Kuhn, 1977a, pp. 320–339.

Kuhn, T. (1979). "Metaphor in Science." In A. Ortony, ed., *Metaphor and Thought*, pp. 409–419, Cambridge: Cambridge University Press.

Kuhn, T. (1983a). "Commensurability, Comparability, Communicability." In P. Asquith and T. Nickles, eds., *PSA 1982, Volume 2*, pp. 669–688, East Lansing: Philosophy of Science Association.

Kuhn, T. (1983b). "Rationality and Theory Choice." *Journal of Philosophy* **80**: 563–570.

Kuhn, T. (1987). "What Are Scientific Revolutions?" In L. Kruger, L. Daston and M. Heidelberger, eds., *The Probabilistic Revolution, Volume 1, Ideas in History*, pp. 7–22, Cambridge, Mass.: MIT Press.

Kuhn, T. (1991). "The Road Since Structure." In A. Fine, M. Forbes and L. Wessels, eds., *PSA 1990, Volume 2*, pp. 2-13, East Lansing: Philosophy of Science Association.

Kuhn, T. (1993). "Afterwords." In P. Horwich, ed., *World Changes: Thomas Kuhn and the Nature of Science*, pp. 311–341, Cambridge, Mass.: MIT Press.

Lakatos, I., and A. Musgrave, eds. (1970).*Criticism and the Growth of Knowledge: Proceedings of the International Colloquium the Philosophy of Science, London, 1965*. Cambridge: Cambridge University Press.

Lakatos, I. (1970). "Falsification and the Methodology of Scientific Research Programmes." In I. Lakatos and A. Musgrave, eds., *Criticism and the Growth of Knowledge: Proceedings of the International Colloquium the Philosophy of Science, London, 1965*, pp. 91–196, Cambridge: Cambridge University Press.

Laudan, L. (1996). *Beyond Positivism and Relativism*. Boulder: Westview Press.

Lewis, D. (1984). "Putnam's Paradox." *Australasian Journal of Philosophy* **62**: 221–236.

Martin, M. (1971). "Referential Variance and Scientific Objectivity." *British Journal for the Philosophy of Science* **22**: 17-26.

Martin, M. (1972). "Ontological Variance and Scientific Objectivity." *British Journal for the Philosophy of Science* **23**: 252-256.

Nola, R. (1980). "Fixing The Reference of Theoretical Terms." *Philosophy of Science* **47**: 505–531.

Nola, R. and H. Sankey, eds. (2000a). *After Popper, Kuhn and Feyerabend: Recent Issues in Theories of Scientific Method*. Dordrecht: Kluwer Academic Publishers.

Nola, R. and H. Sankey. (2000b). "A Selective Survey of Theories of Scientific Method." In R. Nola and H. Sankey, 2000a, pp. 1–65.

Oberheim, E. and P. Hoyningen-Huene. (1997). "Incommensurability, Realism and Meta-incommensurability." *Theoria* **12**: 447–465.
Papineau, D. (1979). *Theory and Meaning*. Oxford: Oxford University Press.
Papineau, D. (1996). "Theory-Dependent Terms." *Philosophy of Science* **63**: 1–20.
Parrini, P. (1998). *Knowledge and Reality. An Essay in Positive Philosophy*. Dordrecht: Kluwer Academic Publishers.
Putnam, H. (1975a). *Mind, Language and Reality. Philosophical Papers, Volume 2*. Cambridge: Cambridge University Press.
Putnam, H. (1975b). "Explanation and Reference." In Putnam, 1975a, pp. 196–214.
Putnam, H. (1975c). "The Meaning of 'Meaning'." In Putnam, 1975a, pp. 215–271.
Reisch, G. (1991). "Did Kuhn Kill Logical Empiricism?" *Philosophy of Science* **58**: 264–277.
Sankey, H. (1991). "Translation Failure Between Theories." *Studies in History and Philosophy of Science* **22**: 223–236.
Sankey, H. (1993). "Kuhn's Changing Concept of Incommensurability." *British Journal for the Philosophy of Science* **44**: 759–774.
Sankey, H. (1994). *The Incommensurability Thesis*. Aldershot: Avebury.
Sankey, H. (1995). "Review of P. Hoyningen-Huene, *Reconstructing Scientific Revolutions: Thomas S. Kuhn's Philosophy of Science*." *Australasian Journal of Philosophy* **73**: 487–489.
Sankey, H. (1996). "Normative Naturalism and the Challenge of Relativism: Laudan versus Worrall on the Justification of Methodological Principles." *International Studies in the Philosophy of Science* **10**: 37–51.
Sankey, H. (1997). "Incommensurability: The Current State of Play." *Theoria* **12**: 425–445.
Sankey, H. (1998). "Taxonomic Incommensurability." *International Studies in the Philosophy of Science* **12**: 7–16.
Sankey, H. (2000). "Methodological Pluralism, Normative Naturalism and the Realist Aim of Science." In R. Nola and H. Sankey, 2000a, pp. 211–229.
Scheffler, I. (1967). *Science and Subjectivity*. Indianapolis: Bobbs–Merrill.
Shapere, D. (1984). *Reason and the Search for Knowledge*. Dordrecht: Reidel.
Siegel, H. (1987). *Relativism Refuted*. Dordrecht: Reidel.
Siegel, H. (1990). "Laudan's Normative Naturalism." *Studies in History and Philosophy of Science* **21**: 295–313.
Sterelny, K. (1983). "Natural Kind Terms." *Pacific Philosophical Quarterly* **64**: 110–125.
Worrall, J. (1988). "The Value of a Fixed Methodology." *British Journal for the Philosophy of Science* **39**: 263–275.

RICHARD N. BOYD

REFERENCE, (IN)COMMENSURABILITY AND MEANINGS

Some (Perhaps) Unanticipated Complexities

Abstract. Received conceptions of the meanings of scientific terms assign to meanings an essentially benign methodological role: the meaning of a term consists of principles or inference rules which are, always or for the most part, (approximately) true or reliable. In fact, many scientific terms have meanings which are malignant: which are mainly false or misleading and which detract from, rather than contribute to, scientific progress. Kuhn's conception of incommensurability can be fruitfully extended to take account of malignant meanings. Malignant meanings are especially implicated in cases, like that of human sociobiology, in which the influence of social ideology on scientific practice is especially profound.

0. OVERVIEW

0.0. Malignant Meanings and the Limits of Commensurability

In the present essay I propose a substantial revision of received naturalistic conceptions of the semantics of scientific (and other) terms and to the received understanding of the methodological role which the meanings of such terms play, especially with respect to questions of commensurability and incommensurability between competing paradigms or traditions of inquiry. Received naturalistic conceptions, in so far as they address questions of meaning at all, treat such meanings as *benign*: they assume that the beliefs and inference rules which constitute the meaning of a scientific term within a linguistic community will, typically, be approximately true or approximately reliable.

Such a conception, I argue, fails to recognize the ways in which the meanings of scientific terms can be profoundly misleading – indeed fundamentally incoherent – even when those terms are referentially unambiguous. The failure to appreciate the *malignant* aspects of meaning contributes, I shall argue, to a failure to appreciate important dimensions of incommensurability between research traditions. I propose an alternative conception of the meanings of scientific terms according to which such meanings are often considerably less *benign* than conceptions of meaning (whether naturalistic or not) usually assume, and I explore its relation to issues of real-life incommensurability between significantly different research traditions which share a common subject matter.

One conclusion I reach bears on the cogency of the view (contested by "post-modern" thinkers) that the practice of scientific methods can ordinarily be expected to contribute to intellectual progress through successively closer approximations to the truth. Here I propose to "split the difference" between modernist optimism and post-

P. Hoyningen-Huene and H. Sankey (eds.). Incommensurability and Related Matters, 1–63.

modernist pessimism. I argue that the grains of truth in traditional naturalistic semantic theories provide a justification for a significant level of optimism regarding progress by approximation, but that an appreciation of the role of *malignant* meaning dictates that we recognize the contrary tendency induced by incommensurability in what we may call *conceptual meaning*. In particular I suggest that when scientific practice is significantly influenced by social ideology – as it routinely is in the biological and social sciences – the (malignant) embedding of ideology in the conceptual meanings of scientific terms is often so substantial that the normal internal workings of scientific methodology prove insufficient to overcome the malignancy or to establish commensurability between mainstream scientific research traditions and those which are informed by ideological critiques.

0.1. The Kuhnian Background and the Standard Rebuttal

Kuhn (1970) introduced the issue of (what we may call) *methodological incommensurability* between competing scientific paradigms, an issue that others have raised about competing non-scientific enterprises, like moral or political conceptions. Kuhn's arguments for methodological incommensurability depend on the by now widely recognized theory-dependence of scientific methods. They depend as well on two much more controversial claims about semantics of scientific language, claims which posit two sorts of what we may call *semantic incommensurability*[1] between competing paradigms. According to Kuhn, terms occurring in competing paradigms are *conceptually incommensurable*: they differ in what we might call *conceptual meaning* so that scientists from competing paradigms necessarily fail to communicate: they talk past one another. A consequence of this conceptual incommensurability, Kuhn says, is that there is *referential incommensurability* between the discourses of the two paradigms: the same terms when deployed in the two paradigms, because they have different meanings, must have different referents.

There is a near consensus in the philosophy of science that Kuhn profoundly exaggerates the extent of methodological incommensurability in the history of science and that his mistakes lie in his approach to the semantics of scientific terminology.[2] What we might call the *standard rebuttal* to Kuhn's arguments for incommensurability deploys a "causal" or "naturalistic" conception of reference to rebut the inference from conceptual incommensurability to referential incommensurability, by showing that terms associated with quite different conceptual resources can share a common referent and that this phenomenon is common in the actual history of science.

Naturalistic conceptions of reference are not, by themselves, sufficient to rebut claims of incommensurability for the sorts of cases Kuhn considers. Even when two conceptually divergent paradigms share a common subject matter (that is: are referentially commensurable) it still might be the case that the differences in conceptual meaning between the terms in the two paradigms are so great as to undermine the prospects for methodological commensurability. Although the standard rebuttal de-emphasizes issues of meaning in favor of issues of reference, it remains true, I'll argue, that all versions of the standard rebuttal are committed to a *benign* conception of the meanings of scientific terms. They reflect the estimate that, ordinarily, the beliefs and inferential and explanatory practices central enough to a paradigm or research tradition to count as features of the conceptual meanings of its terms will be approximately true enough (in the case of beliefs) or reliable (in the case of methodological practices) enough that (a) they will contribute to the establishment of methodological commensurability with other paradigms with the same subject matter, and (b) most of them will be recognizable as insightful approximations in any successor paradigm.

0.2. Overview; Critique and Lessons

It is this benign conception of the meanings of scientific (and other) terms which I propose to criticize. Using examples from the emerging but highly influential tradition of research in "evolutionary psychology," I identify a class of inferential patterns or "scripts" connecting evolutionary and genetic premises to psychological conclusions which have the following properties:

1. They are profoundly unreliable.
2. Their unreliability is a logical consequence of fundamental and explicitly acknowledged theoretical principles in evolutionary psychology.
3. This unreliability is unrecognized.
4. Instead, these inferential scripts are central to the methodological practices of evolutionary psychology to such an extent that
    a. they form the basis for the central explanatory strategies in the tradition,
    b. acceptance of these practices – or an appreciation of a deep theoretical criticism of them – is a prerequisite to understanding the literature in the discipline, and
    c. criticisms of these inferential patterns are, for all practical purposes, unintelligible to practitioners in the tradition.
5. The contrast between these inferential practices and analogous practices in other disciplines which study human social behavior are such as to profoundly limit the possibilities for reciprocal criticism, and thus the prospects for methodological commensurability between the relevant disciplines.
6. Inference from similar past cases suggests that the resulting methodological incommensurability is unlikely to be resolved by developments internal to the relevant traditions. In cases like that of evolutionary psychology where social ideology is instrumental in establishing patterns of inference it is often the case that external political criticism is a prerequisite to the establishment of a context in which commensurability is possible.

I argue that the inferential scripts in question are, in a perfectly good sense of the term – one central to the theory of communication – parts of the *meanings* of the relevant terms in evolutionary psychology. They are, however, malignant rather than benign in so far as commensurability is concerned. I draw from this and other examples five lessons.

Lesson One: Traditional empiricist conceptions of the meanings of general terms implied that their meanings consisted of exactly true beliefs about their referents. We need to go much further in rejecting this conception than we have thus far. In particular, the conceptions which are part of the *conceptual meaning* of a scientific (or other) term need not even reflect a coherent conception of its referent. Philosophers who (as they should) incorporate descriptivist elements into their naturalistic conceptions of reference need to take account of this phenomenon.

Lesson Two: Kuhn was right to hold that features of the meanings of scientific terms can be impediments to commensurability. The considerations which establish this point do not require any rejection of naturalistic conceptions of reference or any claims about referential incommensurability.

Lesson Three: The sort of incommensurability which arises from differences in conceptual meanings is especially likely to arise in cases in which several relatively independent traditions examine the same subject matter, as they do in the case of inquiry into human social behavior and human psychological potential. The prospects for incommensurability are enhanced in such cases by the methodological impact of social ideology.

Lesson Four: For this reason, the sort of meaning driven incommensurability we are considering is especially likely to arise in traditions of moral and political inquiry.

Lesson Five: In many cases, the epistemological conceptions necessary for an understanding of how, and under what circumstances, commensurability can be established will constitute a *political epistemology*: one which assigns an epistemic role to some political developments external to the traditions in question. Even in cases in which issues of social ideology or of politics in the usual sense do not arise, it will characteristically be the case that the establishment of commensurability between initially incommensurable disciplines will be a matter of the internal politics of the relevant scientific communities and of the larger institutional structures within which they are embedded.

1. INTRODUCTION.

1.0. Kuhn on Semantic and Methodological Incommensurability

Kuhn's classic (1970) discussion of (alleged) incommensurability between consecutive paradigms is the *locus classicus* for discussions of the possibility that for some disagreements in science (and, by extension, in other domains) there might not exist methods which are both *rational* and *fair* (to the competing positions) for their resolutions. The possibility of this sort of thing, for the sciences at least, is established by the phenomenon (emphasized by Kuhn and Hanson (1958), but also by many logical empiricists in the 1950's and by scientific realists in the 1960's) of the *theory-dependence* of scientific methods.

Once it is recognized that the dictates of scientific methods are always determined by background theories it becomes clear that it is possible that there should be competing positions with theoretical commitments so different that the methods justifiable by the standards of both their positions would be insufficiently powerful to dictate a resolution of the dispute between them.

Kuhn, of course, asserts that such incommensurability is actual: that, for example, Newtonian mechanics, as it was formulated at the end of the 19th century, was incommensurable with the relativistic conception that replaced it.

Such incommensurability claims in the history of science have suggested to many thinkers that the same sort of incommensurability might obtain between consecutive stages in other domains, like ethics or theology. Importantly, it is frequently suggested that the same sort of incommensurability also obtains with respect to scientific or other "paradigms" when these are not consecutive stages of a single research tradition, but rather different approaches to apparently the same subject matter arising in different cultural or intellectual traditions. Thus Kuhn's work has become integrated with broader traditions of relativism in anthropology and other social sciences. I'll be especially concerned with the possibility of this sort of incommensurability in the present paper, but it will be useful to begin by discussing the issue of the incommensurability of consecutive scientific paradigms.

In Kuhn's discussion of the transition between Newtonian and relativistic mechanics, an important appeal to *semantic incommensurability* enters into his defense of the *methodological incommensurability* of the two paradigms. It seems *prima facie* plausible that experiments and observations acceptable by Newtonian standards confirmed relativistic rather than Newtonian conceptions of physical magnitudes. Consider, for example, precise measurements of the orbit of Mercury or of the energy required in a cyclotron to accelerate particles as their velocity approaches the velocity of light. According to this conception, scientists, using fair methods, discovered that the laws of Newtonian mechanics provide only an approximate description of the relevant magnitudes and that relativistic mechanics provides a better approximation.

Kuhn rejects this understanding of the evidence: he denies, that, for example, Newtonianly acceptable measurements of mass confirm its velocity dependence, and he denies that Newtonian mechanics can be seen as an approximation to relativistic mechanics in the sense in question. Oddly, given his historical concerns, Kuhn's arguments for this position do not turn on subtle considerations about measurement techniques or other methodological practices of early 20th century physicists, or on a subtle discourse analysis of their papers, letters or public debates. Instead, they rest on a highly abstract conception (probably borrowed from Carnap – see e.g., Carnap, 1950) of the semantics of theoretical terms. According to this conception, the most fundamental laws involving such a term constitute its *meaning* (in the sense of its *analytic* definition) and *determine its reference* in the sense of constituting an *exact* definite description of the phenomenon to which it refers.

On these grounds Kuhn claims that the meaning *and the referent* of the term "mass" changed during the transition from the Newtonian to the relativistic paradigm, since in the latter, a fundamental Newtonian law about mass (the conservation law) is not retained. Methodological incommensurability is assured because the competing paradigms involve incompatible meanings for their key terms *and do not even share a common subject matter of which their respective theories may be seen as providing approximate knowledge.*

1.1. Commentary:Semantic Incommensurability and the Dialectics of Inter-"Paradigm" Interaction

It will be important for our purposes to note that there are two different dimensions of semantic incommensurability which, according to Kuhn's arguments, obtain between competing consecutive paradigms and contribute to methodological incommensurability. The key terms in the two paradigms are said to differ in *meaning*, so that there is supposed to be *conceptual incommensurability* between consecutive paradigms. Thus defenders of the competing paradigms will fail to understand the conceptual resources deployed by their opponents.

There is supposed to be as well *referential incommensurability* between such paradigms: the key terms they have in common are supposed to have different referents in the usages of the two paradigms, even when this semantic fact is not evident to the parties to the interparadigm disputes. Thus, according to Kuhn, defenders of two such paradigms are "talking past" each other in two different ways: they fail to understand each other's conceptual resources and they are talking about different things even though they think their uses of the relevant terms are univocal.

Now, Kuhn thinks that conceptual incommensurability of the sort which he believes obtains between competing scientific paradigms entails referential incommensurability, but even if we accept his analysis of scientific paradigms for the sake of argument we can see that the two sorts of incommensurability are, in general, logically distinct. Thus for example when two people are discussing the very same acquaintance regarding whom they have very different experiences and beliefs, the differences in their presuppositions regarding her can be so great as to sharply limit their ability to understand one another, so that there can be conceptual incommensurability with respect to the (name of) the person in question even though there is no referential incommensurability.

Similarly, each of two discussants can be referring to a different one of two very similar people with the same name without being able to discern the ambiguity of the name – indeed without experiencing any conceptual discomfort at all. In such cases, there will be referential incommensurability without conceptual incommensurability.

Two points are important for our purposes about the ways in which these two sorts of incommensurability are related to the dialectics of the interaction between defenders of competing paradigms or research traditions. Consider first the impact of the dialectic of argumentation and persuasion between paradigms or research traditions on the two sorts of semantic incommensurability. Referential incommensurability is permanent: it cannot be eliminated or attenuated as a result of argumentation and persuasion. Two non-coreferential natural kind (magnitude...) terms cannot come to be coreferential without there being a change in the propositions expressed by the laws or generalizations in which they figure.

In the case of conceptual incommensurability, on the other hand, it is by no means obvious that this is so. Conceptual changes within two initially conceptually incommensurable traditions of inquiry – especially changes which result from the dialectics of argumentation between their respective practitioners – might reasonably be expected to establish conceptual commensurability, without any change in subject matter. Only if one accepts quite particular views about the reference of scientific terms like Kuhn's according to which (at least some kinds of) conceptual incommensurability for scientific terms entails referential incommensurability would one be inclined to deny this possibility.

If we turn our attention to the dialectics of methodological incommensurability a similar pattern emerges. When paradigms are referentially incommensurable, their permanent referential incommensurability entails a permanent methodological incommensurability, absent a revision in the referential semantics of the relevant vocabularies.[3] In the case of conceptual incommensurability, methodological incommensurability is entailed – there cannot be methods which accomplish fair adjudication where there is not mutual intelligibility – but, precisely because conceptual incommensurability need not be permanent, neither must the methodological incommensurability it induces be permanent. Again, unless we adopt something like Kuhn's conception of the relation between conceptual and referential incommensurability, there is no reason to exclude the possibility that the dialectic of argumentation and persuasion between practitioners of two competing but conceptually incommensurable paradigms should result in developments within each which establish conceptual commensurability and, thus, eliminate a barrier to methodological commensurability.

It is important to note, of course, that even when conceptual commensurability is thus achieved there remains the apparent logical possibility that the (now semantically commensurable) paradigms might still diverge sufficiently that methodological commensurability is precluded. Still, the picture of initial conceptual (and thus methodological) incommensurability eventually overcome by the insights produced through the dialectics of scientific disputation and persuasion is *prima facie* familiar from, for example, the interactions which take place in the early stages of interdisciplinary investigations.

By contrast, the situation envisioned in Kuhn's treatment of the transition between Newtonian mechanics and special relativity, in which scientists believe that they have understood one another – and even award Nobel prizes for work which establishes new "paradigms" – only to learn from philosophers and historians that they have not even been studying the same phenomena, has, almost everyone agrees, never happened. It is perhaps surprising, therefore, that most philosophical discussions of commensurability and incommensurability in science have focused on the doubtful phenomenon of referential incommensurability rather than on the apparently real phenomena of conceptual incommensurability and the resulting (perhaps nonpermanent) methodological incommensurability. It is this inattention to real life conceptual incommensurability which I shall aim to criticize, and to begin to remedy, here.

1.2. The Standard Reply à la Quine and (Middle Period) Putnam, I: Reference as Primary

Since the publication of Kuhn (1970) a standard response has emerged in the philosophy of science which relies on causal (or "naturalistic") conceptions of reference of the sort initially introduced by Kripke (1971; 1972) and Putnam (1972; 1975a) (but see also Feigl, 1956) and which is influenced by Quinean critiques of analyticity and of related conceptions of meaning. Roughly, the idea behind the standard response is that the real semantic issue raised by Kuhn's examples of alleged incommensurability is one of reference rather than of meaning, and that – once the right theory of reference is employed – Kuhn's arguments for semantic incommensurability, and thus those for methodological incommensurability, fail. There is considerable variability in the details of presentations of what I am calling the standard reply, but I think that it is fair to say that what has emerged in their articulation are two basic ideas.

First, the referential relation between a word and a natural kind, property, magnitude, etc. is constituted by epistemically relevant causal relations between the use of the term in practice and the properties of its referent: relations which serve to establish a tendency for what is predicated of those terms to be approximately true of their referents.[4]

Second, such a causal or naturalistic conception of reference for scientific terms makes a separate theory of meaning for such terms (as such theories are ordinarily understood) either inappropriate altogether or, at any rate, considerably less important for methodological issues like commensurability than has traditionally been thought.

What the versions I'll be interested in of the standard reply have in common include the following more specific claims about the relationship between meanings, true descriptions, and reference:

1. The epistemically relevant causal connections which establish and sustain reference will, at least in the case of theoretical terms in science, always depend in part on the prevalence within the appropriate community (perhaps just the community of experts) of some beliefs which are approximately true of the referents of the terms they use, but

2. Nothing in the causal or naturalistic conception of reference requires that any of these beliefs be *exactly* true.

3. Furthermore, there is no place in the theory of the semantics of scientific terms for any notion of analytic truths about them. It may be possible to reconstruct philosophical intuitions about the "meanings" of scientific terms by treating as parts of the "meaning" of a scientific term at a given time the most fundamental of the laws or principles about it which are accepted in the relevant scientific community at that time. Still, these *law-cluster meanings* play no special *linguistic* or *semantic* role. In particular,

4. When the acceptance by the relevant community of approximately true beliefs is central to the establishment of reference for scientific terms (as it is in all typical cases), the beliefs in the term's law-cluster meaning need not play any special role. Ordinarily, lots of them will be approximately true, but there is no reason why this must be so. Nor is there any reason why approximately true beliefs which are *not* part of the meaning cluster cannot play as important a role in establishing the epistemically relevant causal connections as those in the cluster. Indeed, because of the importance to epistemic access of techniques of,

and low-level beliefs about, measurement and detection, they typically will. Thus,

5. When two different communities use scientific terms (or other terms for which a causal or naturalistic understanding of reference is important) with the same referents, this fact need not be reflected in sameness or significant similarity of the law-cluster "meanings" of the relevant terms in the two communities.

6. There is another semantic or linguistic role which traditional conceptions assign to meaning beyond grounding analytic truths and determining reference *via* exactly true descriptions. The meaning of a term may be thought of as something which a speaker must know in order to be linguistically competent with respect to the term. Law-cluster "meanings" aren't meanings in this sense either. In the first place, causal/naturalistic conceptions of reference all entail the possibility of "borrowing" linguistic and referential competence by deferring to experts or other knowledgeable people, so familiarity with the relevant law-cluster cannot be generally required for linguistic competence.

   If we turn our attention to the referential competence of the relevant experts, it *will* be the case that that competence depends on the acceptance by experts of approximately true beliefs about the phenomena to which reference is achieved *and* that in the typical case many of the beliefs in the relevant law-cluster will be among those which contribute to the establishment or maintenance of reference in this way. Nevertheless, lots of other approximately true beliefs will play a role in maintaining reference and approximately true members of the law-cluster ordinarily play no distinguished role: their contribution to reference is the same as that of other relevantly approximately true and widely held beliefs. To think otherwise is to think that the beliefs which a community judges most central at a given time will always play some special role in establishing the referential semantics of their language. This is just what causal/naturalistic conceptions of reference deny: what determines reference are epistemically relevant causal relations, and the history of science shows that these do not reliably correlate with conceptual centrality as judged by practitioners.[5]

It is important to see that it is no part of the causal or naturalistic conception of reference to deny that *sometimes* the correct explanation for the divergence of conceptions or law-clusters between two paradigms or traditions of inquiry with apparently the same subject matter is that appearance differs from reality and they really lack a common subject matter. The point is, instead, that a divergence of the most central descriptions of that subject matter (a divergence in meaning in the "law cluster" sense of that term) need not dictate such a conclusion. Law-cluster "meanings" are, in an important sense, not linguistic or semantic phenomena: they aren't really *meanings*.

1.3. The Standard Reply, II: How Commensurability is Supposed to Obtain

When we turn to the question of how it is that methodological commensurability obtains (when it does obtain) between consecutive paradigms (and, by extension, in other cases of large-scale disagreement about a common subject matter) the standard reply *does (at least in the hands of Putnam, 1962)* sometimes, perhaps, invoke the law-cluster meanings of scientific terms. Two competing conceptions or theories are commensurable just in case there are sufficiently many beliefs common to the two conceptions, and sufficiently many of them are relevantly approximately true, that they dictate reliable methods for resolving the disagreement between the two conceptions. Those methods thus held in common – when they are grounded in relevantly approximately true beliefs – provide the mechanisms by which actual properties of the common subject matter are indicated: they give the world a voice in the dispute, so to speak. The voice thus accorded to the phenomena in question explains how commensurability is achieved.

Once the view that the most fundamental laws within a paradigm or theoretical tradition are analytic is abandoned (as naturalistic and causal conceptions of reference require) the possibility becomes clear that empirical evidence will require the abandonment of one or more of these laws. Putnam (1962), in a classic (but very early) stage in the articulation of what I am calling the standard view, does however assign to such laws a special methodological status. He introduces the notion of a statement being *analytic at a time* when *at that time* it possesses the immunity from disconfirmation which genuinely analytic statements would *always* have, and he argues that the elements of a law-cluster have this property, even in the presence of what might seem to be disconfirmatory data, until adequate replacements have been proposed. Only when scientists have such replacements in mind do they consider rejecting fundamental laws.

Putnam (1962) assigns a distinctly semantic role to law-cluster meanings: a term retains its meaning/referent through a period of profound theoretical change just in case sufficiently many elements of its law-cluster are preserved. Importantly this semantic interpretation of law-clusters was (I think rightly) abandoned in later articulations of the standard reply (see, e.g., Putnam, 1972; 1975a; Field, 1973; Boyd, 1993; 1999). Reference came to be seen as a matter of the right sort of causal relation between the use of a term and its referent. The special status of elements of the law-cluster came to be recognized as a methodological rather than a semantic or linguistic matter.

Let me explain. In the first place, laws acquire the status which Putnam calls analyticity at a time by being central to the scientific work of that time, This requires that they be laws of quite general scope (at least this is true of the examples Putnam gives) but they must also be laws which are highly well confirmed – otherwise they wouldn't have become so central to practice. Thus, when such a law appears to be challenged by anomalous data ordinary methodological considerations dictate that the challenge be assessed in the light of the very considerable body of evidence which favors it.

Putnam insists, however, that the laws in question cannot be overturned until a suitable replacement is proposed. In this regard, they might be thought to differ from other well confirmed laws or theories. Here too the requirement appears to be methodological rather than linguistic or semantic. In the first place, a well confirmed and important law will fit appropriately into the prevailing conception of what the relevant phenomena in nature are like (that's part of what it is to be well confirmed and important). When experimental data or other observations appear to compromise a law with these features, the alternative to rejecting the law is always to attribute the compromising observations to the operation of not-as-yet-understood factors, rather than to the falsity of the law itself. As anyone who has tested the predictions of Newtonian mechanics in a Freshman physics laboratory can testify, this is often the better choice.

I suggest that the requirement that an alternative be available which is comparable in scope with the law in question, and which can be comparably well integrated into the rest of science is, to some extent, a reflection of rational methodological standards for choosing between the alternatives just mentioned. Only when there is a credible conception of what the relevant phenomena are like which is compatible with the rejection of a law do we ordinarily have a good reason to attribute embarrassing data to its falsity rather than to the operation of unknown factors. This, too, is a methodological constraint rather than a linguistic or semantic one.

Finally, there is the point made by Putnam himself that fundamental and well established laws may be so methodologically central that scientists have no choice but to continue to apply them pending the acceptance of alternative laws, even when they seem to be challenged by recalcitrant data. Of course this is right; in fact often such laws continue to be applied as approximations after than have been disconfirmed and more accurate alternatives have been accepted. What is important for our purposes is that this, too, is a methodological rather than a linguistic or semantic constraint on the use of the terms appearing in such laws.

All these points are rendered more cogent by the observation that the sort of methodological entrenchment which Putnam (1962) attributes to the member of law-clusters obtains in the case of lots of important and well established scientific principles which are in no sense *fundamental* to the disciplines in which they are accepted. [Examples: the chemical formulae for everyday reagents, the classification of the *Felidae* in *Carnivora*, the currently accepted geological history of the Himalayas, the approximate values of physical constants as reported in your *CRC Tables*.] It is hard to see how there could be even a vague boundary separating the elements of law-cluster meanings from other well established and important principles and harder still to see what it could have to do with the semantics of scientific terms.

Thus the conclusion appears vindicated that the special epistemic or methodological role of law-cluster "meanings" is not a matter of their being *meanings* at all, and this is what later articulations of the standard response entail.

1.4. The Standard Response, III: Meanings as Methodologically Benign

I have characterized a standard response to the semantic presuppositions of Kuhn-style incommensurability arguments: one which involves denying that scientific terms have *meanings* in any strictly linguistic or semantic sense of the term and which portrays the possibility of commensurability between paradigms or research traditions as arising from shared *reference* of relevant terms together with an appropriate sort of overlap in the approximations to the truth reflected in the two paradigms or traditions. Since I propose to challenge this response, I want to be sure to represent it fairly. With this aim in mind, I want to articulate a more moderate position on meaning which I think would satisfy many defenders of the standard response. It, too, underwrites the conclusion that issues of meaning are not central to issues about commensurability and incommensurability and it too is, on the view I'll develop here, mistaken.

In the first place, fairness dictates that I acknowledge that philosophers responding to the incommensurability arguments of Kuhn had good reasons for focusing on the question of referential incommensurability. The arguments which Kuhn presented for referential incommensurability posed an extremely important challenge to prevailing conceptions of the referential semantics of theoretical terms. These conceptions were essentially *empiricist in origin and, hence, essentially descriptivist*: they assigned to the most fundamental laws involving a theoretical term precisely the role of uniquely picking out its referent. Thus when Kuhn deployed the same conception to defend the highly implausible conclusion that consecutive paradigms were always referentially incommensurable he produced a crisis in semantic theory requiring, as it turned out, the articulation of distinctly non-empiricist naturalistic conceptions of reference.

By contrast, there was no analogous crisis in the theory of conceptual meaning or, to put it more neutrally, the theory of mutual intelligibility across scientific "revolutions." In the first place the theory of "borrowed reference" which came free, so to speak, with naturalistic theories of reference made it clear that no special knowledge of the meaning of a scientific term is necessary in order for one to achieve mere referential competence with respect to it; it can be "borrowed" by linguistic deference to experts.

With respect to the question of conceptual intelligibility between experts, there was a similar absence of crisis. With respect to cases of "scientific revolutions" discussed by Kuhn – cases in which new "paradigms" are developed by practitioners trained in earlier ones and articulated using (what the founders take to be) conceptual resources shared with the earlier paradigms – almost any even remotely plausible conception of mutual intelligibility for experts, except one like Kuhn's which treats fundamental laws as analytic truths, will underwrite the (correct) conclusion that mutual intelligibility was largely achieved during the revolutions in question. Thus no new theory of conceptual meaning was required in order to articulate a response to positions like those of Kuhn.

In a word, it is just too easy to produce a theory which yields the conclusion that Copernicus, Galileo, Newton, Darwin and Einstein succeeded in communicating with other experts. No revolution in the theory of conceptual meaning is required. What I shall be arguing here is that the situation is quite different with respect to cases in which what is at issue is mutual conceptual intelligibility between relatively independent paradigms or traditions of inquiry with the same subject matter.

What I propose to do now is to articulate a more moderate position regarding the issue of conceptual meaning to which, I believe, many defenders of (versions of) the standard response tacitly subscribe and which, I believe, best captures the rationale for emphasizing issues of reference and de-emphasizing issues of conceptual meaning in the case of scientific terms.

The more moderate position I have in mind agrees with what I have called the standard response on two points. First, the establishment of the referential connection between scientific terms and their referents depends on all sorts of beliefs, methods and practices characteristic of (most) members of the scientific communities which use those terms. Second, although it is possible, in some or all cases, to rank such beliefs, methods and practices as more or less central or as more or less well established – as, for example, law-cluster conceptions of the meanings of scientific terms do – there appears to be no special way in which the most central or most highly confirmed beliefs play a special role in fixing the *referential* semantics of scientific terms of the sort which would distinguish their role as the role of *meanings*. What makes it a more moderate response is that it adds a hedge against the possibility that scientific terms really do have *conceptual* meanings: meanings which must be acknowledged in order to account for communication between scientists. Here are the components of the moderate position I have in mind:

1. Questions of conceptual meaning are questions about the prerequisites for mutually intelligible communication. If scientific terms have conceptual meanings, then the conceptual meaning of a term, t, within a paradigm or tradition of inquiry, P, involves a theoretical or methodological commitment, c, just in case a shared commitment to c is central to the possibility of intelligible communication between participants in P.

2. It is true that mutually intelligible professional communication within a paradigm (or between paradigms for that matter) depends on their being a sufficient level of similarity between the theoretical and methodological commitments of the relevant practitioners. Nevertheless, it is unlikely that scientific terms have conceptual meanings in the sense just indicated. Probably all that is required for two scientists working within a paradigm to communicate is that there be some substantial similarity in their theoretical and methodological commitments, without there being any particular commitments that have the special status required of components of conceptual meanings.

3. Nevertheless, if it should turn out otherwise – if scientific terms do have conceptual meanings – then those meanings can ordinarily be expected to be *methodologically benign* at least in the mature sciences. That is: at least in mature sciences the fact that scientific terms have conceptual meanings can be expected ordinarily to contribute to, rather than to detract from, the prospects for methodological commensurability between paradigms or traditions of inquiry. Here's why.

    a. In the first place, if scientific terms within paradigms or traditions of inquiry do have conceptual meanings, then successful communication between practitioners from different paradigms does not require that the conceptual meanings of the relevant terms be the same in the paradigms in question or even that they be mutually consistent. Conceptual meanings – whatever they may be – are not analytic definitions of scientific terms; scientific terms don't have analytic definitions. All that is required for communication between participants in different paradigms is that their adoptions of the conceptual meanings in their respective paradigms, together with whatever other conceptual resources they may possess, allow them to find each others theories and methods intelligible (even if profoundly mistaken). The dialectic of discussion and persuasion between paradigms can thus establish conceptual commensurability even when conceptual meanings differ.

b. Moreover, whatever conceptual meanings are they will surely reflect the best confirmed theories and best established methods in the relevant paradigms.

c. In mature sciences, at any rate, there is a general – although by no means absolute – tendency for the best confirmed theories to approximate the truth more closely over time and for methods to become more reliable (see, e.g., Boyd, 1989; 1999 for a discussion of the dialectical relation between these two tendencies).

d. Thus it is to be expected that whatever theoretical principles or inferential practices are part of the conceptual meaning of a scientific term in a mature paradigm will tend to be approximately true or approximately reliable.

e. So, it is generally to be expected that when participants in two mature paradigms with the same subject matter subscribe to the conceptual meanings of the terms in their respective paradigms they will each be subscribing to approximately correct conceptions *of the same phenomena.* This will enhance, rather than diminish, the likelihood that their conceptual resources will be similar enough to underwrite conceptual commensurability. Thus conceptual meanings, if there are any, will ordinarily contribute to conceptual commensurability.

f. Similar considerations suggest that conceptual meanings should generally contribute to methodological commensurability in ways that go beyond their contribution to conceptual commensurability. Methodological commensurability obtains just in case there are sufficiently many points of (approximate) agreement between paradigms, among which are sufficiently many approximately true principles or approximately reliable methods, that these principles and methods underwrite fair and reliable methods for adjudicating the differences between them. Since the components of conceptual meanings will generally be instances of such approximate insights, they can generally be expected to make a positive contribution to the prospects for methodological commensurability.

g. Thus, if it should unexpectedly turn out that an account of the semantics of scientific terms *does* require identifying certain beliefs, practices or methods associated with such terms as components of their conceptual meanings, it will ordinarily turn out that the fact that scientists accept such meanings will be *methodologically benign* with respect to issues of commensurability between alternative paradigms or traditions of inquiry. Ordinarily, the beliefs and methods which will be parts of the meanings of scientific terms within a paradigm will reflect real insights into its subject matter of a sort which will usually contribute to the establishment of conceptual and methodological commensurability with other approaches to the same subject matter.

So, differences in the meanings of terms will not ordinarily undermine the prospects for commensurability when two paradigms or traditions of inquiry share a common subject matter. In so far as semantic issues are at stake in matters of commensurability and incommensurability they are issues about reference rather than about meaning. Where referential commensurability obtains between paradigms in mature science we may *prima facie* expect both conceptual and methodological commensurability.

Since I propose to criticize this position I should indicate that I don't understand it to entail that there couldn't be referentially commensurable but conceptually incommensurable paradigms in mature science or that there could not be referentially and conceptually commensurable mature paradigms which resist the development of methods underwriting methodological commensurability. What I intend is for this defense of the *benign meaning thesis* to represent a highly plausible justification for the emphasis placed on referential commensurability when philosophers of science discuss the theses defended by Kuhn.

One other point is relevant here. It is widely agreed, I believe, that the sorts of epistemically relevant casual relations between the use of scientific terms and their referents which constitute the reference relation can obtain only when the relevant community has some significant approximately true beliefs about the referents in question. This is the grain of truth in the descriptivist conception of reference for scientific terms (see Boyd, 1993 for a discussion). I suspect that some philosophers are attracted by the idea that the conceptual meanings of scientific terms will be given by approximately true fundamental principles within the relevant paradigms because they are attracted by a position even closer to descriptivism according to which the principles which are part of the meaning of a scientific term will always be among the approximate truths which help to establish the reference constituting causal relations. I believe that this is false – and indeed the arguments of the present essay are designed to show that it is profoundly false – but note that I have not built this particular sympathy towards descriptivism into the characterization of the benign meaning thesis.

1.5. Meaning and Communication in Practice

Traditional empiricist semantic theory assigns to the analytic definition, or "nominal essence" of a general term two distinct roles: the fixing of reference and the establishment of the possibility of unambiguous communication between speakers. The acceptance of the analytic definition of a general term is what determines linguistic competence with that term, and the descriptions which constitute its analytic definition define its referent or extension.

The new naturalistic conception of semantics which underwrites the standard response undermines both components of this conception, and it extends the critique so that it applies not just to analytic conceptions of meaning but to any accounts which assign to certain widely held beliefs (or other cognitive attitudes) a special role in determining referential competence with, or fixing the reference of, a scientific term. On the one hand, the possibility of "borrowed reference" and of deference to the semantic competence of experts entail that a speaker can use a term coreferentially with the rest of her community even if her beliefs about its referent are highly atypical. No deference to *conceptual meanings* – analytic or not – is required. Likewise, although some of the beliefs about a term's referent widely shared within a community may play a crucial role in determining its reference, these need not be components of its meaning – analytic or otherwise.

I agree with the naturalistic conception just sketched but I deny that it should be seen as supporting a benign conception of the meanings of scientific terms. I agree that research communities (or other communities of inquiry) can share a common subject matter even when their beliefs and other cognitive attitudes are very different, and that the possession by two such communities of a common subject matter ordinarily contributes to the prospects for methodological commensurability. I believe, however, that it happens fairly often that there are features of what we might call the *conceptual meanings* of scientific (or other) terms which differ between research traditions in such a way as to give rise to a quite robust form of methodological incommensurability even when the traditions in question unproblematically share a common subject matter. The sort of incommensurability which results from differences in conceptual meanings is, I shall argue, importantly different from that which might arise just from profound differences in theoretical conceptions. If I am right, what is involved in cases of incommensurability grounded in differences in conceptual meaning is really a matter of limits of mutual intelligibility between traditions – of "talking past" one another, just as Kuhn suggests – even when there is an indisputable commonality of subject matter. I'll argue, moreover, that in many such cases the conceptual meanings in (at least one of) the traditions will prove not to be benign. They will involve beliefs and inferential or explanatory practices which are so far from being true or reliable that they detract from, rather than contribute to, the prospects for establishing commensurability.

Of course the phenomenon of talking past one another comes in degrees and grades. We are all familiar with situations in which two friends discussing a mutual acquaintance talk past one another until they realize that one of them knows and presupposes, whereas the other does not even know, that their acquaintance has recently married. Obviously cases like this do not raise questions about incommensurability.

Let me explain what I have in mind when I talk about conceptual meaning. Philosophers like Kuhn defend a law-cluster conception of the meanings of theoretical terms. As we have seen, they can be thought of as defending two related semantic claims about the role of fundamental laws in determining the semantics of such terms. On the one hand, they can be thought of as making a claim about how the reference of theoretical terms is fixed – as maintaining (mistakenly as it happens) that the referent of a theoretical term must be such that all or most of the central laws about it are (perhaps approximately) true. Alternatively, they can be thought of as making a claim about communication and mutual intelligibility. According to the second sort of claim, intelligible communication between researchers is impossible unless they accept the same law-clusters for the theoretical terms they employ.

I think that this latter claim is false, but I think that something like it is true, and the notion of conceptual meaning is supposed to capture the relevant grain of truth. In order to get a better understanding of what this grain of truth consists in, it is important to see some ways in which the law-cluster conception – understood as a theory of the prerequisites for communication – is *not* refuted. In the first place, it is not refuted by the phenomenon of borrowed reference. Suppose that a law-cluster theorist maintains that the law of conservation of mass-energy is part of the conceptual meaning of the notions of mass and of energy. She will be maintaining that acceptance of this laws is necessary for communication among physicists and others who deploy the terms "mass" and "energy" in doing science. She need not deny, for example, that a physics teacher can understand the error made by a student who misunderstands a thermodynamics lecture and writes on an examination that in closed systems mass is conserved but total energy decreases over time.

More importantly, the law-cluster conception of conceptual meaning is not undermined by the fact that some researchers – historians of science for example – can, by suitably immersing themselves in traditions with different law-clusters, come to understand, and compare, the doxastic and methodological commitments of the two traditions. Provided that this sort of specialist's understanding is not what underwrites the communicative capacities of the overwhelming number of participants in the traditions in question, the law-cluster conception of conceptual meaning would not be compromised.

Let's introduce some terminology here. Let's say that someone *engages with* some features or other of the conceptual or inferential resources of a tradition of inquiry just in case either (a) she completely accepts those conceptual resources and participates in the relevant inferences (let's call this sort of case *uncritical engagement*) or (b) she fully appreciates the relevant conceptual and inferential resources but engages in a serious theoretical or methodological critique of them, either rejecting some or all of them or adopting some intermediate skeptical or agnostic attitude towards them (call this *critical engagement*).

Using this terminology we may say that the law-cluster conception defended by Kuhn holds that the conceptual meanings of scientific terms are such that communication between scientists across research traditions ordinarily depends on their being *uncritically* engaged with the law-cluster associated with the relevant terms in the two traditions, *so that* – except for historically or philosophically sophisticated specialists who are critically engaged with their respective traditions – communication is impossible unless scientists accept the same law-clusters.

Of course this is mistaken, as the case of the transition between Newtonian mechanics and special relativity indicates, and we may use the terminology just introduced to identify the crucial error. Under the circumstances involved in this transition – in which the predecessor paradigm and the successor paradigm are both approximations to the truth and (more importantly) the claims made by the theories central to the two paradigms are *in ways obvious to practitioners in either paradigm* approximations to each other – critical engagement does not require specialized skills or knowledge (historical or philosophical skills for example) beyond those which are available to competent investigators in either paradigm. Thus, although critical engagement on the part of participants in the two paradigms *is* required to establish conceptual commensurability, this is no barrier to communication or to the establishment of methodological commensurability.

Thus Kuhn, for example, greatly exaggerates the conceptual barriers to communication between participants in the scientific "revolutions" he discusses. These "revolutions" were, after all, initiated by participants in the "paradigms" whose replacement they urged – participants who deployed the conceptual resources they shared with other participants in the earlier paradigm to argue for the new one. What I want to argue here is that, in other respects, Kuhn (at least in the key semantic argument on pp. 101–102 of Kuhn, 1970) greatly *underestimated* the range of conceptual resources which are properly counted as components of the conceptual meanings of scientific terms.

1.6. Engagement and Conceptual Meanings

I suggest that the notion of conceptual meaning can be usefully explicated in terms of the notion of engagement we have just deployed. Let's say that a law, doctrine, or inferential practice is part of the conceptual meaning of a term within a paradigm or tradition of inquiry just in case engagement with it (either critical or uncritical) is central to an investigator's capacity to intelligibly appreciate and discuss the findings and methods of the paradigm or tradition in question.

Of course, centrality is a matter of degree, so it would be open to someone to raise the (essentially Quinean) challenge that there is so smooth a "continuum" between the most and the least central doctrines and practices that the notion of conceptual meaning just defined fails to correspond to anything fundamental to the theory of scientific communication. This is an empirical issue, of course, so I'll provide examples of doctrines and practices which, I think the evidence will show, are presupposed in the literature in particular research traditions, and in the dialectic of argumentation and persuasion within them – and between those traditions and rival ones – *and* which are such that only someone who either assimilated those doctrines and methods as a member of the tradition in question or critically engaged with them from an alternative research perspective could make any scientific or methodological sense of the work and findings of the tradition in question. That there may be a continuum along conceptually or methodologically relevant dimensions connecting such cases to principles or practices not thus central to communication and understanding does not mitigate the conceptual centrality of the former.

This conception of conceptual meaning is, it should be noted, broader than that to which Kuhn appeals in arguing for methodological incommensurability. That conception – borrowed as it is from Carnap and other late logical positivists – makes the doctrines which are part of the meaning of scientific terms *analytic truths* with the consequence that obviously non-analytic doctrines (or methodological practices which are obviously not analytically justifiable) are not candidates. On the conception I am proposing, features of "paradigms" as Kuhn understands them – inferential practices, for example, or standards for experimental design, or non-"fundamental" but widely applied truisms – are potential candidates for conceptual meaninghood even when it is obvious that they are not analytic or analytically justified.

One special case of such features will illustrate the sort of conceptual meanings I have in mind and will also aid in our understanding of malignant conceptual meanings when we come to discuss them. Kuhn (1970) remarks that scientists working within a paradigm have "quasi-metaphysical" knowledge which allows them to know in advance, in typical cases, what the form of the solution to a scientific question will be. The phenomenon he has in mind is a matter of *projectibility* judgments in the sense of Goodman (1973). These are paradigm (or tradition, or whatever) dependent judgments of theoretical plausibility and they set the research agendas for paradigms. To a good first approximation, methodology within a paradigm involves (theory dependent) testing of *projectible* answers to scientific questions, subject to the important constraint that the test procedures control for those possibilities for experimental artifacts (or artifacts of observational technology) which are themselves suggested by *projectible* conceptions of the relevant experimental or observational situation (for a systematic treatment see Boyd, 1985a; 1985b; 1990a).

Now, as Kuhn recognizes, coming to be able to make appropriate projectibility judgments is not entirely a matter of having explicit knowledge of the relevant paradigmatic theories. A large part of this sort of methodological sophistication involves tacit knowledge of inferential standards and practices which is acquired in the process of graduate education and through other methods of professional indoctrination.

What is important for our purposes is that engagement with at least the most important (tacit as well as explicit) standards of projectibility within a paradigm is a prerequisite to understanding its published literature or to engaging in scientifically or methodologically fruitful conversations with its practitioners about their work.

To see this, just try reading professional journals in some discipline with whose basic terminology you are familiar but in which you have not been (officially or unofficially) trained, and then share your impressions of the papers you have read with a sophisticated practitioner. You will find – and I am here reporting the experience of everyone in the philosophy, history or social studies of science – that you have missed, or misjudged, most of the important substantive or methodological issues.

The explanation, of course, will be that the authors of the papers in question will have designed their studies and written up the results within the framework of projectibility judgments dictated by the current state of the art in their paradigm or tradition. Their papers will provide evidence which bears on the choice between (what they and their readers will take to be) projectible answers to the relevant questions and they will have incorporated controls for artifacts in ways similarly dependent on projectibility judgments. They will presume that their readers are equally immersed in that paradigm and share basically the same methodological standards, including standards of projectibility.

Partly (but only partly) for that reason, they will not make explicit many of the central methodological judgements (of projectibility, in particular) which inform their studies. The other reason, of course, is that they could not make explicit most of these judgments if they wanted to. I do not mean merely that they could not classify these judgments as, for example, *projectibility* judgments, because they aren't familiar with Goodman's notion of projectibility. I mean instead that (as Kuhn notices – this is part of the importance of *exemplars* in determining the direction of research within a paradigm) most of those judgments are, as a matter of fact, only *implicitly* represented in the theories and practices of the paradigm or tradition. Moreover this is not a state of affairs which practitioners could fully remedy if they wanted to. As historians, sociologists and philosophers of science have discovered, the task of (even partly) explicating the methodological presuppositions of a paradigm or research tradition requires special skills which are not ordinarily part of the intellectual equipment of its practitioners.

Thus when you read the literature in the paradigm in question, and when you discuss scientific and methodological issues with its practitioners you will *always* fail to appreciate many of the fundamental substantive and methodological issues which are at stake unless and until you have (perhaps critically) engaged with those of its doctrines and practices which underwrite projectibility judgments. Thus the most central of these doctrines and practices must be counted as components of the conceptual meaning of the relevant scientific terms within the paradigm or tradition in question.

Similar considerations indicate that certain standard features of experimental design or of instrumentation – ones an understanding of which is presupposed in the relevant literature – will count as components of the conceptual meanings of those terms whose applications, in routine scientific practice, they help to determine. Without (perhaps critical) engagement with these features of paradigmatic practice you would be unable to understand papers in the literature or appreciate the dimensions of methodological or substantive argumentation within the paradigm.

Note that the account of conceptual meaning offered here accommodates exactly Kuhn's understanding of the importance, for a scientist's understanding of scientific concepts, of her immersion in her paradigm's professional practice and of her (partly tacit) appreciation of its exemplary research techniques and findings. The present account differs from Kuhn's more Carnapian conception of meaning for scientific terms, as it is deployed in his argument for referential incommensurability, in that it does not require of components of the conceptual meaning of a scientific term that they be either analytic (in the case of doctrines) or analytically justifiable (in the case of methods), nor does it require of each such component that it play a crucial role in determining the referent of the term. Only their role in underwriting intelligible communication among specialists distinguishes components of conceptual meaning in the sense proposed here from other features of the use of a term.

For this reason, the present conception can recognize as components of a term's meanings features of its deployment within a paradigm which are central to the paradigm's (communicative and experimental) practices even when they are not candidates for components of its meaning in Kuhn's more Carnapian sense. The conception offered here is, therefore, more consonant with what he had to say about conceptual understanding within a paradigm than is the Carnapian conception upon which his own arguments for referential incommensurability rest.

1.7. Conceptual Meanings and Conceptual Commensurability: Conceptual Commensurability as a Cognitive Achievement

Let's consider the question of conceptual commensurability between paradigms and research traditions in the light of the proposed account of conceptual meanings for scientific terms. One important historiographic point made by Kuhn (1970) is that coming to understand the doctrines and methods of a paradigm different from the one in which one has been trained is a significant cognitive achievement, often a very difficult one. It is this achievement which, among other things, permits us to avoid the sort of whiggish history of science which sees earlier scientific theories and practices as irrational in so far as they differ from our own.

The Carnapian conception of meaning upon which Kuhn explicitly relies permits us to see some of the ways in which this is true. In so far as fundamental laws are part of the meanings of scientific terms, then we can see why appreciating the rationality and achievements of paradigms other than one's own requires a special sort of engagement with unfamiliar conceptual material and the deployment of skills which are not ordinarily required of participants in any particular scientific paradigm (and which may be, in some sense, the special province of historians and philosophers of science).

Nevertheless, because Kuhn's Carnapian conception must do "double duty" as an account of analytic reference fixing descriptions, it cannot be deployed to explain another *quite Kuhnian* phenomenon. The achievement of conceptual commensurability between different research traditions is a significant achievement – requiring the deployment of skills and resources not ordinarily required for practice within a research tradition – even in cases in which the most fundamental laws accepted within the traditions are the same, and in which there is no serious prospect of a persuasive argument for either referential *or* methodological incommensurability. It requires, as we have seen engagement (critical or otherwise) with a complex body of partly tacit doctrines and inferential practices.

Of course the notion of conceptual meaning proposed here is designed precisely to explain this phenomenon. Methodologically or substantively significant conversation between research traditions always does require the *achievement* of conceptual commensurability through mutual engagement. When research traditions disagree, or even appear to disagree, in their doctrines or methods methodological commensurability cannot be achieved without the establishment of conceptual commensurability. There is never – or almost never – in such cases *immediate* methodological commensurability, and the reason that this is so *is* that scientists in different traditions use scientific terms with different meanings. Kuhn was right about that and, once this insight is distinguished – as it is here – from an empiricist conception of the relation between conceptual meanings and reference fixing, we can see that Kuhn's insight extends to a far greater range of cases than those which fit his model of "scientific revolutions."

It remains to examine the relation between conceptual meanings, properly understood, and methodological commensurability. Given that differences in the conceptual meanings of scientific terms within a paradigm or tradition will ordinarily be an *initial* barrier to the achievement of conceptual commensurability (and, thus to methodological commensurability) the question remains whether or not the acceptance of those meanings for terms by participants within a research tradition can ordinarily be expected to contribute to their ability to achieve conceptual commensurability with another tradition sharing a common subject matter. According to the *benign meaning thesis* the answer is "yes": ordinarily the components of meaning for a scientific term within a mature paradigm or research tradition will reflect substantial insights into its subject matter and these insights will facilitate a participant's understanding of the (approximately equally) genuine insights reflected in the conceptual meanings of the same terms in another mature tradition.

In the simplest case, on the benign view, participants in two such traditions would be able to rely on the insights of their respective traditions to come to a (perhaps critical) engagement with each other's traditions through exploration of each other's research literature and through serious substantive and methodological discussions. In more complex cases, the differences between the traditions might be such that even the establishment of conceptual commensurability might require experimental observational studies of the relevant phenomena conducted in the light of the special need to clarify points of tension between the traditions. But, according to the benign meaning thesis, even in cases of the latter sort the components of term meaning within either of the traditions can ordinarily be expected to reflect genuine (perhaps partial and approximate) knowledge of the common subject matter, and thus to contribute to the success of the relevant empirical studies.

This thesis is plausible enough if we focus our attention on the examples of "scientific revolutions" discussed by Kuhn. It is also rendered plausible by the reasonable idea that the doctrines and methods which are best established as parts of the conceptual machinery of either one of the two traditions are likely to occupy that position because they are among its best confirmed findings, and are thus likely to reflect real (approximate) knowledge of their common subject matter. It may also, as I suggest earlier, gain some plausibility from the conception that empiricists descriptivist conceptions of reference are right in one respect: that the conceptual meanings of scientific terms ordinarily contribute to the establishment of the reference relation for those terms by reflecting approximate knowledge of the phenomena to which they refer.

It is this benign conception of conceptual meanings that I propose to dispute. I think that there are scientifically (and socially) important cases in which, far from representing approximate knowledge or reliable methods, the conceptual meanings of terms within scientific research traditions are not only profoundly mistaken but, if made explicit, profoundly incoherent. For practitioners in such traditions, their acceptance of such meanings profoundly detracts from – rather than contributes to – the possibility of conceptual commensurability with other traditions sharing the same subject matter.

The existence of such cases, I shall argue, indicates that Kuhn was right to think that divergence in meaning can significantly compromise the prospects for methodological commensurability, even though he was mistaken in thinking that they ordinarily imply divergence in reference. The almost exclusive focus in the post-Kuhnian literature on issues of referential commensurability has thus led to an under appreciation of an important dimension of methodological commensurability.

Equally important are the implications of such cases for theories of reference. Conceptual meanings of scientific terms are, of course, important to the establishment of reference: they are central to the possibility of communication within relevant research traditions (this is how *conceptual meaning* is defined) and, since reference is a matter of *socially coordinated epistemic access* (Boyd, 1993;1999), the establishment of social communication is central to the phenomenon of reference. It is also true that the epistemic access characteristic of reference is ordinarily achieved only when research communities have lots of approximately true beliefs and lots of approximately reliable methods. *But*, the contribution of conceptual meanings to reference need not be the one suggested by venerable descriptivist conceptions of reference: the conceptual meanings of scientific terms within a tradition need not provide an even approximately accurate conception of their reference nor even approximately reliable methods for studying them. Conceptual meanings can be – and in important cases are – utterly misleading and fundamentally incoherent. We now turn our attention to some examples of this phenomenon.

2. EVOLVED BEHAVIORS AND HUMAN NATURE: AN EXAMPLE OF INCOMMENSURABILITY IN THE MAKING

2.0. Cultural Studies (and Cognitive Science) vs. Sociobiology

A wide variety of disciplines in the humanities and social sciences – especially history, anthropology, social psychology, political science, economics and philosophy – have been concerned to provide explanations for large scale features of human social behavior: family structures, social arrangements, cooperation and competition, religion, warfare, divisions of labor, and the like. With the publication of Wilson (1975) there came to be recognized an alternative research tradition, initially called "sociobiology" and then later renamed (to avoid the embarrassment occasioned by obvious political excesses in its earlier stages) "evolutionary psychology." The aim of this new approach is to substantially reduce questions in human social psychology to questions in evolutionary biology and, thereby, to obtain significant insights into the patterns of human social behavior previously in the domain of the humanistic and social scientific disciplines just mentioned. The basic strategy is to deploy what are called "optimality models" in evolutionary biology to make estimates about probable patterns of behavior in early human societies and to infer, from these patterns, generalizations about human motivational structure and social psychology. It is within this research tradition that I propose to identify a family of terms with conceptual meanings which are malignant rather than benign. In particular I shall argue that there are inferential practices at work in contemporary evolutionary biology which:

a. Are profoundly unreliable,

b. reflect – if made explicit – an essentially incoherent conception of the evolution of human psychology, one inconsistent in many points with explicit (and well established) doctrines established within the same research tradition,

c. are parts of the conceptual meanings of the key terms, and

d. render almost unintelligible – for those uncritically engaged with them – deep and correct methodological criticisms of evolutionary psychology arising from the other disciplines mentioned, so that

e. the prospects for methodological commensurability are greatly reduced.

[Wilson (1975) and the more popular Wilson (1978), together with Barash (1979) and Alexander (1979) provide an introduction to the early work in this literature, when the term "sociobiology" was standard. Betzig (1997) provides a sample of the very large contemporary literature. Pinker (1996) provides an example of a recent popularization. Sherman and Reeve (1997) provides a methodological discussion of extraordinary sophistication. The key Journals include *Ethology and Sociobiology, Behavioral Ecology and Sociobiology, Journal of Social and Biological Structures, Human Nature, Trends in Ecology and Evolution.*]

2.1. Anecdotes

My confidence in the position I am going to defend here about conceptual meanings in evolutionary psychology has a history which provides some of the evidence I'll offer for it. For the last decade and a half I have been teaching an undergraduate course, "Science and Human Nature," which examines substantive and methodological controversies about human sociobiology. With the help of colleagues who have taught with me (Professors Nicholas Sturgeon and Karen Jones) I came to see that – as I'll indicate below – certain central inferential practices in human sociobiology are obviously in conflict with extremely secure findings in post-behaviorist psychology, findings which, in other contexts, human sociobiologists all acknowledge.

When I and my colleagues tried to explain this conflict to our students, and to criticize the inferential practices in question, what we found was that – especially in the cases of students who had previously studied evolutionary psychology, but even for those who had only taken introductory courses in evolutionary theory and animal behavior – it was extremely difficult for students to remember the criticisms from one lecture to the next or to apply them to actual cases of sociobiological inferences.

I do not mean merely that not all of our students were convinced by the criticisms we offered. That was to be expected. What I mean is that many of them reported the experience of finding the criticisms plausible sounding at first, but then finding themselves unable even to paraphrase them. When we would assign exercises in which the criticisms were made explicit (in short, logically uncomplicated, sentences) and students were asked to say how someone who accepted these criticisms would respond to very simple sample cases of the relevant sociobiological inferences, they often were unable to do so, and they themselves found this puzzling.

At first, we attributed this phenomenon to the methodological immaturity of undergraduate students, but our experiences, and those of other philosophers who presented the same criticisms to sophisticated professional human sociobiologists, persuaded us that something deeper was going on which rendered the criticisms extremely difficult for those initiated into standard patterns of evolutionary thinking about human psychology to understand, even though (as you shall see) the logical structure of the criticisms is hardly daunting.

The account of conceptual meanings presented here is my attempt to explain the phenomenon just described and to set it in an epistemologically and semantically informative context.

2.2. The Standard Pattern of Extrapolative Sociobiological Inference, I: Inferring from Evolutionary Scenarios to Motives

Terminology first: I distinguish between two sorts of projects in human sociobiology. The first project – let's call it the *explanatory* project – seeks evolutionary explanations for *independently identified* features of human psychology. It is a special case of a standard pattern of explanation in evolutionary theory in which evolutionary explanations are offered for independently identified phenotypic traits of organisms. There are controversies about the methods proper to such explanations (see, e.g., Gould and Lewontin, 1979 for a famous criticism of some standard methods), but I will not be concerned with this project or those criticisms here. What I am concerned to discuss is the project of *extrapolative* human sociobiology (or extrapolative "evolutionary psychology"): the project of using the resources of evolutionary theory to determine which theories of human psychology are likely to be true, rather than to explain independently confirmed psychological theories.

There is an inferential pattern which is more or less characteristic of the extrapolative project just discussed. An evolutionary scenario regarding the evolution of certain behaviors, B, in early humans is presented. Usually, but not always, this scenario is said to have been retrodicted from evolutionary theory according to the dictates of an "optimality model."[6] An underlying motivation, M, for B is then specified (with greater or less precision depending on the extrapolative argument in question) and evolutionary theory is said to "predict" or to "suggest" or otherwise to confer a privileged epistemic status on an hypothesis, H_M to the effect that M is an innate and relatively non-malleable feature of human psychology available for explaining human behavior under conditions different from those in which prevailed in the environment in which human psychology evolved[7] (the *environment of evolutionary adaptation*; henceforth: *EEA*).

There is an important assumption which underwrites this inference pattern and which I will not challenge here. According to the "no one here but us hunter gatherers" thesis, although human populations have been subject to the operation of selective forces at all times from the development of the first complex agricultural societies to the present, since humans stopped living in small hunter gatherer groups there has been no uniformity to the direction of selection with respect to psychological characteristics. Because the environments in which humans have lived since the hunter gatherer experience have been so varied, and have changed so rapidly, psychological traits which were favorable for reproductive fitness under one social arrangement were likely to be unfavorable under subsequent ones and *vice versa*. The effect of this rapid variation in selection pressures is that the basic developmental psychology of humans has not changed significantly since we were hunter gatherers. The EEA for *contemporary* human developmental psychology was the condition of life for human hunter gatherers.

It follows that contemporary humans have the same developmental psychology as did hunter gatherers, so that if evolutionary theory can provide insights into the developmental psychology of hunter gatherers it will thereby provide insight into developmental psychology generally (see Cosmides and Tooby, 1987 for an excellent discussion).

What will be important for our purposes will be the inferential patterns by which the motivation M is identified and the inferential patterns which support the hypothesis H_M that M is an innate and relatively non-malleable feature of human psychology. What I claim is that the identification of M is routinely achieved by a profoundly illegitimate inference in which (something very much like) the *evolutionary role* of a feature of human behavior is taken as the *propositional content* of a (usually unconscious) motive underlying it.

Example: It is often suggested that altruism arose in early humans through *kin selection*: that although acts of altruism reduced the individual fitness of altruists, they increased the fitness of their kin sufficiently that genes which were expressed in altruistic behavior were favored. One way to describe this proposed evolutionary scenario is to say that humans "evolved a tendency to be altruistic towards their kin." Understood as a description of the evolutionary function of altruism among early humans, this is an accurate report of the scenario in question. It is, however, routine to fallaciously infer from scenario descriptions of this sort, that the (perhaps unconscious) motive which underwrites (most?) displays of human altruism is a concern for the altruist's kin (or members of his or her "in group"). For a presentation of this scenario, together with a different but equally fallacious inference to a motivational structure, see Barash, 1979, pp. 132–169.

Example: In a widely cited paper, clearly from the era of "evolutionary psychology" rather than "sociobiology," Cosmides and Tooby (1987) come as close as anyone to making explicit the inferential pattern I am discussing here. They hold that "*The evolutionary function of the human brain is to process information in ways that lead to adaptive behavior*; the mind is a description of the operation of a brain that maps informational input onto behavioral output" (Cosmides and Tooby, 1987, p. 282, emphasis theirs).

They elaborate this doctrine as follows.

> *When applied to behavior, natural selection theory is more closely allied with the cognitive level of explanation than any other level of proximal causation. This is because the cognitive level seeks to specify a psychological mechanism's function, and natural selection theory is a theory of function.* Natural selection theory specifies how an organism should respond to different kinds of information from its environment. It defines adaptive information processing problems that the organism must have some means of solving. Cognitive programs are solutions to information processing problems. (Cosmides and Tooby, 1987, p. 285, emphasis theirs)

In explaining the application of this principle, they examine the implications of the kin selection scenario about altruism we have just discussed. They conclude as follows.

> ... an organism's behavior cannot fall within the bounds of the constraints imposed by the evolutionary process unless it is guided by cognitive processes that can solve certain information processing problems that are very specific. To confer benefits on kin in accordance with the constraints of kin selection theory, the organism must have cognitive programs that allow it to extract certain specific information from its environment: who are its relatives? which kin are close and which distant? what are the costs and benefits of an action to itself? to its kin?. The organism's behavior will be random with respect to the constraints of kin selection theory unless (1) it has some means of extracting information relevant to these questions from its environment, and (2) it has well-defined decision rules that use this information in ways that instantiate the theory's constraints. A cognitive system can generate adaptive behavior only if it can perform specific information processing tasks such as these. (Cosmides and Tooby, 1987, p. 288)

Note that Cosmides and Tooby explicitly link the evolutionary function of adaptive behaviors to the *computational function* (and thus to the propositional content) of the underlying psychological states. What is important is that Cosmides and Tooby propose that one can infer the computational structure of the underlying psychology from the evolutionary function or role of the behavior, rather than merely that one can infer that the underlying computational structure must have been one capable of underwriting the predicted behaviors *in the EEA*. See section 2.4.0. for a discussion of the fallacy in inferences of the first sort.

The credibility of routine fallacious inferences of this form is facilitated, I claim, by the situation that many of the linguistic forms which are deployed in describing evolutionary scenarios or evolutionary hypotheses are either metaphorical uses of forms ordinarily employed to report propositional attitudes like desire, preference or purpose, or are ambiguous between propositional attitude expressing uses and their use in the making of evolutionary claims. The inferences I have in mind obtain their plausibility in part from the fact that they trade on the ambiguities involved.

Example: Sometimes the evolutionary scenario just mentioned is expressed by statements like "Humans are altruistic *in order to benefit kin*." It is characteristic – indeed, I suggest, part of the meaning of such expressions as "in order to" in this literature – for authors of such claims to infer that the (unconscious) motive underwriting human altruism is a concern for kin. Wilson (1978) suggests that this inference would be appropriate, although in that book he dissents from the evolutionary scenario in question.

Another Example: See the discussion of Daly and Wilson (1997) in the next section.

Similarly, some psychological/social terms ("altruism," "nepotism," and "strategy," as in "kin recognition strategy," for example) have been introduced *metaphorically* as technical terms in the study of the evolution of behavior. They retain, I shall argue, *as parts of their meanings in the literature in evolutionary psychology*, fallacious inference patterns appropriate to their literal, as opposed to their metaphorical, uses.

Example: Phenotypic traits are said to be *altruistic* just in case they reduce the individual fitness of organisms that exhibit them while enhancing the fitness of their conspecifics. Some evolutionary psychologists reject the scenario for the evolution of altruism just mentioned in favor of a scenario according to which other-regarding or moral motivations were established by *individual* selection because they enhanced the effectiveness of those who had them in so far as participation in cooperative reciprocal arrangements was concerned (this is the "evolution of reciprocal altruism" *sensu* Trivers, 1971). If some such scenario is correct, then altruism in humans was not *altruistic* in the metaphorical technical sense just discussed. It is routine (and, I suggest, part of the meaning of the term "altruism" in the literature) for defenders of this sort of evolutionary scenario to fallaciously conclude from the scenario in question that the *motivation* for apparently altruistic actions in humans is largely selfish.

In Wilson (1978), Wilson subscribes to this conception of the evolution of apparently altruistic motives. He concludes,on the basis of this scenario, that there probably is not much "hard core altruism." It is clear from the context that he intends this as a finding about *human psychology*. Nowhere in the book does he inform the reader that there is a technical (albeit metaphorical) use of the term "altruism" in evolutionary biology and that his arguments rely on using the term in this sense. Instead, he routinely grounds inferences about the presence or absence of genuine ("hard core," non-selfishly motivated) altruism in human psychology in considerations about the extent to which traits are altruistic in the metaphorical technical sense, without comment. This is precisely what one would expect if one thought that a conflation of the two meanings of "altruism" was part of the conceptual meaning of that term in the sociobiological literature.

Recall that the inferences we are discussing proceed from an evolutionary scenario involving a behavior B, to an alleged motivation M for B, and then to the hypothesis H_M that M is an innate and relatively non-malleable feature of human psychology. I have indicated some of the ways in which M is inferred illegitimately from the evolutionary function posited in the evolutionary scenario for B. It remains to explain how inferences to innateness and non-malleability are facilitated.

2.3. The Standard Pattern of Extrapolative Sociobiological Inference, II: Inferring the Innateness and Non-Malleability of Motives

Here the standard pattern of inference itself is less complex, but the roots of its credibility may be somewhat more complex. The pattern is simply that, once the motive M for B has been arrived at, it is inferred from the evolutionary scenario that M is innate and not very malleable.

Example: In one famous and famously controversial example, the preference for kin or members of one's in-group which is posited as a result of a (fallacious) inference from the kin-selection scenario for human altruism is taken to be a manifestation of an evolved innate and relatively non-malleable xenophobia that, in turn, is offered as an explanation of the persistence of contemporary racism even in the face of reforms aimed at its elimination. This argument appears in Barash (1979) and is hinted at in Wilson (1975). Wilson (1978) rejects its conclusion because he adopts the individual selection scenario for apparent altruism discussed above – and concludes that we are motivationally largely selfish, but happily not innately xenophobic *as we would be if real ("hard core") altruism had been established by kin selection.*

Example: Here is an example from the more recent literature, after "evolutionary psychologists" became less bold in their pronouncements about politically controversial matters than they were during the "sociobiology" period. In a widely cited study, Buss (1989) explores the predictions regarding human mate preference which, he maintains, can be made from evolutionary theory. He considers two different scenarios for the evolution of mate choice in human males, and draws different predictions from them about male mate choice *in general* (i.e., not just in the EEA). According to one scenario, early human males sought short-term mating partners. On this assumption, he expects that natural selection would have favored "a preference for females in their early 20s who show cues that are positively correlated with fertility" (Buss, 1989, p. 177).

According to the other scenario, early human males sought long-term mating partners. On that assumption, Buss takes evolutionary theory to predict a preference for "females in their mid-teens who show cues indicative of high reproductive value" (Buss, 1989, p. 177).

It is plain from the text that Buss, without further argument, takes these predictions to be applicable to human males under conditions other that those of selection. For example, immediately after introducing the two hypotheses, Buss considers the views of other theorists. He describes the position of Williams (1975), without comment, as follows:

> Williams (1975), in contrast, predicts a compromise preference between reproductive value and fertility due to the existence of both long-term mating bonds *and* some possibility of *divorce* and extrapair matings. [First emphasis the author's; second emphasis mine.] (Buss, 1989, p. 177)

Since *divorce* is, presumably, a particular social institution not always present in hunter-gatherer groups, it seems obvious that Buss understands the various evolutionary predictions he is considering as applying to conditions other than those which prevailed in the EEA.

That he understands the predictions of evolutionary theory in just that way is also made clear by his own experimental design. The data he brings to bear on the evolutionary predictions in question consist of the *contemporary* subjects' responses to questions about the parameters they consider important in mate choice. The data are cross cultural, and are obtained from subjects from 37 different *contemporary* cultures.

It is thus clear that Buss understands the various evolutionary scenarios to imply predictions about human motivational structures under a variety of conditions; his cross-cultural tests of the predictions are precisely designed to test for the presence of innate and largely non-malleable (rather than culturally specific) motivational structures. Thus his study reflects a paradigm case of the sort of inference from evolutionary scenarios to innate non-malleable psychological structures we are discussing.

It is worth noting that Buss nowhere defends (or even considers) this particular feature of his inferential practices. This is exactly what one would expect on the assumption defended here that inferences of the sort we are discussing are parts of the conceptual meaning of the relevant terms and can thus be presupposed, by authors and readers in the literature in question.

The plausibility of such inferences appears to have several sources. At least three different sources of plausibility seem to be involved, which deserve separate discussions.

2.3.0. Evolution, "Biological Bases," and Instinct: The "Two Natures" Inference

One source is certainly the (culturally widespread) tendency to contrast "nature" with "nurture" and to identify the former with the "biological," and the latter with rationality and culture. According to this conception, behavioral flexibility is by and large to be identified with the effects of human rationality or human culture, whereas the biological aspects of human nature are identified with a motivational psychology of (relatively non-malleable)impulses or instincts. Roughly, this is the classical conception according to which human nature is composite from our "animal nature" and our "rational nature," with the former nature identified with the biological aspects of human psychology.

This sort of rationalization for inferences from evolutionary scenarios to non-malleable innate motivational structures is, I believe, at work whenever a biological story – evolutionary or not – about some feature of human behavior is summarized by saying that the behavior in question has a "biological basis." The inference that the behavior is rooted in an innate non-malleable motivational structure is essentially automatic – indeed such an inferential tendency is, if I am right, part of the meaning of the expression "biological basis" in the relevant research traditions.

Example: The best illustration of this rationale at work may well be from the work of LeVay (1991; 1993), even though LeVay is not an evolutionary psychologist. LeVay

(1991) reports differences in size of a particular brain structure (one of the cell groups in the interstitial nuclei of the anterior hypothalamus) in heterosexual and homosexual men and then adduces the biological basis thus uncovered for sexual orientation as providing *prima facie* evidence that sexual orientation in human males is largely unlearned and non-malleable. Wilson's (1978) claim that "biology" has "culture" on a short leash reflects the same dichotomy between the biologically based components of human psychology and behavior and the rationally or culturally determined ones.

2.3.1. Free Will, Biological Determinism, and Psychological Compulsion

There is another related source of the plausibility of the inferences we are discussing which is harder to establish by reference to the literature, but which seems to me evident in conversations with students and others. I suggest that many times the inference from an evolutionary scenario regarding a behavior, B, to a conclusion about its non-malleable innate motivation is facilitated by a tacit pattern of fallacious reasoning along the following lines. B is an instance of *evolved* behavior; thus B is *biologically determined* and hence B is not *freely chosen*. Instead, B is the product of a *compulsion* or other difficult to resist urge, of which its author is perhaps unaware. It is this pattern of inference which sometimes emerges explicitly when someone says, for example, that people cannot be blamed (or blamed as much as one might think) for being territorial, or xenophobic or selfish, or whatever, because they *evolved* so as to have the trait in question.

2.3.2. "Evolved" (and thus) "Natural" "Inclinations" or "Tendencies"

Another source of the plausibility of the inferences we are considering lies in systematic ambiguities in the uses of expressions like "evolved inclination," "natural tendency," and the like. In one use of this expression, a *natural tendency* or (somewhat less clearly) *natural inclination* is an *evolved* (and in that sense *natural*) dispositional property of an organism. Example: In some areas, where there is snow in winter but not during the rest of the year, some mammals have a *natural tendency* to change their coat colors with the seasons so as to be protectively colored.

In another perfectly standard use of the expression, a *natural tendency* (or *inclination*) is a feature of motivational psychology which is innate (or is likely to be learned under almost any conditions) and hard to extinguish. Example: People (perhaps) have a *natural tendency* to become angry when threatened.

When, according to an evolutionary scenario, a motivational trait *evolved* (and is thus *natural* in the first sense) it is often tempting (natural?) – indeed, I claim, it is part of the meaning of terms like "evolved" and "natural" in the relevant literature – to infer, fallaciously, that the trait is innate and non-malleable (*natural* in the second sense).

Example: Here there is a famous example in the literature. In a widely cited paper, Daly and Wilson (1997) deploy the resources of evolutionary psychology to provide an explanation for an interesting fact about child neglect and child abuse *in contemporary industrial societies*. When couples marry, (or otherwise form a family) into which they bring their biological children from previous marriages or relationships, each partner is less likely to abuse or neglect his or her biological children than to abuse or neglect the children of his/her partner from the previous relationship. Daly and Wilson offer as an explanation our "usual inclination" towards such asymmetries which they say is predicted by optimality models in evolutionary theory. They do not spell out the evolutionary scenario in any greater detail, but it is evident that what optimality modeling might be taken to predict is a tendency or inclination (in the sense of a dispositional property) to favor one's own biological offspring *under conditions of the sort prevailing in the EEA*. What the explanation they offer for the relevant *contemporary* behaviors requires, on the other hand, is an innate and (relatively) non-malleable *motivation* to prefer one's own offspring – one which is exhibited under conditions quite different from those of the EEA. The fallacious inference from the posited evolved inclination to the relevant "usual" (apparently: innate and relatively non-malleable) motivational state is required for their paper to be intelligible. I emphasize that they do not make this inference explicit. They take it for granted, just as one would expect if such inferences were parts of the meanings of terms like "evolved" and "tendency" in the relevant discourse.

2.3.3. Behaviorism, Behavioral Ecology and the "Scientific" Study of Behavior

It's worth remarking that this last source of plausibility for the fallacious inferences we are discussing – and perhaps the others as well – is enhanced in *its* credibility by the residual impact of behaviorism on research in extrapolative evolutionary psychology, and by the development of *behavioral ecology* as a research framework for the study of non-human animals.

Behaviorism – understood as a methodological approach to the study of behavior in which reference to mental or psychological processes is (allegedly) foregone in favor of purely behavioral descriptions – has had a much more persistent influence in studies of animal behavior than it has in the study of human psychology. In part this has been so because the adoption of behaviorist methods and behaviorist rhetoric has been central to the devices by which students of the behavior of non-human animals have sought to insulate their methodology from the tendency to anthropomorphize non-human animals. It has come to be seen as a mark of the *scientific* study of animal behavior that the researchers adopt a behaviorist perspective.

Of course, as philosophers and students of human cognitive psychology have recognized for several decades, it is not possible to make generalizations about the behavior of animals, human or non-human, without employing, at least tacitly, some conception of the source of those behaviors. One of the points which early critics of behaviorism as a methodology in human psychology emphasized was that when "behaviorist" psychologists deployed taxonomies of human behavior in proposing general theories, they were usually (and usually pretty transparently) importing into their methodology tacit assumptions about mental states and processes. Thus, in practice "behaviorist" human psychology did not represent the abandonment of theorizing about the mental. Instead, behaviorist psychologists, *believing falsely about their own methods that they were without presuppositions about the mental*, left largely unexamined (and uncriticized) the conceptions about mental states and processes which they tacitly adopted.

I suggest that much the same thing currently happens in extrapolative evolutionary psychology. Many of its practitioners are primarily students of (non-human) "animal behavior," or were trained in the behaviorist tradition associated with "animal behavior" studies. Although, as the examples cited above indicate, their methodology certainly involves the tacit positing of innate and relatively non-malleable (perhaps unconscious) motives, many of them (like earlier behaviorists) believe about their own practices that they do not involve positing mental states and processes. They are inclined to understand their "evolutionary" (or "optimality") predictions about human *behavioral dispositions* as being largely independent of psychological theorizing.[8]

In the context of such tacit but unrecognized psychological theorizing it is unsurprising that sometimes descriptions of the evolutionary function of behaviors, or "purely behavioral" characterizations of them, do double duty as descriptions of the propositional content of tacitly posited motivational states. It is equally unsurprising that the profound epistemic problems with the patterns of inference by which these states are (tacitly) posited should go unrecognized in the relevant literature. The deep structure of these inferences is largely invisible to most participants. [Note that Cosmides and Tooby (1987) are clear exceptions.]

There are, I believe, two other factors at work which enhance the effect of residual behaviorism in masking the problems with the inferences in question. In the first place, even the most behaviorist students of (non-human) animal behavior are aware that some (explicit or tacit) assumptions are always at work when researchers frame inductive inferences in terms of some particular way of taxonomizing behaviors. In the discipline of *behavioral ecology*, to which evolutionary psychology is very closely related (arguably as a sub-discipline) a conception of the appropriate taxonomy of behaviors has arisen which *is* largely independent of theorizing about underlying psychological mechanisms. According to that conception, behaviors are to be categorized in terms of what we might call *ecological* parameters: parameters which contribute to determining Darwinian fitness like, e.g., effects on food gathering efficiency, predator avoidance, access to mates, etc.. Evolutionary theory is then taken to predict that organisms will exhibit behaviors which are near optimal with respect to fitness. Thus, for example, it might be predicted that territorial defense behaviors would increase during the breeding season in those bird species in whose environments suitable nesting sites are uncommon.

Most behavioral ecologists and most evolutionary psychologists accept roughly this conception of the proper parameters for descriptions of behaviors. If (and only if) the broadly "adaptationist" optimality modeling approach to evolutionary theory which informs the literature is correct (an assumption I do not challenge here) then ecological parameters are the right ones *for predicting behaviors in the EEA*.

The apparent successes such ecological parameters in predicting the behaviors of non-human animals in the wild has apparently led many evolutionary psychologists to expect that human evolutionary psychology can be successfully carried out using the same sort of ecological parameters in characterizing human behaviors. In particular, they are inclined to believe not only that this would be possible in principle but also that current inferential practices in human evolutionary psychology generally conform to this standard: that psychological terms (like "altruism") used in evolutionary psychology are, in the final analysis, just metaphorical ways of describing behaviors and their impact(s) on parameters relevant to fitness, and not descriptions of posited mental states or processes.[9]

We have just seen, in the examples cited above, that this is not so. On any plausible conception of mental states (in particular, on any functionalist conception) the inferences involved in those examples do posit (perhaps unconscious) motivational states.

Still, it is easy to see how researchers whose main training and/or research is in the study of non-human animals could misunderstand their own methods when they turn their attention to the human case. The use of ecological parameters in categorizing behaviors is appropriate in the study of non-human animals *precisely because* it is the main aim of researchers to characterize their behaviors in the wild – that is, in conditions which are, or duplicate, the EEA for their psychological/behavioral dispositions and capacities. It is a peculiar feature of the project of extrapolative evolutionary psychology that it aims to predict/understand the behaviors of humans in environments quite different from the EEA for human psychology. Many evolutionary psychologists fail to appreciate the fact that this new project places very different demands on the resources they use in taxonomizing behaviors, with the result that they make tacit assumptions about the underlying psychology of human behavior *while believing about their own practice that it conforms to behaviorist strictures*. Thus the *existence* of the inferential patterns we have been discussing has been largely (but not entirely; see Cosmides and Tooby, 1987) unrecognized.

The other factor which, I believe, operates to enhance the influence of residual behaviorism is the peculiar ideological setting of reductionist approaches to the study of human behavior discussed in Section 4.3. It is part of a widespread conception of "objectivity" in such studies that consideration of psychological (as opposed to "biological"!) mechanisms in the study of human behavior represents a departure from scientific rigor and objectivity. That such a position is, in the final analysis, incoherent (since, according to the materialist approach presupposed in the literature, psychological mechanisms *are* biological mechanisms in the nervous system) does not preclude its having a profound methodological influence.

2.4. Reality Check

I have been arguing that there are fallacious inference patterns which are parts of the meanings of words and phrases like "evolved," "natural," "tendency," "altruism," "... in order to ...," "biological basis," "strategy," and the like. Indeed, I propose to argue that these inference patterns are sufficiently fallacious, given the explicit findings of evolutionary theory and related disciplines, that the principles and inferential practices which are parts of the meanings of such terms present an essentially incoherent picture of evolved behaviors, rather than the approximately true picture posited by a benign conception of scientific meanings. Of course, my critique of the benign conception is fundamentally mistaken if the inferences in question aren't seriously fallacious. In case their fallaciousness isn't obvious, I'll argue for it here.

2.4.0. Behaviorism, Functionalism and How Not to Infer Motives

Let's consider first the inferences from evolutionary scenarios to underlying motives which proceed by deploying (something like) the relevant evolutionary function description as a description of the propositional content of the underlying motivational structure. As an example let's consider the inference which takes as a premise

(1) Early humans evolved to be altruistic towards their kin, and reaches the conclusion that

(2) The (perhaps unconscious) motive for altruism was (or is – but we'll come to issues of innateness and non-malleability later) concern for one's kin.

We have seen that an (unsuccessful) attempt to follow the strictures of behaviorism and (thus) to speak only about behaviors rather than about their psychological bases can lead one to tacitly make inferences of this sort without recognizing that one has done so. As it happens, an appreciation of why behaviorism failed can help us to see why inferences like the one above are fallacious.

The least controversial aspect of the standard functionalist critique of behaviorism is that it is impossible to mirror, in behaviorist terms, the valuable explanatory richness which we get by offering explanations of behavior in terms of underlying psychological states because there is no simple correspondence between behaviors (described in purely behavioral terms) and underlying psychological states. The mapping from behaviors to motives is many-many.

In particular, it is central to the critique of behaviorism that, for any given pattern of behavior under a narrow range of stimulus conditions, there will be a large number of quite different scientifically reasonable explanations in terms of underlying motives, beliefs and other psychological states. That's why, even as first approximations, "operational definitions" for psychological states are inadequate.

Let's apply this most basic anti-behaviorist insight to the cases of inferences like that from (1) to (2). Call two psychological theories *behaviorally equivalent under conditions, C,* just in case they predict the same behaviors for humans under C. We can state the basis for the critique of behaviorism just discussed this way: for any narrow range of conditions, C, and any scientifically plausible general theory of human psychology, T, there will always be several scientifically plausible competitors to T which are behaviorally equivalent to T *under C* but which differ from T, and from each other, in the psychology they posit and the behaviors they predict under conditions *different from C.*

Now, suppose that we have accepted some evolutionary scenario concerning certain human behaviors in the EEA (that is: under the special conditions of life of hunter gatherer tribes). Let T be the hypothesis about the underlying psychology of those behaviors which is suggested by the sort of inference we are discussing. In so far as T is scientifically plausible, it will have several different scientifically plausible competitors which are behaviorally equivalent to it the under conditions which prevailed in the EEA, but which yield different predictions about human psychology and behavior under different conditions. Each of these theories will be equally compatible with the evolutionary scenario, so accepting that scenario provides no scientifically justified reason to prefer T over these competitors, and the inference to T is thus fallacious.

Thus, for example, premise (1) is compatible with the psychological theory affirmed in (2), but it is equally compatible with any plausible theory of the development of human altruism which, like it, would predict that *in early tribal societies* altruistic behaviors would mainly benefit the kin of altruists. Given the plausible view that small tribal groups were mainly groups of kin, a plausible theory behaviorally equivalent to T might not have to posit any special psychological mechanism to explain the kin bias in early human altruistic behaviors. An alternative which did posit such a mechanism could posit any plausible mechanism such that – under the conditions of early tribal life – the factors which conduced to displays of altruism towards a person were *correlated with* kinship, like, e.g., familiarity.

This result is quite general (as the reader is invited to check for herself): for any evolutionary scenario of the sort popular among evolutionary psychologists there will be lots of importantly different scientifically plausible psychological theories behaviorally equivalent under conditions prevailing in the EEA to the one posited as a result of the standard pattern of inference from evolutionary functions descriptions to descriptions of the propositional content of motives. Thus such inferences are scientifically unjustified.

2.4.1. Dualism, Free Will, and the Scope of Tendencies: How not to Identify Instincts.

We have just seen that the standard pattern of sociobiological inference from evolutionary scenarios to the propositional contents of motivational states is fallacious. It remains to examine the equally standard pattern of inferring that the motivational states (however identified) which underlie the behaviors specified in such scenarios are innate and non-malleable. Let's again focus on the sample inference involving altruism and kin-regard, this time with the commitment to innate and non-malleable "instinct" made explicit:

(1) Early humans evolved to be altruistic towards their kin, thus

(2) The (perhaps unconscious) motive for altruism is (non-malleably) a concern for one's kin or for members of one's in-group.

Of course, the anti-behaviorist functionalist considerations just rehearsed show that this inference is fallacious. In particular they show *both* that (even if the evolutionary scenario is right) the motivation for altruism in early humans need not have been a concern for their kin (or for their in-groups) *and* that, even if it were, that motivation need not have been innate or non-malleable. What I want to emphasize here is that the various considerations, rehearsed above, which might initially be thought to lend plausibility to the inference in question do not in fact provide any rational support for it.

Consider first the rationale which identifies evolved motives as *biologically based* and then infers that they are instinctual rather than being subject to rational or cultural influence. From a suitable dualist perspective – one in which our "animal natures" are aspects of the physical world whereas our "rational natures" are not – this inference might be justified.

As it is, however, from the tacitly materialist perspective regarding human psychology which (properly, I believe) underwrites any evolutionary study of human psychology, the opposite conclusion follows. Evolved features of our psychology have, of course, a biological (and thus a material) basis. Indeed *all* features of our psychology have a biological basis, since our mental life is a feature of the activity of our nervous systems.

It follows, however, that the dualist picture which assigns instinctual motives to the biological features of our psychology and rationally or culturally malleable motives to our non-material minds *can't* be right: we *have no* non-material mental life. Thus from the fact that a feature of our motivation "has a biological basis" nothing whatsoever follows about the extent to which it is innate or culturally or rationally malleable. The innate and non-malleable and the learned and malleable (and the innate and malleable and the learned and non-malleable, for that matter) are on an ontological par: they all have a "biological basis."

Similar considerations show that the rationale which rests on considerations about free will makes a fundamental mistake about the relationship between ontological and psychological questions. Suppose, for the sake of argument, that incompatiblism is true, so that the physical determination of behaviors makes them unfree in whatever sense is relevant to the question of free will. It follows from the assumption that the motivational basis for a pattern of behavior evolved that the relevant motivational structure is biological, and thus physical. So, assuming incompatiblism, it follows that the relevant behaviors are unfree (surely whatever level of quantum uncertainty there is in biological systems will not restore free will if free will is really compromised by physical determinism).

So, it follows that evolved behavioral patterns are unfree. What does not follow is *anything* about the psychology of those patterns. Incompatiblism, which we have assumed for the sake of argument, is *not* the view that physically determined actions are always the product of instinctual desires or compulsions, nor the view that physically determined action patterns are not subject to rational deliberation or cultural influence. Instead, it is the view that – whether they are innate or learned, rationally calculated or instinctual, culturally malleable or not – physically determined actions are unfree is some metaphysically important sense. Physical determination which proceeds *via* rational deliberation or *via* the causal influence of culture is, according to incompatiblism, just as much a source of unfreedom as determination *via* the operation of innate and non-malleable motivational structures.

Thus a thinker with incompatiblist views who infers that behaviors with a "biological basis" must derive from an innate and non-malleable motivational structure because they are unfree engages in fallacious reasoning.

Finally, let us turn to the inferential rationale involving appeals to evolved behavioral tendencies. Consider again the inference involved in Daly and Wilson (1997) from an evolutionary scenario involving special concern for one's own offspring to the conclusion that humans have a "usual inclination" to care preferentially for their own biological offspring. From the evolutionary scenario Daly and Wilson must have in mind (they do not state it explicitly) one can rationally infer that early humans had an *inclination* (in the sense of a dispositional property) which, *in the EEA*, resulted in their caring for children who were their own biological offspring, and did not (often) result in their caring for other children. This is the only "usual inclination" which one can infer from the evolutionary scenario: in fact the assertion of the existence of this "inclination" simply restates the scenario in question.

The point just made – that only the most limited sort of "inclination" or tendency can warrantedly be inferred from the evolutionary scenario in question – follows, of course, from the functionalist and anti-behaviorist critique presented earlier. All that the evolutionary scenario allows us to infer is that the psychology of early humans was such that, *coming to maturity in the EEA* they had some adult psychology or other which led *in the EEA* to the predicted pattern of child care. Nothing about innateness or non-malleability can be inferred, nor can it even be inferred that a special concern for one's biological children *because they are one's biological children* characterized the adult psychology of early humans.

It is important to see how conflating the notion of an inclination as a dispositional property exhibited under specific conditions (like conditions of the EEA) with the notion of a usual or natural inclination in the psychological sense can lead one to make the fallacious inference which Daly and Wilson make. In the *psychological* sense of "inclination," to attribute to humans a usual or natural inclination to do something *is* to attribute to humans a motivational state which could reasonably be expected to be exhibited under a wide range of conditions; that's what "usual inclination" means. So, if one mistakenly believes that the evolutionary scenario Daly and Wilson have in mind predicts an *inclination*, in that sense, to care preferentially for one's own offspring, then one will mistakenly conclude that the scenario leads to the prediction they advance.

2.4.2. Incoherence

I intend to argue that the inference patterns we have been discussing are parts of the meanings of the key terms in evolutionary psychology and that, as such, they are counterexamples to the initially plausible idea that meanings in mature scientific disciplines are generally *benign*: that such meanings tend to reflect genuine insights into the relevant subject matter of the sort which would ordinarily be expected to contribute to the establishment of methodological commensurability between different research traditions with the same subject matter. It is thus important for me to argue that the inferences in question are very far from reflecting insights into the evolution of human psychology. In fact, what I propose to show is that these inference patterns, when taken together with basic evolutionary principles which are also candidates for components of meanings of the relevant terms if any things are, give rise to a conception of human psychology and its evolution which is essentially incoherent. It is this last claim which I propose to defend in the present section.

We have already seen that the inference patterns in question are seriously fallacious. Might it still be the case that the picture of the evolution of human psychology prevailing in evolutionary psychology is basically insightful, despite the prevalence of these inference patterns?

That the answer is "no" is indicated by two different considerations. In the first place, the inference patterns we have been discussing are not peripheral to evolutionary psychology. In fact, they are what defines the discipline and distinguishes it from other approaches to human psychology. Secondly, if made explicit, the inference patterns in question are profoundly inconsistent with the most basic components of evolutionary theory – with, that is, conceptions which are presupposed in any scientific discipline in which evolutionary theory is applied.

Consider first the question of the centrality of the inferences in question to evolutionary psychology (or "sociobiology"). It is important to see that what is distinctive about evolutionary psychology as a disciplinary project or research strategy is not simply the idea that human psychology is the product of evolution. That presupposition is fundamental, for example, to comparative psychological studies generally and to comparative neuropsychology in particular. It is a broadly evolutionary conception of the origins of human and non-human psychological structures which justifies the assumption that comparative studies will be informative with respect to the human case.

What *is* distinctive of evolutionary psychology is precisely the research strategy of deploying the resources of optimality models and associated evolutionary scenarios as sources of insight into human psychological structures, especially motivational structures. But, the inferential patterns we have been discussing are precisely the methodological tools by which the alleged insights are obtained. In so far as they are seriously fallacious the central methodological practices of the enterprise of evolutionary psychology are seriously compromised.

In that regard, then, the conception of human psychology and its evolution reflected in these aspects of (what I claim are) the meanings of key terms in evolutionary psychology are anything but insightful. Worse, there just isn't any even remotely coherent (much less correct) conception of such matters reflected in the meanings of such terms. I assume here that the most fundamental explicit principles of evolutionary theory are components of the conceptual meanings of the relevant terms: they are certainly presupposed in the literature in such a way that someone who had not engaged with them would find it unintelligible.

So, the question of the coherence of the conception of the evolution of human psychology reflected in the meanings of the key terms in the literature of evolutionary psychology amounts to the question of whether or not the fallacious inferences we have been discussing could be part of a coherent (even if false) conception which also included the most basic principles of evolutionary theory. The answer is "no."

To see this, first consider the inferences from evolutionary scenarios which proceed by confusing the evolutionary function description of a posited pattern of behavior with a description of the propositional content of the motivational state which causes it. What such inferences require is that when expressions which are metaphorical uses of psychological descriptions (like "... in order to ..." or "altruism") or terms ambiguous between psychological and non-psychological uses (like "inclination") are used to describe the adaptation of behaviors to the environments in which they were displayed one takes those expressions as expressions of genuine motives. Now one of the things which is explained in every basic course in evolutionary biology is precisely that Darwin showed that it is *never* appropriate to interpret the purposive or teleological language which we find it so natural to use in describing adaptations as literally describing motivational or purposive or teleological structures.

Thus the fallaciousness of the inference rules we are considering follows trivially from absolutely central principles of evolutionary theory. There is no coherent story which incorporates the appropriateness of those inference rules and the central principles in question.

The same is true of the inferences from evolutionary scenarios to the conclusion that the underlying motivations for the relevant behaviors are innate and non-malleable. One of the points explicitly acknowledged by evolutionary psychologists even if they tacitly ignore it (see e.g., Lumsden and Wilson, 1981 and Cosmides and Tooby, 1987) is also central to any conception of the evolution of learning capacities in humans or any other animals. It is that natural selection can operate to favor *learned* patterns of behavior. If this were not so, there would be no way in which natural selection could favor the establishment of learning mechanisms themselves.

Let me explain. Of course, it is true that when selection favors learned behaviors (for example, behaviors which depend on learned motives) it must also favor innate learning mechanisms whose operation underwrite the relevant learning, but neither the behaviors themselves nor the motives for them need be innate or non-malleable. In fact, if, when natural selection favored behavioral patterns those patterns were always unlearned and grounded in innate and non-malleable psychological states, there could not be any evolutionary explanation for the evolution of the capacities involved in learning itself. There would be no evolutionary function for learning mechanisms to play.

This point follows easily from the most basic principles of evolutionary theory and is (as I mentioned) explicitly recognized even by evolutionary psychologists who routinely ignore it in their inferential practices. Thus there is no coherent conception (much less an insightful one) which incorporates this second feature of the standard pattern of evolutionary psychological inference and acknowledges the most basic principles of evolutionary biology. If the inferential principles in question and the most basic evolutionary principles are parts of the meanings of the key terms in evolutionary psychology, then we have an actual and important case in which meaning of terms in a mature scientific research tradition are anything but benign.

2.4.3. Meanings?

It remains to examine the question of whether or not the inferential practices we have been discussing – and the basic principles of evolutionary theory – are constituents of the meanings of the key terms in evolutionary psychology. This is, of course, an empirical question about the sorts of conceptual engagements which are required for someone to understand the relevant literature and to appreciate methodological and theoretical discussions in evolutionary psychology. I'll present evidence for the claim that these components of the conceptual resources of evolutionary biology are constituents of meaning in the required sense but, of course, the reader may choose to conduct her own investigations into this question.

The key reason for thinking that the inferential principles we are discussing – as well as the most basic principles of evolutionary theory – are constituents of meaning is that, in the literature in question and in professional discussions in evolutionary psychology they are (a) tacitly presupposed and (b) central to the argumentative strategies of the papers and discussions in question. Since they are rarely made explicit (which is hardly surprising given the conflict with fundamental tenets of evolutionary theory) the inferential practices in question must be presupposed as appropriate (in the case of uncritical engagement) or explicitly identified by the critical reader/listener as central to the relevant literature (in the case of critical engagement) in order for the reasoning engaged in by professional evolutionary psychologists to be intelligible. This is the key mark of components of conceptual meanings.

Other factors also indicate the appropriateness of thinking of the inferential practices in question as constituents of meaning. If engagement with a particular conceptual resource within a research tradition is centrally important to the intelligibility of that tradition, and if the standard mode of engagement within the tradition is uncritical engagement, then one would expect that the typical participant in the tradition would have difficulty understanding claims which denied the cogency of the conceptual resource in question.

I now report – as a summary of 20 years of experience teaching a course on methodological issues in sociobiology/evolutionary psychology – that this is indeed the case. Especially for students who have been exposed to this research tradition in biology courses it is extremely difficult to even remember the broad outlines of the sorts of criticisms of the inferences in question which I have presented here.

I want to emphasize that I am not reporting merely that such students fail to be convinced by those criticisms. That would be unsurprising in a context in which teachers they regard as *prima facie* authoritative disagree about fundamental methodological matters. What I am reporting is that students who have been made familiar with the literature in question have very great difficulty even paraphrasing the criticisms in question.

Similar results obtain, I now report, when philosophical critics of evolutionary psychology encounter able practitioners of that discipline in methodological discussions. The response of those practitioners is precisely what one would expect if the inferential practices in dispute were, for the evolutionary psychologists, components of the very meanings of the relevant terms/concepts. In my experience, evolutionary psychologists respond to criticisms of the sort we have been discussing by acknowledging the logical possibility that the conclusions of the relevant inferences might be false, but by expressing incredulity that the critic would take this fact to underwrite a serious methodological criticism. That is, they respond to the criticisms in question just as one should if responding to someone who rejected a fundamental principle of inductive inference on the grounds that it was not deductively valid. What they – like their students – are unable to do is to appreciate the fact that the criticisms in question are not exercises in philosophical skepticism but ordinary scientific critiques. That, I suggest, is true precisely because, for practitioners who are uncritically engaged with the inferential practices in question, those practices are components of the meanings of the key disciplinary terms and thus define the limits of relevant scientific criticism.

Here's another way to think about the same question. No one who works in evolutionary psychology would report her view that some motivational feature of human psychology was established by individual selection by saying that it was not *altruistic* unless she intended to convey to her listeners/readers that the motivation for that feature was not (or was not mainly) other regarding, *even though* the term "altruism" has, in the literature in question, the technical meaning "contributes to the reproductive fitness of conspecifics while reducing the fitness of the organism that exhibits it." The inferential pattern which involves (in effect) conflating this technical sense of "altruism" with the standard psychological sense is so much a standard of reasoning within the discipline that it is an essential component of communicative practices.

Still, someone might say that the inference in question was not part of the *meaning* of the technical term "altruism" but rather that the fact about communicative practices I have just mentioned is simply a matter of conversational *implicature*. Here's the rebuttal. This "implicature" has the property that it takes a semester long graduate seminar to cancel it, and the cancellation will not be successful with respect to some members of the seminar. They will not be able to remember from day to day that there are two senses of "altruism" involved. Similar results obtain for the other standard inferential patterns in evolutionary psychology. If one is interested in having a theory of communication, "implicatures" which are that hard to cancel *are* constituents of meaning.

2.4.4. Malignancy?

We need also to address the question of just how far from benign the meanings of the key terms in evolutionary psychology are. Recall that the benign meaning thesis is not the thesis that the meanings of scientific terms will ordinarily be exactly true (if they are doctrines) or perfectly reliable (in the case of methods) but only that the meanings of scientific terms in mature sciences can be expected to reflect sufficient insight into the relevant subject matter that, when two traditions share a common subject matter, the meanings of the relevant terms in the two traditions can be expected, in general, to contribute to – rather than to detract from – the establishment of methodological commensurability between them.

This is pretty plainly not the contribution which the meanings of the key terms in evolutionary psychology make to the prospects of commensurability between that research tradition and those (like traditional anthropology, sociology, political economy and social psychology) which address many of the same issues about human social behavior. The key inferential patterns which are among the relevant meanings in evolutionary psychology make it extremely difficult for its practitioners to appreciate the plausible methodological criticisms arising from competing traditions as anything other than (a) a desperate retreat into skepticism in the face of the new science of evolutionary psychology or (b) a reflection of a failure to appreciate the implications of evolutionary biology. The consequence is that the prospects for fruitful dialogue upon which the prospects for commensurability depends are greatly reduced. Meanings in evolutionary psychology are genuinely malignant.

3. LESSONS, I: MEANINGS, REFERENCE AND SCRIPTS

3.0. Meanings

We now need to extract some philosophical lessons from our examination of meanings of theoretical terms in sociobiology/evolutionary psychology. In the first place, we have seen that scientific terms do sometimes have conceptual meanings. There may be domains of discourse or research traditions about which the Quinean dictum is true that there is no real distinction to be drawn between those principles or inferential patterns which are so central to practice that they count as components of meaning and those whose centrality is insufficient for them to qualify for that honor. Nevertheless this is by no means always true. In the case of sociobiology/evolutionary psychology, for example, there *are* inferential practices such that engagement with them is essential to any understanding of the most basic inferences and arguments in the literature and these *are* components of the conceptual meanings of the terms and phrases involved. There is a place in the semantic theory of scientific terms for a notion of conceptual meaning.

Second, although it may often happen that doctrines or inferential patterns become established as components of conceptual meaning because they reflect important insights into the relevant subject matter(s), this is not generally true even for well established scientific research traditions. *The* central inferential practices which are components of meaning in sociobiology/evolutionary psychology are deeply fallacious.

Indeed, the components of the meanings of the key terms in a scientific research tradition need not provide anything like a coherent conception of its subject matter: the meanings of key terms in sociobiology/evolutionary psychology, for example, do not.

For this reason, the meanings of terms in a scientific research tradition can be barriers to the establishment of methodological commensurability with respect to other traditions with the same subject matter, just as Kuhn suggested. The meanings of scientific terms, even in a well established research tradition, can be *malignant* rather than *benign*.

3.1. Meanings and Reference

Once it is acknowledged that scientific terms sometimes have meanings it is natural to suppose that they figure in the establishment of reference for such terms in approximately the way suggested by traditional descriptivist conceptions of reference: that a scientific term will refer to whatever property, magnitude, or whatever its meanings are *approximately* true of (in the case of doctrines which are components of meaning) or approximately reliable about (in the case of inferential practices). At least it is plausible to think that a constraint of this sort is part of what determines the reference of a scientific term, with the rest of the work being done by relevant causal relations between instances of the referent in question and use of the word in question.

Descriptivism isn't even that close to being a correct account of reference. As the case of sociobiology/evolutionary psychology shows, the meanings of a scientific term within a research tradition need not be such that they could be approximately true of, or insightful about, *anything*. They can provide a fundamentally incoherent picture of the subject matter in question.

It is true, of course, that reference is an epistemic success phenomenon (Boyd, 1989; 1993; 1999). In all but the most contrived cases a term, t, refers to an entity, e, only if inferential (and perceptual) practices in the relevant community are such that there is some explanatorily important tendency for what is predicated of t to be true of e's. What the examples from sociobiology/evolutionary psychology indicate are two points about the epistemic success which is thus central to reference. First, the practices whose epistemic reliability underwrite reference need not be the practices engagement with which is central to the possibility of successful communication: they need not include the components of the conceptual meanings of the terms in questions.

Secondly, even when scientific terms are used in mature scientific research traditions with all the trappings of "paradigms" in Kuhn's sense – graduate programs, refereed journals, professional societies, and (most importantly) shared standards of evidence – it need not be the case that the epistemic successes which underwrite the referential connections between the vocabulary of the "paradigm" and features of the world are mainly achieved by practices within the paradigm. The term "altruism," as it is used within sociobiology/evolutionary psychology refers (some of the time) to a real other-regarding aspect of human motivation. This can only be so because the use of the term "altruism" is to some extent determined by inferential practices which are epistemically reliable with respect to that other-regarding motivational state.

As it happens, sadly, the methods of sociobiology/evolutionary psychology are not among those practices: they are very poor methods for finding out about altruism. So, in effect, evolutionary psychologists who use the term "altruism" to refer to an other regarding motivational state are *borrowing* the reference of this term from more reliable practices which are situated in other psychological research traditions and in everyday "folk psychological" practice.

3.2. Reference and Progress

Part of the import of Kuhn (1970) was to undermine the Whig conception that the operation of scientific methodology can, almost always, be expected to lead to progress, so that, when scientists fail to make such progress the appropriate diagnosis of their situation is, at least ordinarily, that they failed to apply scientific methods correctly.

Part of the Kuhnian challenge to that conception is available just as a consequence of the recognition of the theory-dependence of scientific methods: whether or not progress is made with respect to a family of scientific questions depends, not (just) on whether or not practitioners conscientiously practice scientific methods, but on whether or not the background theories and conceptions which determine their methods happen to be relevantly approximately true. Since theory-independent inductive methods do not exist, the expectation of progress at any point in the history of any scientific discipline must rest on the profoundly *a posteriori* estimate that the background theories in question are relevantly approximately true. So the expectation of progress in science needs to be significantly qualified.

There is, of course, an additional dimension to Kuhn's critique of the Whig conception. According to Kuhn there is no progress – in the sense of successively closer approximations to the truth – during "scientific revolutions." Kuhn's reasons for denying that there is this sort of progress have – as we have seen – a semantic component: the key terms in scientific paradigms are supposed to change meanings *and referents* during "revolutions," so that it *can't be* that there is progress towards a more accurate understanding of the common subject matter of the earlier and latter paradigms *because they don't have a common subject matter*!

Naturalistic rebuttals to Kuhn's conception of reference for scientific terms refute this last claim and establish that terms in quite different research traditions can share a common subject matter, so that *Kuhn's* semantic argument against the expectation of progress across revolutions fails.

Since reference is a matter of some degree of epistemic success, it might be reasonable to conclude that when the key terms in a scientific research tradition refer there is reason to expect that the research methods within that tradition will be epistemically reliable enough to underwrite a *prima facie* expectation of progress.

What we have just seen is that, perhaps surprisingly, a semantic argument closely related to those offered by Kuhn shows that this expectation is not generally speaking justified. In at least some important cases the methods, and the doctrines which determine the methods, within a research tradition – including those whose centrality is such that they properly count as components of the meanings of the key terms – are so misleading that they cannot be expected to underwrite progress (or progress towards commensurability with other traditions sharing the same subject matter). In such cases there will, of course, be epistemic successes upon which the reference relations for the various key terms supervene, but they need not lie within the purview of the tradition in question: they can be almost entirely borrowed.

Thus the causal or naturalistic theory of reference for scientific terms underwrites a general expectation of scientific progress only in the most highly qualified sense: scientific terms used within a research tradition have determinate referents only if *somewhere or other* perceptual or inferential practices which govern the uses of those terms produce *some sort of* approximations to the truth. But, these practices need not lie within the domain of the research tradition in question, *or of any scientific research tradition at all*. It is fully compatible with a causal/naturalistic conception of reference for the theoretical terms of some scientific disciplines or research traditions that most of the relevant epistemic successes which underwrite reference are achieved outside of scientific practice. Whether or not things are ever this bad, the expectations of scientific progress which follow from just causal theories of reference are very limited indeed.

3.3. Quinean Scruples Revisited

We may restate some of the points just made in a way which illuminates the ways in which Quinean criticisms of the notion of meaning fail to apply to the questions about conceptual meanings which we are addressing here. According to the traditional empiricist conception of meanings which Quine challenges in rejecting the analytic synthetic distinction the meaning of a general term (like a theoretical term in science) is provided by a set of conceptual entities – statements, criteria of application, or inferential principles – which:

a. are analytic or analytically grounded,

b. are thus rationally unrevisable,

c. determine the most basic rational norms for inferences involving the terms in question, and

d. are such that acceptance of them is definitive of linguistic/conceptual/ communicative competence with respect to the terms in question.

What the Quinean critique of the analytic synthetic distinction suggests is that there are (almost) no terms, and certainly no theoretical terms in science, which are associated with conceptual entities satisfying a.–d. because (almost) no terms of any sort are associated with conceptual entities satisfying a. and b.

We may think of a.–c. as specifying that the sorts of meanings anticipated by the empiricist tradition are supposed to be *methodologically normative* in an especially strong sense. They are supposed to be (analytically) methodologically immune from revision and to (analytically) set the most basic inferential standards for the terms in question.

d., by contrast, requires of meanings only that they be *communicatively normative*: that they underwrite the conditions for mutual intelligibility of statements and arguments.

Now Quine's arguments that there are (at least for almost all words) no conceptual entities satisfying a.–c. do not address the question of whether or not there are such entities satisfying d., alone. All they establish is that whatever *communicatively normative* conceptual entities there might be associated with a word (in the sense of satisfying d. with respect to it) they will never (or almost never) satisfy the standards which a.–c. set for *methodological normativity* (because nothing will).

Similarly, when Quine argues against the prospect that a clear distinction can be drawn between conceptual entities central enough to the use of a term to count as parts of its meaning and those conceptual entities which are merely well established, his arguments are entirely directed towards questions of methodological normativity: he argues essentially that there is a continuum of levels of immunity from revision for the conceptual entities associated with a word at a time, but that, because no conceptual entities rise to the analytic level of immunity, there is no non-arbitrary level of immunity along this continuum for a given term above which such entities are methodologically normative enough to be counted as components of meaning.

I suppose that this is entirely right, but it does not address the central question of the present essay regarding communicatively normative meanings. What I have been arguing is that for lots of theoretical terms in science there *are* conceptual entities which are unproblematically components of their conceptual meanings – which are communicatively normative for those terms. I have not argued that there is a sharp boundary between such entities and those which are not components of conceptual meaning, but only that there are (at least in some cases) clear cut cases of such components.

3.4. "Conceptual Roles" and Two Kinds of Normativity

The points just made depend on the fact that, once the analytic-synthetic distinction is abandoned for scientific terms, there are two distinct sorts of normativity which can be attributed to conceptual entities. On the one hand, they may be said to be normative in a methodological sense, such that someone who affirms that they are normative in that sense is making a methodological judgment regarding the (approximate) truth, or reliability of the conceptual entity in question. To say of a doctrine that it is normative in this sense is to say that it reflects some important truth or approximate truth about the relevant subject matter; to say that an inferential pattern is methodologically normative is to affirm its reliability and justifiability with respect that subject matter.

By contrast, to describe some conceptual entity as communicatively normative is *not* (as it would be on the standard empiricist conception of the semantics of scientific terms) to endorse it in this way,but merely to indicate that engagement with it (uncritical *or critical*) is centrally important to one's ability to understand the argumentation and reasoning in the scientific tradition. As the treatment of the standard inferential practices in sociobiology/evolutionary psychology in the present essay indicates, it is possible to argue that conceptual entities are communicatively normative while insisting that they are very, very, far from being methodologically normative.

The distinction between the two sorts of normativity is related to corresponding distinctions between conceptions of the *conceptual role* of a scientific term and between conceptions of the *rationality of inferences* in science. Conceptual role first. The conception of conceptual meaning I have offered here is a species of the genus of conceptual role accounts of term meanings. In the first place, I have insisted that certain inferential roles (which are the paradigm cases of "conceptual roles") are components of the meanings of scientific terms. Secondly, in identifying doctrines as components of the meanings of such terms I have emphasized the question of whether or not engagement with them is crucial to an understanding of the central inferences and arguments within the relevant research tradition (again a question of conceptual role).

Where the proposal offered here differs from that reflected in more traditional empiricist conceptual role semantic theories is that, for those theories, the conceptual roles which are parts of the meanings of scientific terms are analytically justified, so that to identify a conceptual entity as a component of the conceptual role of a scientific term is to provide for it the strongest possible justification. On this sort of conception, conceptual role meanings are as benign as conceptual entities could possibly be.

By contrast, on the conception proposed here the conceptual roles in terms of which the meanings of scientific terms are defined are those which would be identified by the true theory of communicative practices within the relevant research tradition rather than by the (analytic components of) the true theory of the relevant subject matter. So one need not be endorsing a conceptual role in identifying it as a central component of the meaning of a term. Scientific terms can – and do – often have conceptual meanings that they shouldn't have.

It is not part of my thesis here that we cannot have a different sense of conceptual role according to which the conceptual role of a theoretical term is the conceptual role which it ought to have. My only claim is that, if we pay attention to those the insights of traditional conceptual role semantics which survive the critique of the analytic synthetic distinction, we will see that conceptual roles in this latter sense play no significant role in theories of communication or commensurability in science.

All of the distinctions we have examined in the present section rest on the rejection of the analytic-synthetic distinction for theoretical terms. The rejection of that distinction is closely related (just how closely we need not figure out for present purposes) to the rejection of the conception that there are *a priori* justifiable, and thus theory-neutral methods in the sciences. As we have seen it is the fact that there are no such methods which makes it possible in the first place that methodological incommensurability should obtain between competing scientific research traditions or "paradigms," even when their practitioners are scientifically rational.

Recognizing that the methods of science are theory-dependent in this way permits us to draw one more important distinction. Recall that I maintained that the components of the conceptual meanings of theoretical terms are *communicatively* but not *methodologically* normative: in identifying a conceptual entity as a component of the conceptual meaning of a theoretical term one need not *endorse* the methodological or theoretical role it plays.

There is, however, a sense in which the components of the conceptual meanings of theoretical terms must be methodologically normative: no conceptual entity associated with a term in a research community is going to be such that engagement with it is crucial for understanding inferences and arguments unless *by the standards prevailing in that research community* the entity in question plays a very important methodologically normative role. Once we see this, we can see that it is a reflection of an important phenomenon of the *relativity of scientific rationality* to the doctrines and practices prevailing within a research tradition at a time. The relationship between scientific rationality and epistemic success is complex. Roughly, a scientist or scientific community acts rationally just in so far as they conscientiously apply the methods dictated by (what their careful and thoughtful judgment takes to be) the best available theories. Practices which are rational in this sense are epistemically reliable *only* to the extent that these best available theories are relevantly approximately true.

So, scientific rationality is (in the sense specified) relative to a research tradition. There need be no failure of individual or collective scientific rationality when rational scientific practices (in unfortunately situated research traditions) fail to reliably resolve scientific issues or (as in the sorts of cases we are considering here) fail to establish commensurability between competing research traditions.[10]

Thus, for investigations of the relationship between the semantics of scientific language and of the phenomenon of incommensurability have uncovered two respects in which Kuhn was right: there are sometimes barriers to methodological commensurability arising from differences in the meanings of the same terms in different research traditions *and* this fact is crucially related to the phenomenon of tradition-relativity of scientific rationality. Note, however, that nowhere have we had any occasion to invoke a relativity of truth or of "reality" or of the "worlds" scientists study.

3.5. Insulated Scripts and Incoherent Meanings

I have argued that the conceptual meanings of key terms in sociobiology/evolutionary psychology are essentially incoherent. A natural question arises of how it is possible that a system of inferential practices and substantive doctrines which, taken together, provide no coherent picture of the relevant phenomena whatsoever could be sustained as methodological norms in a serious scientific discipline practiced by many intelligent and sophisticated people. Why isn't the incoherence fairly readily detectable – and thus correctable?

Part of the answer, of course, is that the inferential practices are not made explicit. Roughly, the relevant "rule" of inference is this: From a premise which attributes a particular evolutionary function, F, to a class of human behaviors in the EEA, infer that an innate non-malleable motive to accomplish F is part of human psychology and underwrites analogous behaviors in conditions other than those prevailing in the EEA *unless* this conclusion is implausible in the light of what is already known about human learning mechanisms. In the latter case, infer that there is an innate and non-malleable psychological mechanism at work and posit for it a structure as closely related to an innate and non-malleable motive to do F as you can without sounding silly.

Now no one would subscribe to this inferential principle if it were spelled out explicitly. It is not (but see the discussion of Cosmides and Tooby, 1987, in section 2.2.). Instead, I suggest, what happens is that students who are acquiring the inferential standards prevailing in sociobiology/evolutionary psychology become familiar with, and learn to deploy, what we might call *inferential scripts* reflecting the standard applications of the standard sociobiological/evolutionary psychological inferential strategy. They acquire, through practice and through reading the literature, a "feel" for the sorts of extensions of those scripts which are treated as legitimate innovations within the research tradition. This tacit knowledge of the standards prevailing in the tradition constitutes their (uncritical) engagement with the inferential practices in question.

This is, I believe, a special case of a phenomenon recognized by Kuhn which helps to underwrite his choice of the term "paradigm." According to Kuhn assimilation into the community of researchers whose work constitutes a paradigm crucially involves mastering certain key "exemplars" of appropriate research and acquiring (largely) tacit understanding of the prevailing standards for innovations within the framework represented by those exemplars. In the case of the standard inferential practices in sociobiology/evolutionary psychology what is involved is tacit knowledge of the prevailing standards for the acceptability of new variations on the received applications of the standard inferential practice.

One further point will help us to understand how a basically incoherent conception can set the inferential standards for a scientific research tradition. In my (perhaps caricatured) account of what the explicit form of the standard inferential practice would be I characterized the general inferential pattern and then added a clause to the "rule" which instructs the inferer to depart from the standard practice when necessary in order to avoid scientific embarrassment. Without this "loophole" the rule would dictate inferences which would be profoundly – and obviously – fallacious in the light of cases in which there is selection favoring learned behaviors, and thus selection for learning mechanisms connected with motivations and behaviors rather than for innate motives and behaviors.

In practice, of course, practitioners do not apply an explicit (and methodologically peculiar) rule which contains a clause recommending avoiding embarrassment. Instead, the tacit understanding of the methodological standards prevailing within the research tradition which they have acquired includes a tacit understanding of when the more general standard inferential strategy is to be invoked and when it is not. This tacit understanding *insulates* the general inferential strategy from the embarrassment which would result from its consistent application. This phenomenon of insulation of inferential practices helps to explain why the incoherence of the conceptual resources of the research tradition remains invisible to practitioners.

The exact psychology and sociology of the establishment of tacit inference principles and of their insulation are empirical issues which go beyond the scope of this paper. For philosophers however, some appreciation of how the relevant mechanisms work can probably be gleaned from an appreciation of the role of skeptical arguments in philosophy, especially within the empiricist tradition. It is routine for philosophers to deploy skeptical arguments against metaphysical or normative doctrines to which they are unsympathetic, even when those arguments are so powerful that, *if they were consistently applied*, they would result in the rejection of knowledge claims to which the philosophers in question are committed (for a discussion of this sort of selective skepticism in empiricist philosophy of science see Boyd, 1983; 1985a; for a discussion of the same phenomenon in metaethics, see Sturgeon, 1984).

In such cases, the philosophers have acquired – through their training in philosophy or their understanding of the relevant literature – a tacit understanding of the skeptical inferential practices in question: an understanding which embody a suitable (tacit) insulation of those practices from potentially embarrassing applications. Something like the same sort of tacit understanding operates to establish insulation in cases of incoherent meanings in scientific research traditions.

4. LESSONS, II: DISCIPLINARY BOUNDARIES, IDEOLOGY AND THE PROSPECTS FOR COMMENSURABILITY

4.0. How Widespread is the Phenomenon?

We have seen that scientific terms have conceptual meanings, and that their meanings can be malignant with respect to commensurability, in much the way Kuhn thought, even though his arguments for referential incommensurability fail. We've also learned that a certain plausible "descriptivist" proposal within a broadly naturalistic conception of reference – that when scientific terms have conceptual meanings they can be expected to be approximately true/reliable – is subject to important counterexamples.

An important question obviously remains: how widespread is the phenomenon of malignant meaning? This is, of course, an empirical question whose resolution would require investigations that go beyond the scope of the present paper. What I propose to do here is to identify some features of the situation of sociobiology/evolutionary psychology which I think may plausibly have contributed to the emergence of malignant meanings, and to suggest that in addressing the question of how widespread malignancy of meaning is it will be important to examine other research traditions or other frameworks of inquiry which share these features.

I suggest that among the key factors responsible for the emergence of meaning malignancy in the case of sociobiology/evolutionary psychology are three interrelated phenomena: the *disciplinary isolation* of specialized research traditions like sociobiology, the *ideological setting* within which its research takes place, and the (ideological) *prestige of reductive science*. I'll explore these factors in the following sections and suggest, on the basis of my exploration, that the prospects for the eventual achievement of commensurability in the face of malignant meanings may, in some important cases, depend on social and political factors external to the institutional practices of science.

4.1. (Sub)Disciplinary Isolation

One of the ways in which it is possible to see that the standard inferential patterns in sociobiology/evolutionary psychology are fallacious is to recognize that they rest on tacit premises and methods regarding human (and non-human) psychology which have been rejected for good reasons in the critiques of behaviorism which underwrote the development of contemporary "cognitive science" approaches to human and non-human psychology. I have suggested that the plausibility of many (perhaps all) of the fallacious inferential practices in question rests in part on the influence which those behaviorist premises and methods still have in programs in "animal behavior" within which much of the research in sociobiology/evolutionary psychology is conducted.

What is important for our purposes here is that, roughly speaking, it is possible to become a professional evolutionary psychologist without becoming familiar enough with the more "cognitive science" literature to anticipate or appreciate the tensions between one's own (sub)disciplinary inferential strategies and those of researchers with a keener appreciation of the problems with behaviorism. Arguably, at least, the way in which this (sub)disciplinary isolation insulates the inferential practices in evolutionary psychology from anti-behaviorist criticisms helps to explain how the fallacious inferential practices in question could become sufficiently normative *within sociobiology/evolutionary psychology* that they emerged as components of the conceptual meanings of its key terms.

In so far as this is true one might expect to find (some of) the preconditions for malignant meanings whenever subdisciplines develop their own journals, graduate programs, professional organizations and the like largely independently of such institutions within the relevant broader disciplinary context. At least arguably, a tendency towards this sort of subdisciplinary isolation is a consequence of very widespread tendencies towards specialization, and of the institutional arrangements which grow out of competition between subdisciplinary approaches for funding and prestige. In this regard, then, the prospects for malignant meanings seem unfortunately favorable.

4.2. The Opposite of Isolation: Ideological Embedding

I am inclined to think, however, that the primary contributor to the emergence of malignant meanings in sociobiology/evolutionary psychology lies in a way in which inferences in that research tradition are *not* isolated from the conceptual elements of other approaches to the same subject matter. What I have in mind is the ways in which the standard inferential pattern in sociobiology/evolutionary psychology is ratified by prevailing ideology in the culture at large.[11]

I mean to speak here of *ideology* in both senses of the term: in the sense in which it refers to conceptual resources which are central to a particular cultural setting at a time *and* in the sense in which it refers to those aspects of ideology in the first sense whose prevalence is explained by their role in the ratification of existing patterns of power and wealth.

It is important to note that the general conceptions (or misconceptions) of rational and animal natures, of behaviors with a "biological basis," of "nature" and "nurture," of "free-will," etc. which partly explain the plausibility of the standard inferential strategy in sociobiology/evolutionary psychology are paradigm examples of prevailing ideology in the first of these senses, so much so, in fact, that some of the fallacious inferences we have been discussing might plausibly be said to be constitutive of the meanings of the terms mentioned above in everyday intellectual discourse. That many of the fallacious inferences in the research tradition are thus underwritten by elements of the broader culture surely serves to enhance their immunity from the criticisms that one might otherwise expect to be forthcoming within the discipline itself.

What seems especially important in the case of sociobiology/evolutionary psychology is that both its inferential strategies and its findings are embedded as well in an ideological setting of the second, more distinctly political sort. It is a characteristic feature of even much of recent (and more "moderate") work in evolutionary psychology that its inferential principles and its conclusions underwrite and are underwritten by the sort of cynical conception of human nature and of human potential which have been the premises and the conclusions of social Darwinism since Darwin. They thus carry with them the special credibility which, as the history of science indicates, attaches in any modern historical period to the scientific rationalizations of (the least agreeable and least justifiable of) its social structures and relations.

The combination of these two aspects of ideological embedding constitute, I conjecture, the primary explanation for the fact that an essentially incoherent combination of doctrines and inferential practices has come to constitute the conceptual meanings of the key terms of sociobiology/evolutionary psychology. There is, however, an additional ideological factor which, I conjecture, is also important.

4.3. The Prestige of Reductive Science

Another factor which tends to insulate the inferential practices of sociobiology/evolutionary psychology from the scrutiny which might undermine their credibility – and which thus helps to explain how they maintain their status as components of conceptual meaning – must surely be the epistemic and methodological prestige which currently attaches to reductive explanatory approaches in the sciences. My own experience teaching about methodological issues in evolutionary psychology indicates that the very fact that the approach in this research tradition involves a "reductionist" research strategy serves to lend its methods and findings a credibility quite independently of the details of the arguments presented in the relevant literature.

This, too, is a bit of prevailing ideology. Its roots are, of course complex. It seems evident that, to some extent, the spectacular success of biochemical approaches to issues in genetics, physiology, neurochemistry and the like have contributed to this prestige. So has the greater prestige of the "hard" sciences (of which biology is now one) in comparison with the "soft" sciences from which alternatives to sociobiological/evolutionary psychological explanations are likely to arise. At least arguably, these factors are ideological mainly in the first of the two senses we are discussing. [I actually think that they are ideological in the second sense as well, and importantly so, but it is beyond the scope of this essay to develop this point.]

What I want to emphasize here (because my experiences in teaching this material and in arguing with practitioners of evolutionary psychology suggests to me that it is very important) is an aspect of the prestige of reductionist explanations in the human sciences which is plainly ideological in the political sense. To a remarkable extent there is an association between (1) the conception of an objective scientific opinion on matters human, (2) a reductionist approach to such matters, and (3) "toughmindedness" regarding them, where the latter carries simultaneously the connotation of unbiased investigation and the implication that only a cynical conception of such matters can possess the appropriate level of objectivity.

The connection between (2) and (3) is partially explained by the fact that reductionist explanations for unfortunate features of human social practice *will* characteristically result in a cynical conception of human nature, because what characterizes reductionist explanations of that sort is that they attribute the unfortunate features of human social relations to properties of individual human psychology rather than to, e.g., features of social or economic organizations. But the general association of objectivity with reductionism and with toughminded cynicism is a much deeper feature of prevailing social ideology. It reflects a gendered (masculine) conception of scientific objectivity and underwrites cynicism about the prospects for social progress in general.

This association of the authority of (alleged) scientific objectivity with the defense of prevailing relations of power and wealth is, of course, the paradigm case of the political ideological role of appeals to scientific authority in modern societies. It would be unwise to underestimate the extent to which this association undermines the prospects for the sorts of criticisms of the (cynical) standard inferential strategy in sociobiology/evolutionary psychology which would be required to overcome the malignant meanings and make possible methodological commensurability with other approaches to human nature.

The examination of these ideological effects suggests that malignant meanings might be more likely to be found in cases in which (sub)disciplinary isolation is associated with ideological embedding, especially of the political sort. If this guess is correct, then such (sub)disciplines as psychometrics, applied game theory and the psychology of individual differences might be places to look.

4.4. The Prospects for Commensurability: The Politics of Criticism

The question I want to address now is what the prospects are for eventually achieving commensurability between competing research traditions when the conceptual meanings of their shared terms are malignant in one or more of them. Recall that, as I understand the term, *commensurability* obtains between two research traditions only if there are methodological resources which are fair to both traditions *and epistemically reliable*, and which are adequate to adjudicate the differences between them. For example, in the case of the establishment of commensurability between sociobiology/evolutionary psychology and other approaches to human psychology it would be necessary for practitioners of the former research tradition to come to appreciate the deep problems with those standard inferential practices which are currently components of the conceptual meanings of their key theoretical terms.

There is surely no informative generalization which applies in all cases of malignant meanings, but I think that an important lesson can be drawn regarding cases in which there is important ideological embedding underwriting the sustenance of malignant meanings. In the case of sociobiology, as it was called at first, there was a period (roughly from the publication of Wilson, 1975, until shortly after the publication of Kitcher, 1985) during which a dialogue of sorts emerged in which practitioners of sociobiology and their critics (especially Lewontin, 1976 and Gould and Lewontin, 1979) engaged in exchanges which were accessible to substantial parts of the intellectual community and which had the effect of producing a *quite temporary* pause in the growth of the plausibility which sociobiological theorizing enjoyed within the scientific community generally.

What has happened subsequently has been the re-establishment of the credibility of evolutionary psychology (as it is now called), *with the standard (fallacious) inferential practices largely unchanged.* The effect of the criticisms by Gould, Lewontin, Kitcher and others was *not* to undermine those inferential practices, even though the critics showed (along lines roughly like those explored here) that those practices are deeply fallacious and indicated their ideological roots. Instead, the only significant and lasting effect of these (quite cogent) criticisms was to reduce the frequency of sociobiological (I mean "evolutionary psychological") speculations about racism and war and to replace these "hot" topics with reproductive strategies, sex differences, cooperation, and altruism. To a limited extent this took some of the political "edge" off the research traditions, but it left the inferential practices – and their ideological roots – largely untouched.

No doubt the story of why the important criticisms raised during the "sociobiology" period had such limited effects is quite complicated, but one factor seems definitely to have been important. The criticisms in question emerged during a period of radical critiques of political ideology in intellectual life. The very possibility of the articulation of those criticisms – and of their intelligibility in the intellectual community generally – rested on political developments (the emergence of militant anti-racist, anti-imperialist and anti-sexist movements) largely external to institutional science.

Had these movements not emerged, I suggest, the ideological embedding of the standard sociobiological inference practices would have been sufficient to make challenges to them hard to invent and even harder to present persuasively. The return to almost unchallenged status (and thus to the role of components of conceptual meanings) of those practices is largely explained, I suggest, by an equally external phenomenon: the worldwide defeat of leftist movements of the sort which helped to underwrite the critiques in question. Whatever the explanation may be for that defeat, it was certainly not largely a matter of considerations internal to institutional science or to rational application of scientific methodology.

What this suggests, and what I now propose, is that in cases in which malignant meanings are sustained by embedding in social ideology, the prospects for the establishment of commensurability may depend (much) more on the emergence of political movements embodying critiques of social ideology than on any developments internal to the research traditions in question.

4.5. Politics and the Relativism of Rationality

I should emphasize two ways in which the point just made about the relationship between social ideology and malignant meanings illustrates important dimensions of the relativity of scientific rationality. In the first place, consider the situations of researchers who are pursuing a malignantly flawed research program as a result of ideologically sustained commitments to mistaken doctrines or fallacious methods. In the absence of a change in *political* circumstances which reduces the conceptual and cognitive impact of the relevant bits of social ideology, even the exercise of perfect (theory-dependent) scientific rationality need not suffice to permit them to recognize the flaws in their methods or doctrines *even if* there exist scientific arguments against these doctrines and methods which (a) are actually presented somewhere in the literature and (b) would be persuasive were it not for the ideologically sustained malignant errors in question. The question of the possibility of error correction and progress under such circumstances is, in an important sense, a political question.

This so because there is *always* an important political dimension to the reliability or unreliability of scientific methods. Because they are ineliminably theory-dependent, scientific methods are systematically reliable only when, and to the extent that, there are available background theories which are themselves relevantly approximately true. But, this is not a sufficient condition for their reliability. Appropriate social, economic and political arrangements must exist for relevantly accurate theories to be disseminated and for work properly grounded in them to be supported and its results assigned an appropriate level of credibility (see Section 5.1).

This, in turn, happens within a research tradition – under ordinary circumstances – *only* when powerful economic and social interests are served by the promulgation of approximately true answers to the relevant questions. When powerful interests would be ill served by the truth, the various mechanisms of social ideology operate in such a way that – under ordinary conditions – the products of institutional science serve those interests at the expense of the truth.

This pattern is obscured when we direct our attention only to the evident reliability of the methods of the sciences in those areas of the physical sciences and biology where the most important achievements of contemporary science have emerged. If we fail to recognize that these are precisely the areas with respect to which there are industrial and military interests which are served by the discovery of (approximate) truths, we can fail to recognize the role of economic and political interests in determining the institutional and ideological structures in which these successes are made possible.

Examining cases like that of sociobiology/evolutionary psychology, in which those very factors conduced to the systematic scientific ratification of ideology rather than to uncovering the truth, helps us to see that the prospects, such as they are, for progress towards the truth depend crucially on political factors in both sorts of cases.

The lesson of the present discussion is that there are important cases in which the operation of scientific methodology – even when practiced conscientiously and rationally by smart people – is inadequate, in the absence of political transformation, to overcome the ways in which social ideology introduces malignancy into findings, methods, and even concepts and meanings in the sciences.

5. BROADER IMPLICATIONS

5.0. Commensurability and Incommensurability in Ethics

Standard relativist arguments against moral realism are similar in structure to Kuhn's arguments for referential incommensurability: they identify conceptual differences between different traditions of moral inquiry and practice, treat these as representative of differences in meaning, and conclude that the traditions in question cannot share a common subject matter. I have elsewhere (Boyd, 1988) maintained that these arguments are not decisive against moral realism and that an appreciation of the semantics and epistemology of scientific inquiry can help us to see that moral realism is a serious possibility in metaethics.

What I want to do here is to indicate ways in which – even if moral realism is correct (as I believe it is) – there may be insights as well as errors in the relativist treatment of metaethical issues. If we think of moral reasoning not primarily as the province of professional philosophers but as something engaged in by members of various different social, religious, national, and professional groups then, like sociobiology, moral inquiry within many of these groups will have the property that those who engage in it are largely insulated from critiques from inquirers in other groups. Moral inquiry certainly shares with inquiry in the human sciences the property that it is embedded (deeply embedded!) in social ideology.

Thus moral inquiry shares the properties which, I conjecture, are especially likely to give rise to malignant meanings and consequent incommensurability. I want to explore briefly some further consequences of these properties of moral inquiry.

First, it is important to see that the embedding of moral inquiry within political ideology is manifest along several different dimensions. One is obvious: there is a systematic tendency for the prevalent moral views at a time to be such that their acceptance serves the needs of the powerful. Indeed, so strong is this tendency that the debunking critique of morality attributed to Marx has much to recommend it (see Boyd,

1988; 1995; Miller, 1985; Wood, 1972; 1979; 1984; Gilbert, 1981; 1982; 1984, for discussions).

Secondly, although morality need not (I think should not) be thought of as having a religious foundation, there is in fact an intimate connection between moral inquiry and religious institutions and practices. In so far as these are instruments of political ideology, their ideological embedding will extend to moral inquiry as well.

Moreover, whatever else moral inquiry may be about, it certainly involves inquiry about human nature and about the effects of possible social policies on human well being; so it is about human social structures as well as about human nature. Thus moral inquiry centrally includes topics which are, along with religion, the paradigm *loci* of the influence of political ideology.

Finally, consider the relative isolation of communities of moral inquirers from each other which, I am suggesting, can be expected to contribute to the emergence of malignant meanings and incommensurability. This isolation is itself partly a phenomenon of ideology in the political sense. A central social function (*the* social function, if you're inclined to be pessimistic) of moral discourse and practices is to protect the interests of the powerful. [I think that Marxists are right to think of these interests as the interests of ruling classes, but people with radically different conceptions about how political and economic power work can agree that morality serves the powerful.]

The capacity of morality to serve this ideological function is, in turn, greatly enhanced by the phenomenon of isolation of communities of moral inquirers. Consider, for example, cases in which racist anger is directed at the allegedly immoral behaviors or character traits of isolated minorities in order to deflect criticisms from unpopular economic policies, or in which chauvinistic moral condemnation of the "enemy" is used to rationalize unjust wars. Neither sort of ideological deployment of moral sentiments would serve its function of social control with nearly the same effectiveness in situations in which there was an appreciation, on the part of members of the target group for the propaganda, of the moral perspective and outlook of the scapegoated group or of the "enemy." It is thus unsurprising that ordinarily social mechanisms are in place to create and rationalize the required isolation between moral communities. Such isolation is a component of the (political) ideological role of appeals to morality.

I am thus inclined to think that the battery of ideological forces arrayed against the prospects for progress in moral understanding, and thus against the establishment of commensurability between moral communities, is much more substantial even than those arrayed against progress and commensurability in many of the human sciences (not, perhaps, more substantial than in: military and diplomatic history, the economics of poverty, the psychology of gender differences, or the genetics of intelligence). It follows that, even on the moral realist assumption (which I endorse) that it has a real subject matter, moral inquiry may be especially vulnerable to the phenomenon of incommensurability and that overcoming incommensurability may crucially involve political struggle against those interests whose power prevailing ideology rationalizes.

5.1. Concluding Radical Postscript

In a certain sense the question of commensurability and incommensurability with respect to competing paradigms or research traditions is the question of the possibility of *objectively* resolving the differences between them. In the present essay I have followed standard scientific realist practice in arguing that Kuhn's arguments for referential incommensurability fail to demonstrate permanent incommensurability for the cases he considers and I have offered a conception of the possibility of the emergence of commensurability as a result of suitable dialectical interactions between practitioners in competing paradigms between which there are not initially the methodological resources for commensurability.

Nevertheless I have made a gesture towards Kuhnian arguments, and towards the sorts of postmodernist relativism they are often taken to support, by agreeing with Kuhn, and with many postmodern thinkers that:

1. Scientific (and other) terms have conceptual meaning,
2. there can be conflicts of meaning between competing paradigms or research traditions, such that
3. they seriously compromise the near term prospects for commensurability.

I have argued as well that

4. The phenomenon of malignant meanings and of the incommensurability they can occasion are indicative of a sort of relativism about scientific rationality.

I have also agreed with a point more often made by postmodern thinkers than by Kuhnian ones, namely, that

5. Features of social ideology can be parts of the conceptual meanings of scientific (or other) terms.

I want to explore the relationship between the points made in the present essay and the proper understanding of the notion of objectivity. It is part of the stereotype of objectivity that objective methods are theory-neural and thus immune to social ideology (and to individual idiosyncracy as well). I have suggested, in my discussion of the scientific prestige of reductive explanations, that there is a complex ideological association between scientific objectivity, reductionist explanations and "toughminded" cynical conceptions in the human sciences. These features of the prevailing conception of objectivity are, of course, matters of political ideology: the associations with reductionist toughmindedness are part of a broader pattern of scientific rationalization for the prevailing relations of power and wealth, and the conception that science is "objective," and *thus immune from ideological influence* itself serves the *ideological* function of enhancing the credibility of the political rationalizations produced by the sciences.

There is, however, such a phenomenon as scientific objectivity and – just as our stereotypical conception suggests – it serves to explain those cases in which scientific methodology is systematically and often spectacularly epistemically reliable. Scientific objectivity is a matter of the reliable operation of scientific methods to systematically produce real knowledge, and scientific methods *do* exhibit this property under a variety of historically important circumstances. Where our stereotypical conception is wrong is that it gets the identification of those circumstances essentially backwards.

Scientific methods, when they are reliable, are reliable because they are theory-dependent *and* because the background theories upon which their applications depend are relevantly approximately true, *but* even the availability of relevantly accurate background theories will not render the methods of science reliable *unless* the scientific disciplinary practices in question are situated appropriately with respect to sources of political and economic power.

In ideological contexts like the present one, in which scientific findings serve to ratify profound inequities in power relations, ordinary institutional science can function in an epistemically reliable way with respect to questions about 'human nature' *only* when political developments largely external to institutional science create temporary shifts in the balance of political influences within scientific institutions. It is not an accident, for example, that critiques of biological determinist cynicism about the prospects for social and political progress – in so far as they have any significant impact on mainstream professional life in the human sciences – do so only in the context of large scale and (largely) external political struggles for the progress in question. This was, for example, the case with respect to critiques of sociobiological methods and with respect to related critiques of the professional literature on the genetics of intelligence (see, e.g., Block and Dworkin, 1976).

In so far as scientific objectivity regarding politically controversial issues in the human sciences is possible in the absence of such large scale movements, it is always within the context of oppositional movements which substitute a politically less misleading context for that which obtains in mainstream scientific institutions. The role which scholars close to the Communist Party played in the U.S. in the 1930's and 1940's in keeping alive a tradition of objective study of racism and its effects provides a useful example here.

Thus, far from depending on insulation from politics, objectivity about politically controversial matters in the human sciences (and elsewhere) is, when it obtains, *always* a political achievement. Consider now cases in which powerful interests sustain scientific institutions in which methods reliably lead to real knowledge, when that sort of epistemic success is in their interests. The establishment and maintenance of such institutions, and of their epistemic reliability (when it is wanted), is *also* a political achievement of sorts, even in those cases – military research for example – where epistemic successes are not especially to be admired. Scientific objectivity, when it obtains, is thus *always* a (partly) political achievement.

So, scientific objectivity is a theory-dependent, politics-and-power dependent phenomenon, and the view that it arises under conditions of theory-neutrality and political immunity is itself a piece of ideology in the political sense of the term.

Now many postmodernist critics of science have maintained that the notion of scientific objectivity is an essentially political or ideological notion, and that it is a mistake for critics of the *status quo* or of the ideologies which support it to deploy the notion of scientific objectivity. They often hold that the notion of objectivity (and, for example, the notions of truth and of knowledge) are political tools of oppression, inappropriate for use in the context of ideology critique.

Obviously I dissent from this view, but I think that the resources of the present essay allow us to appreciate an important grain of truth in it. I am inclined to think that the ideological conception of objectivity we have been discussing is, in contemporary educated circles (at least in the U.S. – I haven't sampled adequately elsewhere), constitutive of the conceptual meaning of the term "objectivity." This is, I believe, one of those cases in which ideologically determined elements are pretty clearly parts of the meaning in academic discourse of a politically important term.

If this is so then it is true, as critics of objectivity maintain, that the *concept* of objectivity, as that concept is manifest in contemporary intellectual discourse, is a tool of oppressors, at least in so far as its applications in politics and in the human sciences and related areas of inquiry are concerned.

On the other hand, the *phenomenon* of objectivity – the epistemically reliable deployment of scientific methods – is essential to projects of social criticism and ideology critique. In the human sciences and related domains, scientific objectivity cannot – absent wholesale political and economic change – be achieved within the normal working of institutional science, except in rare contexts of sustained political struggle. It can be obtained outside the normal workings of institutional science only in the contexts provided by oppositional movements.

It is probably also true that only in the contexts provided by political struggle can the concept of objectivity be freed from its malignant meanings. In almost every respect the critics of scientific objectivity are thus correct about the depth of the ideological association between the concept of scientific objectivity and processes of economic and political oppression. They are right as well to think that the situation can only be remedied politically. But they need not – and indeed must not, if they are to succeed – abandon the projects of criticism to which concepts of objectivity, knowledge, and truth are essential.

In an era in which we are routinely called upon to celebrate the defeat of the left and the emergence of a globalized capitalist economy, it is a good idea to reflect that the points just made about the special epistemic role of political struggle and of oppositional political institutions are part of the legacy of Marxist (and in particular of Leninist) political theory. They have not outlived their usefulness.

Cornell University

ACKNOWLEDGEMENTS

I wish to thank Eric Hiddleston, Karen Jones, Barbara Koslowski, Jeff Roland, Susanna Siegel, Janson Stanley, Nicholas Sturgeon, Christopher Sturr, Zoltan Gendler-Szabo, and Aaron Zimmerman for many helpful comments and criticisms.

NOTES

[1] In the Editors' Introduction the editors distinguish between the *semantic incommensurability thesis*, according to which alternative scientific theories may be incommensurable due to *semantic variance* in their terms and the *methodological incommensurability thesis*, according to which they may be *incommensurable* because of the absence of *common standards of theory appraisal*. To a good first approximation, what they mean by "incommensurability" is what I here call "methodological incommensurability." The semantic incommensurability thesis is the thesis that theories may be incommensurable (in their sense) due to what

I call "semantic incommensurability," whereas the methodological incommensurability thesis is the thesis that theories may be incommensurable due to differences in the theory-dependent evidential standards accepted by their defenders.

2 I am here adopting (not just for the sake of argument) what I think of as the standard interpretation of the distinctly philosophical arguments for incommensurability that are presented in Kuhn (1970). I think that the semantic theory apparently relied on in Kuhn's discussion of Newtonian and Einsteinian mechanics (p. 102) should be seen as absolutely central to the his general defense of incommensurability because, even though this point is not explicitly recognized by Kuhn, only by arguing that there are changes in the physical referents of scientific terms during scientific revolutions could he defend the strong thesis of methodological incommensurability which he appears to advance. If, for example, terms like "mass" are understood to be coreferential in Newtonian and Einsteinian mechanics then there are lots of experiments, *prima facie* acceptable by both Newtonian and Einsteinian standards, which provide evidence for the velocity dependence of mass.

Other interpretations of Kuhn are possible which attribute to him less dramatic conceptions of semantic and methodological incommensurability (see, e.g., Hoyningen-Huene, 1993). It is not my aim here to dispute them. Instead, I am concerned to argue that, even if Kuhn's technical arguments for incommensurability are as unpersuasive as I believe they are, the phenomena of semantic and methodological incommensurability to which he drew attention are even more widespread and important than he knew. I do not, however, advance my theses about malignant meanings as an exegesis of Kuhn's own views. To my knowledge he did not explore examples of the sort I discuss.

3 What I have in mind is this. There might be two paradigms or research traditions *with the same subject matter* but which were such that they employed the same natural kind (magnitude,...) terms to refer to different phenomena. Imagine, for example, two chemical paradigms which employed the same terminology for the elements but assigned different names to the same element, one using the term "hydrogen," for example, for the element the other called "oxygen." Obviously it might be possible for practitioners in the two paradigms to figure this out and to agree on some new standard terminology – which would have to be different from the terminology of at least one of the paradigms. Once this redefinition of terms had taken place the *redefined* paradigms would no longer be referentially incommensurable and that barrier to methodological incommensurability would have been overcome. Absent this sort of redefinition however, even if practitioners in two referentially incommensurable paradigms believed that they had identified fair methods for resolving the differences between their positions they would be mistaken, since those methods would inevitably be unreliable in the light of the (undetected) ambiguity in the terms they jointly employed.

4 For a more detailed account of reference which makes such connections definitive both of reference and, in a sense, of kinds, properties, magnitudes, etc., see Boyd (1989; 1990a; 1991; 1993; 1999).

5 In thus characterizing the standard reply, and in offering Putnam (1975a) as one of its primary sources, I am placing diminished emphasis on Putnam's claim in that paper that a knowledge of *stereotypes* is a component of expert level knowledge of meaning for natural kind terms. This proposal does, of course, assign a semantic role to a conceptual aspect of "meaning" and it has affinities with the law-cluster account. In Section 1.4 I explain why, in my view, it has been deservedly uninfluential. For the present it suffices to remark that the considerations which establish it that law clusters do not play any essential descriptive role in reference fixing also establish this point with respect to stereotypes.

6 Optimality models involve calculations to determine what behaviors (or other features of organisms' phenotypes) would represent the reproductively optimal solution – under the conditions in which selection took place – to the evolutionary "problems" facing the organisms in question. Those who deploy such models then retrodict that evolutionary scenarios eventuating in the optimal "solutions" probably took place. It is this feature of sociobiological argumentation, and of many other sorts of evolutionary explanation which is the special target of Gould and Lewontin (1979). Note that none of the criticisms of extrapolative socio-biological inferences discussed here reflect any criticisms of optimality modeling. I do not have a settled view on this matter.

7 There is another pattern of inference also represented in the evolutionary psychology literature. Some authors hold that evolutionary theory predicts that humans will exhibit reproductively optimal behaviors not only in the EEA but all subsequent environments, including the present ones, as well. I have focused on the alternative inference pattern from optimal behaviors in the EEA to innate, perhaps unconscious and (relatively) non-malleable motives for three reasons. First, those authors who deploy the "optimality everywhere" strategy, like those who deploy the strategy discussed here, are tacitly and unjustifiably inferring innate and non-malleable (usually unconscious) motivational whose propositional content is given by the

alleged evolutionary function of the relevant behaviors. Second, many authors do not clearly distinguish between these inferential strategies. So important is the role of standard "inferential scripts" (see Section 3.5) in this literature that sometimes the mere mention of "optimality modeling" suffices to address the evolutionary questions at issue, even though this does not distinguish between the two inferential strategies; this is true, for example, in Daly and Wilson (1997), discussed below.

Finally, it is my aim here to argue that standard inference patterns in evolutionary psychology are *malignant* components of the meanings of relevant terms, in the sense discussed above. I did not want to stack the cards in my favor by focusing on an inferential strategy which is even less credible that the one I discuss at length. It is widely recognized that there are deep problems with positing optimality everywhere (see Cosmides and Tooby, 1987). In particular, if it turned out that animals routinely exhibited near optimal adaptations to environments quite different from those in which they evolved, then Darwinian evolutionary theory would have to be abandoned in favor of (something like) a theory of adaptation by design! [Sherman and Reeve (1997) offer an even more sophisticated taxonomy of inference patterns in evolutionary psychology. They consider applications of the "optimality everywhere" strategy in those special cases in which a pattern of behavior, characterized in what I have been calling *ecological parameters* (see Section 2.3.3), might have been subject to constant positive natural selection up until the very recent past. The considerations rehearsed in Section 2.3.3 raise doubts about whether constant selection for an *ecologically characterized* pattern of behavior must be reflected in the sort of constancy of selection at the *psychological* level which would be presupposed by the "optimality everywhere" inferential strategy in such cases, but this is an issue beyond the scope of the present paper.]

8 The tendency in evolutionary psychology to fail to appreciate the tacit psychological presuppositions at work in inferences from evolutionary scenarios to conclusions about human behaviors in environments different from those characteristic of the EEA is reflected in (and encouraged by) the widespread view that evolutionary explanations for behaviors and explanations in terms of proximate neurological or psychological mechanisms are always non-competitors because they operate at different "levels of analysis," so that evolutionary psychology can be conducted largely independently of psychological or neurological research. [For applications of this view see, e.g., Alexander (1987), Sherman and Reeve (1997), Thornhill and Thornhill (1990).]

Of course, the *correct* account of a phenotypic trait (whether psychological, neurological, or whatever) will be compatible with the *correct* account of its evolutionary history. Thus when evolutionary psychologists adopt an *explanatory* approach – when they seek to provide evolutionary explanations for aspects of human psychological phenotypes (like, e.g., innate and non-malleable motives) which have been *independently identified by psychological research* – they can reasonably expect that the results of their research will complement, rather than conflict with, the findings of psychological research. In the case of *extrapolative* human evolutionary psychology however the aim of research is to *do psychological research*: to identify general features of human psychology or behavior which persist under conditions very different from those in the EEA. Here there could not possibly be methodological independence from proximal psychology, since the inferences evolutionary psychologists make from evolutionary scenarios are proposed as a way of *doing proximal psychology* – of identifying enduring *proximal* dispositional or motivational features of human psychology, suitable for predicting or explaining behaviors outside the EEA. Nevertheless there is a widespread conviction among practitioners that evolutionary psychology and the study of proximal psychological mechanisms are methodologically independent.

9 The conception that characterizations of behaviors in terms of ecological parameters, together with optimality modeling, can somehow underwrite the study of human behavior in environments quite different from the EEA helps to explain the attractiveness of the "optimal everywhere" inferential strategy mentioned in footnote 7.

10 For the record, I think that there is another perfectly good sense of *rationality* in which conscientious scientific practice counts as rational only in so far as it is associated with some level of systematic epistemic success (see, e.g., Boyd, 1992; 1990b). This is not, however, the sense which is relevant when we address questions about rational methodological commensurability between competing research traditions or "paradigms."

11 For an excellent discussion of the ways in which elements of ideology can be conceptually normative, and a much more sophisticated treatment of the phenomenon of ideology than that offered here, see Sturr (1998).

REFERENCES

Alexander, R. (1979). *Darwinism and Human Affairs*. Seattle: University of Washington Press.
Alexander, R. (1987). *The Biology of Moral Systems*. New York: Aldyne de Gruyter.
Barash, D. (1979). *The Whisperings Within*. New York: Harper and Row.
Betzig, L., ed. (1997). *Human Nature: A Critical Reader*. New York: Oxford University Press.
Block, N. and G. Dworkin, eds. (1976). *The IQ Controversy*. New York: Pantheon.
Boyd, R. (1983). "On the Current Status of the Issue of Scientific Realism." *Erkenntnis* **19**: 45-90.
Boyd, R. (1985a). "Lex Orendi est Lex Credendi." In P.Churchland and Hooker, eds., *Images of Science: Scientific Realism Versus Constructive Empiricism*. Chicago: University of Chicago Press.
Boyd, R. (1985b). "Observations, Explanatory Power, and Simplicity." In P. Achinstein and O. Hannaway, eds., *Observation, Experiment, and Hypothesis In Modern Physical Science*. Cambridge: MIT Press.
Boyd, R. (1988). "How to be a Moral Realist." In G. McCord, ed., *Moral Realism*. Ithaca: Cornell University Press.
Boyd, R. (1989). "What Realism Implies and What It Does Not." *Dialectica* **43**: 5–29.
Boyd, R. (1990a). "Realism, Approximate Truth and Philosophical Method." In W. Savage, ed., *Scientific Theories, Minnesota Studies in the Philosophy of Science. Volume 14*. Minneapolis: University of Minnesota Press.
Boyd, R. (1990b). "Realism, Conventionality, and 'Realism About'." In G. Boolos, ed., *Meaning and Method*. Cambridge: Cambridge University Press.
Boyd, R. (1991). "Realism, Anti-Foundationalism and the Enthusiasm for Natural Kinds." *Philosophical Studies* **61**: 127–148.
Boyd, R. (1992). "Constructivism, Realism, and Philosophical Method." In J. Earman, ed., *Inference, Explanation and Other Philosophical Frustrations*. Berkeley: University of California Press.
Boyd, R. (1993). "Metaphor and Theory Change." (second version) In A. Ortony, ed., *Metaphor and Thought*. 2nd edition. New York: Cambridge University Press.
Boyd, R. (1995). "Postscript" to "How to be a Moral Realist." In P. Moser and J.Trout, eds., *Contemporary Materialism: A Reader*. New York: Routledge.
Boyd, R. (1999). "Kinds as the 'Workmanship of Men': Realism, Constructivism, and Natural Kinds." *Proceedings of the Third International Congress, Gesellschaft für Analytische Philosophie*. Berlin: de Gruyter.
Buss, D. (1989). "Sex Differences in Human Mate Preferences: Evolutionary Hypotheses Tested in 37 Cultures." *Behavior and Brain Sciences* **12**: 1–14, as reprinted in Betzig 1997.
Carnap, R. (1950). "Empiricism, Semantics and Ontology." *Revue internationale de philosophie*, 4th year.
Cosmides, L. and J. Tooby. (1987). "From Evolution to Behavior: Evolutionary Psychology as the Missing Link." In J. Dupre, ed., *The Latest on the Best: Essays on Evolution and Optimality*, pp. 277–306, Cambridge: MIT Press.
Daly, M. and M.Wilson. (1997). "Child Abuse and Other Risks of not Living with Both Parents." In L. Betzig, ed., *Human Nature: A Critical Reader*. New York: Oxford University Press.
Feigl, H. (1956). "Some Major Issues and Developments in the Philosophy of Science of Logical Empiricism." In H. Feigl and M. Scriven, eds., *Minnesota Studies in the Philosophy of Science, Volume 1*. Minneapolis: University of Minnesota Press.
Field, H. (1973). "Theory Change and the Indeterminacy of Reference." *Journal of Philosophy* **70**: 462-481.
Gilbert, A. (1981). "Historical Theory and the Structure of Moral Argument in Marx." *Political Theory* **9**: 173–205.
Gilbert, A. (1982). "An Ambiguity in Marx's and Engel's Account of Justice and Equality." *American Political Science Review* **76**: 328–346.
Gilbert, A. (1984). "Marx's Moral Realism: Eudaimonism and Moral Progress." In J. Farr and T. Ball, eds., *After Marx*. Cambridge: Cambridge University Press.
Goodman, N. (1973). *Fact, Fiction and Forecast*. 3rd edition. Indianapolis and New York: Bobbs-Merrill.
Gould, S. and R. Lewontin (1979). "The Spandrels of San Marco and the Panglossian Paradigm: A Critique of the Adaptationist Program." *Proc. Roy. Soc. Lon. (B)* **205**: 581–598.
Hanson, N. (1958). *Patterns of Discovery*. Cambridge: Cambridge University Press.
Hoyningen-Huene, P. (1993). *Reconstructing Scientific Revolutions: Thomas S. Kuhn's Philosophy of Science*, trans. A. Levine. Chicago: University of Chicago Press.

Kitcher, P. (1985). *Vaulting Ambition: Sociobiology and the Quest for Human Nature*. Cambridge, Mass.: MIT Press.
Kripke, S. (1971). "Identity and Necessity." In M. Munitz, ed., *Identity and Individuation*. New York: New York University Press.
Kripke, S. (1972). "Naming and Necessity." In D. Davidson and G. Harman, eds., *The Semantics of Natural Language*. Dordrecht: Reidel.
Kuhn, T. (1970). *The Structure of Scientific Revolutions*. 2nd edition. Chicago: University of Chicago Press.
LeVay, S. (1991). "A Difference in Hypothalamic Structure Between Heterosexual and Homosexual Men." *Science* **253**: 1034–1037.
LeVay, S. (1993). Untitled Short Editorial Piece, *The Nation*, July 5, 1993.
Lewontin, R. (1976). "Sociobiology – a Caricature of Darwinism." In *PSA 1976 Volume 2*, pp. 22–31.
Lumsden, C. and E. Wilson. (1981). *Genes Mind and Culture: The Coevolutionary Process*. Cambridge, Mass.: Harvard University Press.
Miller, R. (1985). "Ways of Moral Learning." *Philosophical Review* **94**: 507–556.
Pinker, S. (1996). *How the Mind Works*. Cambridge: MIT Press.
Putnam , H. (1962). "The Analytic and the Synthetic." In H. Feigl and G. Maxwell, eds., *Minnesota Studies in the Philosophy of Science, Volume3*. Minneapolis: University of Minnesota Press.
Putnam, H. (1972). "Explanation and Reference." In G. Pearce and P. Maynard, eds., Dordrecht: Reidel.
Putnam, H. (1975a). *Mind, Language and Reality. Philosophical Papers, Volume 2*. Cambridge: Cambridge University Press.
Sherman, P. and H. Reeve. (1997). "Forward and Backward: Alternative Approaches to Studying Human Social Evolution." In L. Betzig, ed., *Human Nature: A Critical Reader*. New York: Oxford University Press.
Sturgeon, N. (1984). "Moral Explanations." In D. Copp and D. Zimmerman, eds., *Morality, Reason and Truth*. Totowa, N.J.: Rowman and Allanheld.
Sturr, C. (1998). *Ideology., Discursive Norms and Rationality*. Ph.D. Dissertation, Ithaca, New York: Cornell University.
Thornhill, R. and N. Thornhill. (1990). "An Evolutionary Analysis of Psychological Pain Following Rape: I. The Effects of Victim's and Marital Status." *Ethology and Sociobiology* **11**: 155–176.
Trivers, R. (1971). "The Evolution of Reciprocal Altruism." *Quarterly Review of Biology* **46**: 35–39, 45–47.
Williams, G. (1975). *Sex and Evolution*. Princeton: Princeton University Press.
Wilson, E. (1975). *Sociobiology: The New Synthesis*. Cambridge, Mass.: Harvard University Press.
Wilson, E. (1978). *On Human Nature*. Cambridge, Mass.: Harvard University Press.
Wood, A. (1972). "The Marxian Critique of Justice." *Philosophy and Public Affairs* **1**: 244–282.
Wood, A. (1979). "Marx on Right and Justice: A Reply to Husami." *Philosophy and Public Affairs* **8**: 267–295.
Wood, A. (1984). "A Marxian Approach 'to the Problem of Justice.'" *Philosophica* **33**: 9–32.

MARTIN CARRIER

CHANGING LAWS AND SHIFTING CONCEPTS

On the Nature and Impact of Incommensurability

Abstract. "Semantic incommensurability", i.e., non-translatability of concepts taken from different theories, is at the focus of the argument. I attempt to give a rational reconstruction of the notion underlying the writings of Feyerabend and the later Kuhn. I claim that such a coherent notion can be identified and that relevant instances exist. Incommensurability is brought about by theoretical incompatibility. The translation failure between incommensurable concepts arises from the impossibility of jointly fulfilling two conditions of adequacy that the context theory of meaning places on translations. Potential conceptual analogs either fail to preserve the conditions of application or to reproduce the relevant inferential relations. This feature turns out to be correlated with a cross-classification of the pertinent scientific kinds. These relations between incommensurable concepts are sufficient for making an empirical comparison of the claims couched in these concepts possible.

1. INTRODUCTION

Incommensurability is among the catchwords of later 20th century philosophy of science. The notion of incommensurability in the non-geometrical sense relevant here was simultaneously introduced by Thomas S. Kuhn and Paul K. Feyerabend in 1962 (Kuhn, 1962, p. 103; Feyerabend, 1962, p. 58). Kuhn conceived of incommensurability as a contrast between paradigms or comprehensive theoretical traditions that transcends mere incompatibility. The adoption of a new paradigm entails the restructuring, as it were, of the relevant universe of discourse; the adherents of the two paradigms tend to talk past one another. In particular, incommensurability is intended to express that, first, disparate concepts are employed in each of the theories at hand, second, distinct problems are tackled, third, the suggested problem solutions are evaluated according to different standards, and finally, perceptions are structured differently (Kuhn, 1962, pp. 103–110, 148–150). Feyerabend, by contrast, focused on the "inexplicability", that is, the non-translatability of a term taken from one theory into the conceptual framework of another one incompatible with the first.

While the initial use of the term "incommensurability" varied significantly, it was subsequently restricted to denote the non-translatability of statements from different, strongly contrasting theories. In the following I exclusively address this more limited notion of incommensurability that is sometimes called "semantic incommensurability". In his later years, Kuhn attempted to trace incommensurability back to changing assumptions about what is alike or what is of the same kind. Incommensurability was supposed to represent a translation failure resulting from conflicting structures of kinds; scientific, natural, or otherwise.

P. Hoyningen-Huene and H. Sankey (eds.). Incommensurability and Related Matters, 65–90.

My aim is to give a systematic reconstruction of the nature and impact of incommensurability. The relevant notion is the one entertained by Feyerabend and the later Kuhn. Underlying the conception of incommensurability is the theoretical context account of meaning which both authors adopted for explaining the meaning of scientific terms. My "rational reconstruction" of incommensurability proceeds from this context account as the central premise. For reasons to be explained later, I do not follow Kuhn in regarding incompatible structures of scientific kinds as the origin of incommensurability. Rather, shifting taxonomies are construed as a derivative feature that ultimately arises from a substantial alteration of the relevant theories. The adoption of a system of laws entails assumptions as to what is alike and what is not.

In particular, I try to show, first, that incommensurability can be reconstructed coherently on the basis of the context account. Incommensurability is presented as a consequence of this semantic theory along with the historical observation that substantial theoretical revisions occur indeed. That is, incommensurability qualifies as a sensible notion. Second, I defend the coherence of the notion of incommensurability by presenting relevant examples. My chief case concerns the non-translatability of the concept of "phlogiston" into the conceptual framework of the succeeding oxygen theory. This presentation is intended to further buttress the claim that incommensurability is real and instantiated. Third, I explore the impact of incommensurability on empirical comparability. Incommensurability was perceived in some quarters as a major threat to the possibility of rational theory evaluation. My aim is to dispel such worries. Incommensurable theories admit of empirical comparison. The argument proceeds on the basis of the semantic principles adopted by Kuhn and Feyerabend. That is, given their own linguistic approach, incommensurability does not issue in a breakdown of comparing empirical achievements of the theories in question. Empirical comparison does not require translation and remains largely unaffected by incommensurability.[1]

2. MEANING, INFERENTIAL RELATIONSHIP AND THEORETICAL CONTEXT

The context theory of meaning or inferential role semantics has grown out of aphoristic remarks of the later Wittgenstein to the effect that the *use* of a concept determines its meaning (Wittgenstein, 1953, § 43). This use is fixed in turn by the relations the concept exhibits to other concepts, that is, by its conceptual integration. To specify the meaning of an expression is to sketch the role it plays in the relevant linguistic community. Wilfrid Sellars and Norwood Hanson were the key figures in transforming Wittgenstein's aphorisms into a philosophical theory. Following Wittgenstein, Sellars assumes that the linguistic role of an expression is determined by rules; meaning is established by a system of rules. Sellars' innovation concerns the invocation of "inferences" for characterizing meaning. Such "inference rules" license the transitions between the application of predicates or the acceptance of sentences. The meaning of a linguistic expression is captured by the entirety of those relations to other such expressions that are deemed legitimate in the pertinent linguistic community.

For instance, it is part of the meaning of color terms that the predicate "*x* is red" is implied by "*x* is scarlet", and implies "*x* is colored" and "*x* is not green". Likewise, application of the predicate "*x* is jealous" implies "*x* is human" and "*x* is sexually active", whereas the predicate "x was founded in 1214" (which is assumed to truthfully apply to the city of Bielefeld) implies "x is not sexually active" (which indeed holds of the just mentioned entity). These examples show that the relevant inferences are not generally part of formal logic. Rather, informal, content guided connections are at the primary focus. In the case of scientific concepts, nomological, i.e., factual, relations contribute to their meaning. The transition from the predicate "x is a magnet" to the predicate "x aligns itself in the north-south direction if suspended freely" is justified by laws of nature (along with assumptions concerning boundary conditions). Considerations of this sort show that the theoretical context of a term is not neatly divided into semantic and factual components (as suggested by the traditional analytic-synthetic distinction). Rather, the context theory is tied up with semantic holism (to variable degrees, depending on the version under consideration). In sum, laws and theories supply a concept with a network of relations to other concepts; they provide the context of inferential rules that fixes the use of a linguistic item and determines its meaning. Meaning is dependent on theory or on the pertinent system of beliefs (Hanson, 1958, chapters II–III, Sellars, 1979, pp. 121–122; Churchland, 1979, pp. 46–49; Marras, 1992, pp. 713–715).

Kuhn favors a semantic approach of this sort. He claims that "knowing what a word means is knowing how to use it for communication" (Kuhn, 1989, p. 12) and stresses the holistic nature of meaning. Words usually do not have meaning separately; consequently, changing the usage of a term has ramifications as to the meaning of associated concepts (ibid.). In addition, the rules governing the use of scientific concepts are the laws of the pertinent theories. The meaning of the term "mass" is acquired along with Newton's Second Law or, alternatively, with the law of gravitation. The concepts are learned along with the theory itself (see Kuhn, 1989, pp. 15–20; 1983, pp. 576–577; 1987, p. 8; 1989, pp. 15–20; see Irzik & Grünberg, 1995, pp. 297–298).

Feyerabend likewise commits himself to the context account. He endorses the view that "the meaning of every term we use depends upon the theoretical context in which it occurs. Words do not 'mean' something in isolation; they obtain their meanings by being part of a theoretical system." (Feyerabend, 1965a, p. 180) Theories thus circumscribe guidelines for the use of the concepts featuring in their own principles and theorems. These lawful relations contribute to establishing the meaning of the relevant concepts. Nomological generalizations add to the meaning of the terms employed to express the content of these generalizations (Feyerabend, 1962, pp. 76–81; see also Papineau, 1979, pp. 36–45; Sankey, 1994, pp. 6–10).

3. MEANING VARIANCE AND INCOMMENSURABILITY

The context theory entails that a theoretical change in science brings meaning variance of the affected concepts in its train. One of Kuhn's most prominent historical claims is that science indeed develops sometimes through stages of significant and deep reaching conceptual and theoretical alteration. This view finds its most prominent expression in Kuhn's characterization of scientific revolutions. Kuhnian revolutions are conceived as non-cumulative transitions. They do not involve the sustained elaboration and expansion of an accepted conceptual framework. On the contrary, the framework is abandoned and replaced by a disparate one. In the course of a revolution, accepted problem solutions and alleged theoretical achievements are taken back and supplanted by a theoretical treatment of a different kind. A scientific revolution à la Kuhn involves the revocation of fundamental principles of a discipline and their replacement by disparate ones. Furthermore, the disparity between pre- and post-revolutionary principles prohibits any smooth integration of the former into the framework of the latter. As a result of the fundamental divergence between them, the pre-revolutionary theory cannot be reconstructed as the limiting case of the post-revolutionary one (Kuhn, 1962, chapters VIII–X). In the same vein, Feyerabend emphasizes the deep reaching rupture as to concepts and content between earlier and later theories of the same domain. Preceding accounts are not rectified in detail but preserved in principle; rather, they are typically completely set aside and supplanted by their successors. Scientific change involves profound shifts in formalism and ontology (Feyerabend, 1962, pp. 44–45).

Meaning variance, as generated by theoretical alteration, leads naturally to the idea of untranslatability. Feyerabend's example is the failure to fit the notion of "impetus" into Newtonian mechanics. The conception of impetus was part of late medieval physical theory. The underlying idea was that all bodies by their nature tend to assume a state of rest and that, consequently, every movement requires the enduring action of a force. This force need not be external, it may as well be "impressed" in the body. The thrown stone continues to pursue its path even after the contact with the hand of the thrower is lost and no other external force is effective in the direction of motion. In such cases, it is the impressed force or impetus that acts as a kind of internal motor which propels the body against its intrinsic resistance to motion. The impetus thereby increasingly exhausts itself so that it is insufficient for continuing the motion undiminishedly and only retards the body's slowing down. The intensity of a body's impetus was estimated by the product of its weight and velocity. This empirical measure of impetus is conspicuously close to that of the Newtonian concept of momentum (i.e., mass times velocity) which appears to make the two natural candidates for a translation.

However, as Feyerabend argues, this perfunctory similarity veils a fundamental divergence. Its root is the Newtonian law of inertia which conflicts with the privilege impetus physics confers on the state of rest. The Newtonian law construes uniform-rectilinear motion, along with rest, as force free states of motion. On the impetus theoretical approach, impetus effects the continuation of all motion – including uniform-rectilinear motion. However, by Newtonian lights this latter type of motion has no cause. The required force does not exist, and in particular, it is by no means to be identified with momentum (which is at most an effect of motion but certainly does not bring it about). It follows that the concept of impetus cannot be imported into classical mechanics. Finding a conceptual counterpart of "impetus" in Newtonian physics is ruled out by the incompatibility between the tendency to rest and principle of inertia. As a result of this substantive contrast, appending the concept of impetus to Newtonian theory would create an inconsistency: it would amount to stating that the preservation of all motion (including uniform-rectilinear motion) requires the action of some force, while at the same time denying this claim in virtue of the principle of inertia. Due to this inconsistency the concept of impetus cannot be translated into its *prima facie* Newtonian analog, i.e., momentum. The two concepts are incommensurable (Feyerabend, 1962, pp. 52–62).

Accordingly, incommensurability is understood as non-translatability of a concept of one theory into a superficially analogous concept of another theory, whereby the reason of this translation failure is the inconsistency of the theories involved. Feyerabend later blurred this view under critical pressure. Dudley Shapere obejcted that statements lacking shared meaning can neither contradict nor agree with one another. Feyerabend conceded the point and attempted to characterize incommensurability as incompatibility without inconsistency, drawing for this purpose on notions like "the structure of the ontologies" and the "grammatical habits" associated with the pertinent accounts (Sankey, 1994, pp. 13–15).

However, there is no need to withdraw to such a frail position, being in constant danger of collapsing into either compatibility or inconsistency. The gist of Feyerabend's original view could be saved by realizing that the inconsistency only arises within the language of one of the theories involved, but not, strictly speaking; between these theories. The inconsistency is created by any attempt to import a given concept of one of the theories into the conceptual framework of the other. The contradiction is produced by attempted translations. One might object to this construal that if meaning variance is assumed, even potential conceptual analogs could not be identified so that the adequacy of potential translations could not be assessed. Consequently, as the objection might proceed, it is never justified to claim that each potential translation has failed. But this argument is mistaken. As Kuhn is eager to stress, it is possible to understand a concept without being able to translate it into another language (Kuhn, 1983, pp. 671–673; 1993, pp. 320, 324; see section 10). Actually, it follows from the context theory that each of the relevant theories can be learned by acquainting oneself with the pertinent network of conceptual relations. In such cases, understanding comes from within, as it were, not from anchoring the new content in a framework of previously accepted beliefs. The conceptual gap between two theories may be clearly identifiable without at the same time being able to bridge that gap through translation. In the following, I leave Feyerabend's flimsy later views out of consideration and focus on his more tangible earlier approach.

While Feyerabend confined the application of the term "incommensurability" to problems of translation all along, Kuhn started with a much wider concept that included perceptual, methodological, linguistic, and ontological relations. But later, Kuhn approached Feyerabend's notion in singling out linguistic divergence as the crucial feature of incommensurability. Non-translatability is shifted to center stage. Characteristic of the later Kuhn's position in this question is the focus on the relation between scientific or natural kinds in the theories at issue. A theory is thought to comprise a "lexicon" which contains the kind-terms employed by the relevant scientific community. These kind-terms indicate what is taken as being of the same kind; they represent expected similarity relations among objects or processes. Kuhn recognizes that the structure of kinds is tied up with the generalizations that are constitutive of the corresponding account. That is, kind-terms are connected to the laws of the concomitant theory; they represent the relations of similarity or sameness in kind that are implicitly circumscribed by the relevant laws (Kuhn, 1983, pp. 680–684; 1987, pp. 19–21; 1990, p. 5; 1993, p. 316; see Carrier, 1994, pp. 7–8).[2]

Kuhn's claim is that the adequacy of translation may be judged in terms of the preservation of the kind-structure. Translation succeeds if the lexical structure or taxonomy is retained; conversely, incommensurability arises in cases of divergent lexical structures. Discrepancy among such taxonomies is the chief obstacle to translation (Kuhn, 1983, p. 683; 1993, pp. 325–326; see Hoyningen-Huene, 1993, pp. 217–218). Kuhn attempted to make this condition of kind-preservation more precise by appeal to the "principle of no overlap". The only taxonomic relation that doesn't thwart translation is class inclusion: "no two kind-terms ... may overlap in their referents unless they are related as species to genus." (Kuhn, 1990, p. 4) If class inclusion obtains the old categories are imported as intact wholes into the new taxonomy. If, by contrast, the principle is violated, the overlap among the referents of the affected concepts is only partial. In such cases, scientific kinds are torn into pieces, and the debris is reassembled to form novel, disparate kinds. What was formerly considered as being of the same nature may be regarded as heterogeneous afterward. Conversely, what was thought to be different in kind may be taken as conceptually unified in the new theory. It follows that two items which fall under the same category on one account are possibly denoted using different concepts on the other account. And two items labeled distinctly in one approach could be addressed uniformly in the other approach. Cross-classification of this sort vitiates translation; incommensurability is the result (Kuhn, 1990, pp. 4–5; 1993, pp. 318–319; see Irzik & Grünberg, 1995, p. 299; 1998, pp. 211–212).[3]

The situation can be illustrated using one of Kuhn's own examples. The Copernican Revolution induced a change in meaning and reference of the concept "planet". Geocentrically speaking, "planet" means celestial body rotating around the earth. Mars, Sun and Moon equally qualify as planets in this sense while the Earth does not. Heliocentrically speaking, "planet" means celestial body revolving around the Sun. According to this changed understanding, Earth and Mars pass as planets, while neither Sun nor Moon do. These changes in meaning are intertwined with a shift in the taxonomy of the relevant bodies. Geocentrically, Mars, Sun and Moon form part of the same scientific kind, whereas the Earth was subsumed under a different heading. After the revolution, Earth and Mars were equal in kind, while Sun and Moon shifted in kind in becoming a star and a satellite, respectively. The members of the geocentric kind "planet" were scattered into distinct taxa and placed alongside entities which were formerly taken to be of heterogeneous nature. The no-overlap principle is violated with the result that the geocentric concept "planet" proves untranslatable into its alleged heliocentric counterpart. We are faced with an example of incommensurable concepts (Kuhn, 1962, pp. 115, 128–129; 1987, p. 8; Sankey, 1997, pp. 434, 442; Irzik & Grünberg, 1998, p. 211).

Plausible as it may appear at first sight, though, Kuhn's distinction of the relationship of the relevant kind-structures as the pivot of translatabiltiy is mistaken. Neither does taxonomic agreement vouch for translatability, nor does taxonomic divergence rule out translatability. The first point was successfully argued by Howard Sankey by laying emphasis on the fact that a taxonomic structure à la Kuhn is a purely extensional feature. A taxonomy comprises ordered sets of entities which are presumed to be of the same kind, while the criteria determining membership in these sets are held to be variable and intentionally left out of consideration (Kuhn, 1983, p. 683). A scientific kind is a class of entities assumed to be alike – regardless of the respect in which they are thought to agree. However, as Sankey points out, this is not enough for translation. The general counterexample is co-referring, non-synonymous expressions (Sankey, 1994, pp. 98–100). Think of the venerable case of featherless bipeds and rational animals. The corresponding expressions co-refer; they equally denote the class of all humans. But their linguistic role and argumentative context do not coincide; after all, plucked chickens are inferentially connected exclusively to the first term. Consequently, both expressions differ in meaning and thus fail as translations.[4]

Second, mere taxonomic difference does not suffice to thwart translation. It has to be ruled out, in addition, that the taxonomies in question can be amended and adapted to one another. Taxonomic divergence that is supposed to bear on the notion of incommensurability has to be a matter of principle. Consider the following fictitious case of taxonomic discrepancy in botany. Let one language differentiate between trees and bushes and another one distinguish between deciduous plants and conifers. Each of these languages is more precise in one respect and less precise in another. As a result, one encounters a partial overlap among the relevant equivalence classes; e.g., some conifers are trees and some are bushes. So, the principle of no-overlap is violated and translation should fail. In fact, one is at a loss to identify a conceptual analog to "tree", say. "Conifer" is too narrow since there are deciduous trees as well. "Conifer or deciduous plant" is too broad since this would include gorse bushes which are definitely not trees. However, this notion of untranslatability is far too weak to do justice to the intuitions linked with incommensurability. For this taxonomic gap is so shallow that it can be filled without much ado. Nothing prevents us from simply conjoining the two distinctions, thereby creating a four-fold taxonomy of deciduous trees, coniferic bushes and so on. As to incommensurable concepts, such simple conjunction should be of no avail. In this case, mutual adaptation of the taxonomies has to be ruled out by some contrasting and conflicting beliefs about the relevant realm.

The upshot of both these considerations is that as regards accounting for incommensurability, laws and theories are prior to scientific kinds. Taxonomy is not enough; the criteria for collecting individuals in the respective kinds are of crucial importance as well (Sankey, 1994, pp. 79–81). Concerning incommensurable concepts, there are reasons for setting up the taxonomy in a particular fashion and for resisting attempts to remodel and adjust it to some other structure of kinds. There are reasons underlying discrepant taxonomies which also provide the basis for their sustained mismatch. This consideration brings the issue of theoretical incompatibility back into the picture. Theory and rival explanation is primary, taxonomic divergence is derivative. Discrepancies of lexical structures that bear on the issue of incommensurability originate from contrasting views of the domain in question.

To be sure, Kuhn clearly recognizes that taxonomy is dependent on theory and that, consequently, taxonomic discrepancy testifies to an underlying theoretical divergence. This is conspicuous in Kuhn's comment on the mentioned reshuffling of the astronomic equivalence classes in the course of the Copernican Revolution.

> Changes of that sort were not simply corrections of individual mistakes embedded in the Ptolemaic system. Like the transition to Newton's laws of motion, they involved not only changes in the laws of nature but also changes in the criteria by which some terms in those

> laws are attached to nature. These criteria, furthermore, were in part dependent upon the theory with which they were introduced. (Kuhn, 1987, p. 8)

Kuhn's claim is that theories entail criteria for applying concepts within the realm of phenomena and that these criteria in turn imply structures of scientific kinds. What Kuhn denies is that changes of such criteria are sufficient for blocking translation (Kuhn, 1987, p. 19). Referential shifts are essential for translation failure to arise. In the same vein, Kuhn stresses the importance of reasons underlying a divergence in kind-structure. A purely conventional discrepancy would not last but rather be defused by conceptual adaptation. Taxonomic divergence relevant to the incommensurability issue is not simply semantic but involves a disagreement about things. The choice of names "is no longer about linguistic conventions but rather about matters of evidence and fact." (Kuhn, 1993, p. 318) That is, Kuhn takes contrasting theoretical assumptions as underlying relevantly conflicting lexical structures, on the one hand, but still identifies incommensurability with taxonomic clash. The preceding arguments were intended to make plausible that the adequacy of translation cannot be judged exclusively by the agreement among kind-structures. Criteria employed for establishing kinds are essential as well, and this distinguishes theoretical incompatibility as the foundation of incommensurability.

What emerges from these considerations is the following "rational reconstruction" of the "Kuhn-Feyerabendian" notion of incommensurability. Concepts are incommensurable if they fail to be translatable owing to an underlying theoretical contrast. One of the salient obstacles to translation is cross-classification which is tantamount to the violation of the principle of no overlap. But cross-classification leads to incommensurability only on the condition that it is produced by conflicting views and premises. That is, not each violation of the principle of no overlap engenders incommensurability. Emergence of the latter demands theoretical incompatibility as an essential element.

4. TRANSLATION IN THE CONTEXT THEORY OF MEANING

On the basis of the just given rational reconstruction, cross-classification ceases to be the defining feature of incommensurability. Rather, translation failure becomes a consequence of conflicting theoretical approaches whose clash may in turn issue in deviant kind-structures. We are faced with a sort of common cause scenario in which discrepant principles bring about both untranslatability and cross-classification. It is plain, on the other hand, that not each theoretical conflict has such striking effects. In order to get a firmer grip on the sort of theoretical incompatibility that underlies incommensurability, the nature of the relevant obstacles to translation need to be studied in greater detail. So let's hark back to the context theory of meaning and ask what requirements are to be placed on adequate translations within this framework.

Translation in the sense relevant here concerns the concepts of different theories. Translation requires the coordination of a linguistic item with another one taken from a different theoretical framework but possessing the same meaning. The understanding underlying the entire discussion of the incommensurability thesis is that translation needs to be precise (clumsy paraphrases don't suffice) and to provide a one-one correlation between expressions. The latter condition does not demand that one word is assigned to exactly one word; rather, coordinating strings of words with one another is quite legitimate. The issue is not about words but about semantic resources. It is not about terminology but about what can be said within a conceptual framework (Sankey, 1994, pp. 76–77).

Adequate translation needs to preserve the meaning of the relevant concepts or statements. The question is how this requirement of unchanged meaning is to be spelled out within the context theory. One of the insights popularized by Hilary Putnam's "Twin-Earth" thought experiment (Putnam, 1975, pp. 220–224) is that linguistic relations alone are insufficient for establishing meaning. At some stage a connection between linguistic items and extra-linguistic circumstances is unavoidable. One of the most prominent attempts to forge such a link is "Sellars' triad" of perception, inference, and action. The linguistic role of an expression is captured by its application to perceptual states, the associated network of inference rules, and the actions tied to its utterance (Sellars, 1979, p. 121; Brandom, 1994, pp. 131–134).[5]

These three features can be condensed into two, one referring to the empirical application and the other to the theoretical integration of a term. Sellars highlights the conditions of application as determinants of the meaning of predicates and explains that sameness of meaning demands sameness of the role accorded to the predicate in the "linguistic economy" of the pertinent community (Sellars, 1963, § 31). This can be elaborated to the effect that two demands are to be fulfilled by translations. First, theoretical integration should coincide for the two items at issue. This applies, in particular, to the reproduction of standing inferential relations among the predicates or the sentences formed by using these predicates. Such relations provide the context relevant to the ascription of meaning. For example, the sentence "the tree over there loses its foliage" implies: "there is a deciduous tree." Analogously, "Wilfried is a bachelor" entails "Wilfried is not divorced" (see section 2). The network of such relations supplies predicates with their content; consequently, these relations should be preserved among supposedly synonymous expressions. Second, the conditions of application of concepts should remain unaltered. One of the reasons why the German predicate "*x hat schwarze Haare*" is disqualified as a translation of "*x* is a bachelor" is that their conditions of application differ wildly. In fact, these conditions exhibit hardly any correlation. The two predicates are only accidentally applied to the same objects; they differ in meaning for this reason.

On the whole, then, the theoretical context account recognizes two chief determinants of the meaning of concepts. First, the inferential integration of a concept which is specified by its relations to other concepts. The integration of scientific concepts, in particular, is supplied by the concomitant laws or theories. Second, the conditions of application are determined by the set of situations to which a concept is thought to apply (or not to apply, respectively).[6] To these two sources of meaning correspond two constraints on adequate translations. Rendering a concept appropriately demands, first, the preservation of the relevant inferential relations, and second, the retention of the conditions of application.

5. SHIFTING THEORETICAL GROUND: THE EXAMPLE OF THE CHEMICAL REVOLUTION

Let me turn to a concrete case from the history of science and examine if these two conditions can be satisfied. An often quoted instance of a Kuhnian revolution is the Chemical Revolution which involved the substitution of the phlogiston theory by the oxygen theory (Hoyningen-Huene, 1998, pp. 492–496). This is a revolutionary change by any measure so that the presence of incommensurable linguistic items can safely be expected – if incommensurability is supposed to be instantiated at all.

The Chemical Revolution roughly extends from 1775 to 1790 and was chiefly brought about by Antoine de Lavoisier. Pre-revolutionary chemistry primarily addressed the problem of explaining the properties of chemical substances and the changes of these properties in chemical reactions. The approach underlying these explanations was the so-called chemistry of principles. In its framework abstract bearers of properties were assumed which were supposed to confer their properties on the usual substances they enter. Changes of properties in reactions were attributed to the transfer of such principles. Relevant properties are solidity, volatility, or combustibility. Principles correspond to such properties; they are not identical to chemical substances. Rather, principles are thought to be more fundamental than substances and to provide the basis for an explanation of the features of the latter. On the basis of this approach it makes no sense to require that principles be detectable in the laboratory. Such a requirement would have seriously distorted the deductive structure of the theory. It would have created a circularity by claiming, in effect, that the properties of chemical substances are to be accounted for by the properties of chemical substances. Further, each principle is associated with a number of properties. It is clear that the attachment of one principle to each chemical property would have made the theory rather pointless. But in all its versions, the chemistry of principles introduced but a small number of principles. The challenge was to trace back the variety of empirically accessible properties to only a few property bearing principles.

The phlogiston theory is a specific variant of the principles approach and was developed by Georg Ernst Stahl late in the 17th century. Stahl's chief objective was to give a more unified account of combustion processes. His claim was that there is only one such principle involved in combustion processes from all kingdoms of nature (whereas a number of related sub-principles had been assumed before). This distinguished unique principle of combustibility was called "phlogiston" (which derives from the Greek word for "combustible"). This means that the combustibility of chemical substances is caused by the fact that they contain phlogiston. This principle is released from the burning body during combustion. Fire and flames make it obvious that something is set free in burning; and the residue, i.e., the ash, has lost the property of combustibility. Stahl's unified account entailed that the "calcination" or roasting of metals (their "oxidation" in modern terms) was also produced by phlogiston escape. All combustion processes and all calcinations are brought about by the same mechanism of phlogiston release. Stahl intended to support this claim by inverting the process of calcination, that is, to produce a metal from its calx, by supplying phlogiston from *non-metallic* substances. For instance, Stahl managed to prepare metallic lead out of lead calx (PbO) by means of heating it with glowing charcoal (whose phlogiston is set free in burning). Charcoal is of organic origin and yet it is suitable for restoring the phlogiston to the lead calx and for transforming the latter into the metal. The conclusion was that phlogiston is alike in all the kingdoms of nature; there is only one principle of combustibility (Ströker, 1982, pp. 93–94; Carrier, 1993, pp. 401–402).

This sketch illustrates that phlogistic reasoning centered on properties and their transfer. Lavoisier's oxygen theory, by contrast, represents an essentially modern approach to chemistry. It involved, first, the introduction of a different mechanism of combustion, and second, the abandoning of the entire chemistry of principles (although residues of the principles can still be identified in Lavoisier's thought). Combustion was no longer interpreted as the release of a property bearing principle but as a combination with a particular chemical substance, namely oxygen. More generally, the subject matter of chemistry was restricted to substances that can be isolated in the laboratory. Thus, the explanatory basis of chemistry changed which went along with a corresponding alteration of the explanatory objectives. The aim was no longer to account for the properties of substances. The explanation of chemical reactions was now judged according to the ability to accommodate the reaction weights which had been almost completely ignored earlier and were thought rather to belong to physics.

The upshot is that the Chemical Revolution involved the substitution of a theory by a conceptually disparate one. The assumed entities or the explanatory basis changed drastically: abstract, property bearing principles were supplanted by ordinary chemical substances. The supposed mechanism was exchanged: escape of a constituent was abandoned in favor of compound formation. Finally, the research agenda was overturned: properties of substances shifted into the background, whereas reaction weights entered the realm of chemical problems.

6. THEORETICAL CHANGE, CONCEPTUAL DISPARITY AND TRANSLATION FAILURE

The issue is whether or in virtue of which mechanism such a drastic theoretical change might produce a conceptual disparity that could defeat translation. It is clear, to begin with, that the term "phlogiston" is not part of the oxygen theory; but its lack need not have detrimental effects on the conceptual comparability of both theories. After all, "Buch" is not a word of the English language, and yet this omission in no way creates a conceptual incomparability. It might still be possible to introduce the concept of phlogiston, or some analog, into the framework of the oxygen theory. In order to find out if there are any such options, the two conditions of adequacy developed earlier are to be invoked. First, the theoretical integration or the inferential relations among concepts or predicates need to be preserved and, second, their conditions of application should be retained (see section 2). Let's see how *prima facie* translations fare in light of these conditions.

The first try is to translate "phlogiston" by drawing on its opposite functional role as compared to oxygen. This amounts to rendering "phlogiston escape" as "oxygen bonding". This suggestion is buttressed by the fact that in most cases in which partisans of the phlogiston theory thought it legitimate to apply the predicate "phlogiston escape", adherents of the oxygen theory would speak of "oxygen bonding". It follows that the conditions of application for both terms roughly coincide. However, the drawback is that the inferential relations fail to be preserved. A case in point is that the presence of phlogiston is connected with the color and the nature of the pertinent substance. For instance, a substance containing a large amount of phlogiston is assumed to be of an "oily-fatty nature". Consequently, combustibility and oily-fatty nature are both taken as empirical indicators of the same theoretical state, namely, a high proportion of phlogiston. This correlation provides a basis for inferences of the following kind: resin is combustible because it is of oily nature. But nothing of this sort follows from the oxygen theory if "high proportion of phlogiston" is replaced by "capacity to combine with large amounts of oxygen", as it would have to on the above translation rule.[7] The phlogiston theory specifies relations among properties which are not part of the oxygen theory. Oxides aren't distinguished by their joint characteristics. In fact, matters are even worse. Since the oxygen theory fails to address substance properties, correlations of this sort cannot be added to the theory either. The oxygen theory does not simply happen to remain silent on substance properties. Rather, abandoning the principles approach is tantamount to waiving explanations on the basis of property correlations.[8] Thus, such correlations are conceptually alien to oxygen theory; appending them to the theory would create a serious incoherence. It follows that the acceptance of a translation rule of this sort preserves the conditions of application, to be sure, but fails to reproduce the relevant inferential relations. It falls short of underwriting adequate translations for this reason.

The second try involves the attempt to preserve these inferential relations. The pursuit of this line amounts to explicating "phlogiston" the way I did earlier in sketching the content of the phlogiston theory (see section 5). One could roughly say that phlogiston is conceived as a non-material bearer of properties whose presence generates the combustibility and oily-fatty nature of the relevant substance. However, this translation rule entails a significant alteration of the conditions of application of the concept. In fact, proceeding along these lines involves the loss of all such conditions. There is no such thing as phlogiston by the oxygen theory's lights. If the phlogistic concept is simply grafted onto the oxygen theory, the concept becomes empty; it is no longer legitimately applied to any phenomenon. The attempt to retain the inferential relations is pursued at the expense of losing the conditions of application. It does not provide an appropriate translation for this reason.

The upshot is that the translation of concepts from disparate theories leaves one with the stark choice between two equally unacceptable alternatives. The first one is to translate according to the relevant conditions of application. That is, the two predicates are applied to the same observable circumstances. The catch is that the predicates (or the claims expressed with their help) do not exhibit the same inferential relations. Adherents of the oxygen theory refuse to accept that substance properties are connected by the causal influence of a non-material constituent. The second option is to translate in such a way that the inferential relations are retained. This amounts to giving a general description of the idea of property bearing entities and their transfer in chemical reactions. But the concept specified in this fashion is empty so that the conditions of application are not preserved. It deserves emphasis that the absence of an appropriate counterpart to the concept of phlogiston is in no way a simple gap within the oxygen theory. This concept cannot be introduced into the framework of the latter theory. Dropping the principles is tantamount to abandoning property bearing entities; appending the latter to the theory as an afterthought creates an incoherence. This means that there is no way of retaining the conditions of application while at the same time remaining faithful to the idea of phlogiston and the inferential relations tied to it. By pursuing this second option one is bound to lose all instances of legitimate application. There is no route to an adequate translation of "phlogiston" or related concepts into the terms of the oxygen theory. This is why "phlogiston" is incommensurable with the concepts of the latter theory.[9]

Let's have a quick glance at the other two examples of incommensurable concepts mentioned so as to explore if this conclusion can be generalized. The first example concerns "impetus" and "momentum" (see section 3). Judged by the requirement to retain the conditions of application, "impetus" and "momentum" are conceptual analogs. Both quantities are estimated by the product of a body's velocity and its weight or mass. However, this rendering did violence to the disparate inferential roles of the two expressions in their respective theories. Impetus acts as a cause of motion of the relevant body while momentum doesn't. Drawing instead on the theoretical context, one might explicate "impetus" as the impressed force that contributes to maintaining motion. However, classical mechanics fails to recognize such a force. The concept rendered in this fashion becomes vacuous.

A similar line of argument applies to the relation between the concepts "geocentric planet" and "heliocentric planet". Term coordination via conditions of application can be achieved by listing the relevant objects. Geocentric planets comprehend Moon, Mercury, Venus, Sun, Mars, Jupiter, Saturn. Viewed from the heliocentric perspective, the theoretical context of these entities is heterogeneous. We are dealing with heliocentric planets, stars, and satellites, each of which is accorded a distinct theoretical status. Conversely, trying to preserve the inferential role by invoking the defining feature of geocentric planets, namely, revolution around the Earth, leaves one with the Moon which fails to qualify as heliocentric planet in the first place. So, the conditions of application come out garbled.

These examples suggest that the type of conflict emerging here is a general feature of incommensurability.[10] The translation failure of incommensurable concepts arises from the impossibility to jointly fulfill the two conditions of adequacy that the context theory places on translations. Would be conceptual analogs either fail to maintain the conditions of application or to reproduce the concomitant inferential relations. This is how theoretical incompatibility brings about translation failure.

7. INCOMMENSURABILITY RESULTING FROM A CLASH BETWEEN WORLDVIEWS

Incommensurability is not confined to scientific concepts – as Feyerabend was eager to stress (see, e.g., Feyerabend, 1972, pp. 304–306). Rather, metaphysical positions or worldviews are analogous to scientific theories in the relevant conceptual respect. The alteration of such comprehensive approaches in the course of history also creates incommensurability. The concept of a witch is a case in point. As a result of cultural change, this concept is untranslatable into any concept taken from the contemporary, science based conception. In early modern times, the concept of witch was used to denote roughly the following conjunction of properties. A witch was supposed to be a woman of ugly appearance with unusual and inexplicable capabilities (such as healing or predicting the future) and strange behavior (eccentric or insane appearance). She owes these traits to a pact with the devil in which she traded her salvation for the capacity of performing feats suitable to impress ignorant and untutored people. She enjoys broom riding and lascivious dancing in order to please the Malefactor.

No concept of present-day, science based language is apt to capture this collection of properties. To be sure, one may explicate the concept in contemporary terms and elucidate in this way the inferential context and the pertinent semantic field (which I just tried to do). But this procedure would make the concept empty; it is no longer rightly applied under any circumstances. Each statement containing an ascription of the concept to a person is considered wrong. To our dismay, this was different in the early modern period. Consequently, this explication fails to reproduce the conditions of application and is disqualified as a translation for this reason.

Conversely, one might restrict the rendering of the term "witch" to the observable indications traditionally associated with it. Witches were empirically identified by their unexpected capacities and their strange behavior, and one could use these criteria as clues for a translation into the language of present-day views. Proceeding in this way and dropping, consequently, any reference to the interference of the evil spirit manages to preserve the concept's conditions of application. One might use the term "pseudo-witch" to denote persons exhibiting traits that would have made them appear to be witches in those dark ages. Consequently, the truth value of statements like the following can be retained: "Witches are miraculously successful in healing diseases." One only has to replace "witch" with its modern empirical equivalent, the "pseudo-witch".

However, the conceptual integration of this *ersatz* concept differs wildly from that of its alleged counterpart. While statements about witches possess metaphysical import, sentences referring to pseudo-witches are related to religious mania, psychopathological disorder, and placebo effects. The conceptual tie to salvation is severed, and a connection to hysterical reactions is forged instead. Theology as the relevant discipline is replaced by psychology, or psychopathology for that matter. The theoretical context of the original concept and its assumed modern day counterpart are completely disparate. This is why this procedure fails as a translation as well.

Again, the attempt to retain the conditions of application overturns the inferential relations of the concept; and the preservation of these relations entails the alteration, or the outright loss, of the conditions of application of this concept. Neither procedure leads to adequate translation. I conclude that incommensurability has a wider impact and is not restricted to scientific concepts in the narrow sense.

8. NON-TRANSLATABILITY, CROSS-CLASSIFICATION AND REFERENCE SHIFTS

Let me quickly recapitulate the conceptual situation. I argued earlier that it is inappropriate to accord scientific kinds the principal role in the emergence of incommensurability. Rather, it is the incompatibility of theoretical premises that generates incommensurability in the first place (see section 3). In the framework of the context theory, theoretical alterations induce meaning changes. If the discrepancy between two concepts taken from different theories is such that either the inferential relations or the conditions of application can be preserved, whereas their joint satisfaction cannot be accomplished, these two concepts are incommensurable. I mentioned before that this account of incommensurable concepts removes scientific kinds from the central position Kuhn assigns to them (see section 4). Still, kinds remain definitely part of the picture. I argued that Kuhn connects kinds with laws and theories. Laws contribute to establishing what is equal in kind. A scientific kind is a class of objects or processes which are governed by the same laws or theories (see section 3).

Consequently, the account developed here suggests a correlation between these two features. Incommensurability is expected to result from both the particular sort of translation failure I tried to expound and the cross-classification of scientific kinds that Kuhn took to be pivotal. Allow me to make a quick inspection as to whether the discussed instances of translation failure also exhibit the Kuhnian realignment of kinds, i.e., the dissection of kinds and the formation of new ones in their place (see section 3).

The adoption of the oxygen theory was indeed accompanied by a considerable shift in the taxonomy of kinds. The first relevant relation is the splitting up of kinds: what was formerly considered as being of the same nature is regarded as different afterward. In the phlogiston theory, but not in the oxygen theory, combustibility is conceptually tied to other tangible features of the relevant body (such as color and oily-fatty constitution). Phlogiston is taken as the common cause underlying such correlations. In the oxygen theory, or rather its present day successor, combustibility involves the capacity to combine with oxygen whereas a body's oily-fatty nature has to do with the presence of a large proportion of electronic double bonds. That is, what was conceptually united before the revolution crumbled into separate pieces afterward.

Conversely, the oxygen theory removed the chief conceptual division of the entire principles approach; the distinction, namely, between ordinary substances and non-material property bearers. Chemistry is confined to substances that are identifiable in the laboratory. This move involved a major conceptual unification which essentially supplied chemistry with the outline of its modern ontology. The impact of this unification extended well into the realm of chemical phenomena. It entailed a coherent treatment of distinctions among substances that were thought to go back to the presence of different principles, for one, and other differences that were attributed to the variety of material bases to which the principles were asssumed to combine, for another.

The impetus-momentum example likewise exhibits the Kuhnian cross-classification of kinds. The relevant dividing line runs between forced motion and force free motion. Impetus physics singles out rest as the only force free state; all motion is forced.[11] Classical mechanics collects rest together with uniform rectilinear motion into the unified class of inertial motion and contrasts it with accelerated motion and rotation. Consequently, this newly formed scientific kind of inertial motion cross-classifies the previous distinction between rest and forced motion. Frictionless and otherwise unimpeded uniform rectilinear motion does not fit either one of the older categories. Such motion is connected to resting carriages and separated from flying arrows and spinning tops.

The realignment of equivalence classes can similarly be identified in the conceptual changes produced by cultural development. The loss of the concept of a witch is accompanied by the division of a set of properties that were formerly jointly indicative of the state in question. Miraculous healing was connected then, but not now, with riding brooms and enjoying diabolic company. Conversely, the concept of the pseudo-witch forges novel links of these same empirical indications to contexts like religious mania and psychopathological disorder.

Since the taxonomic conflict going along with the conceptual shift of the notion of planet was explained earlier (see section 3), this quick survey tentatively confirms the earlier expectation: whenever two theories clash such that the rendering of a concept from one theory in the terms of the other preserves either the empirical features or the theoretical integration, but not both, this concept and their would be analogs denote scientific kinds that cross-classify each other. In such cases the underlying theoretical conflict jointly bottoms out as untranslatability and as cross-classification of the nomologically established equivalence classes.

This consideration suggests that incommensurability is tied up with reference shifts. One of the reasons for translation failure to occur is the inability to preserve the conditions of application – which involves a reference shift. Moreover, partial overlap of taxonomic classes is likewise tantamount to a change in denotation. The relevant realm of phenomena is carved up in a divergent fashion. Incommensurable concepts do not just mean different things, they also refer to different things. Actually, a linkage between meaning variance and reference shift is to be expected in the context theory of meaning independent of these more specific considerations of translation failure and taxonomic divergence. For the context theory involves a descriptive account of reference determination according to which properties serve to single out the scope of a concept. The attribute of being a rational animal fixes the corresponding class of denoted objects; humankind emerges as the set of entities satisfying this description. If the assigned property is altered to, say, rational machines, the pertinent set of referents changes as well and now includes some, but by no means all, computers. It is true, it may happen that concepts with different meaning pick out the same referents (as "rational animals" and "featherless bipeds") (see section 3). But in the large majority of cases such concepts refer to distinct objects or processes. Consequently, incommensurable concepts are likely to differ in reference.

It is only due to this wider impact that incommensurability could be viewed as a threat to scientific rationality. On the face of it, there is no reason to be worried about untranslatability. On the contrary, it appears quite plausible that the concepts of mistaken theories cannot be rendered in the framework of their more correct successors. These ill-conceived concepts are dropped as the pertinent theories are superseded by improved approaches. The occurrence of a translation failure of the kind in question indicates that science progresses profoundly. For instance, it was realized by oxygen theorists that there is no such thing as phlogiston. Therefore, it is only natural that a concept which was discovered to be misleading and empty cannot be integrated into the superior theory. However, if theory change goes along with reference change, the successor theory says different things about different objects – rather than different things about the same objects. This militates against a cumulative view of scientific progress. Progress cannot be regarded as being tantamount to understanding more and more aspects of the same entities.

It is true, alternative approaches in the philosophy of language sever this tie between meaning and reference. The causal theory of reference accommodates reference by drawing on conceptions such as "original name-giving ceremony" or "initial baptism". In virtue of the independence of these mechanisms from the meaning of the concepts introduced in this fashion, referential stability may obtain in spite of meaning change. However, establishing the mere possibility of invariant denotation is not enough. Feyerabend and Kuhn stressed that reference shifts are actually to be found in cases of incommensurability. Denotational changes are not merely derived from linguistic principles such that they could be discarded by switching the principles; such changes can rather be demonstrated in the historical record (Sankey, 1994, p. 44; Sankey, 1997, p. 429). The same lesson accrues from the discussed examples. The reference of "phlogiston" definitely fails to be preserved in the Chemical Revolution, and the transition to heliocentric theory doubtlessly entails a profound alteration in the domain of application of the term "planet". There is nothing to be gained here with referential stability. The cumulative character of science cannot be captured by assuming a sustained invariance of meaning and reference of theoretical concepts.

I refrain from discussing the options of accommodating the problems posed by incommensurability within an amended causal account.[12] There are two reasons why the treatment of the issue within the context theory recommends itself. First, historically speaking, the incommensurability problem underlay Putnam's adoption of the causal theory. Putnam argued that this theory was apt to defuse the threat that meaning variance posed to the rationality of scientific change (Putnam, 1973, pp. 196–202). It deserves to be examined more closely, then, whether scientific rationality is really endangered by a context theoretical account or whether the issue can satisfactorily be treated within the account that gave rise to it. Second, systematically speaking, it transpires from the discussion of the causal theory in the literature that the pure causal approach is unsuitable for accommodating ontological change and that it has to be supplemented with descriptive elements. In this "causal-descriptive" theory, "post-baptismal use" is granted a role in reference determination (Sankey, 1994, p. 57; 1997, p. 431). This move amounts to combining elements from the context theory and the pure causal theory. However, if the context theory alone proved sufficient for coping with the rationality of scientific change, the treatment would be more coherent and unified than the one afforded by hybrid, two-pronged causal-descriptivism. So I take the context theory to be the first choice in this matter. Let's see if rationality can be saved within its framework.

9. EMPIRICAL COMPARISON OF THEORIES WITH INCOMMENSURABLE CONCEPTS

A more thorough look at the situation reveals that two issues are advanced as a threat to scientific rationality. The first concerns the cumulative view of the history of science, the second bears on the comparability of the empirical achievements of rivaling theories. The first is to be granted outright. I take it that incommensurability crucially undermines the tenet that the history of science is cumulative throughout. Reference shifts do occur; the attempt to save the traditional conception of progress as unabated pile up of truths is a lost cause.[13] So let me turn to empirical comparability without further ado.

The strand of reasoning leading from translation failure to the exclusion of empirical comparison roughly looks as follows. Non-translatability implies that the claims of one theory cannot be expressed within the framework of the other and vice versa. It follows that the content of one theory cannot be captured by the other. But if it remains opaque what the allegedly rival theory says, there is no way to judge if it agrees or contrasts with one's own theoretical assumptions. Actually, it is not even clear whether the theories at hand are competing with one another. A competition requires some realm of shared reference: competing theories entail deviant predictions about the same objects. But in the case of incommensurable theories, as the argument runs, shared reference can never be ascertained. After all, the supposedly alternative theory is said to be incomprehensible. Consequently, no empirical comparison between the claims at issue seems feasible. On this interpretation, incommensurability would not just thwart the intertranslation of the cognitive content of the theories at hand but also vitiate their comparative empirical evaluation.

Note that this argument in no way rules out empirical examinations of a theory. It would still be quite possible to detect that a theory contradicted empirical findings couched in the terms of this same theory. Feyerabend admits that the presence of incommensurability does not interfere with the option of undermining a theory using data interpreted in its own conceptual framework (Feyerabend, 1970, p. 226; 1975, p. 282).

Thus, the hard problem is not empirical test but empirical comparison. The challenge is to reconstruct the possibility that the *same* piece of evidence underwrites one theory and undermines another one incommensurable with the first. Feyerabend tends to deny this possibility and moves toward the following position – without, however, fully embracing it eventually (Feyerabend, 1970, pp. 220–222, 226; see the more poignant German formulation Feyerabend, 1978, pp. 184–185, 190; 1975, pp. 282–283). Faced with two incommensurable concepts, the relevant data need to be described differently in either theory. But in view of the incommensurability of the concepts employed it appears illicit to maintain that both descriptions refer to the same state of affairs and to suggest that it is the same piece of evidence that, say, was in accordance with one of the relevant theories and militated against with the rival one. To be sure, one may point to the relevant apparatus or setup and say that the findings emerging from "this" device had the effect mentioned. But pointing to an apparatus falls short of characterizing an experiment or observation. The reason is that experiments or observations are *types* of activity; otherwise it would make no sense to speak of a repetition of the same experiment. This implies that experiments or observations need to be identified through a descriptive characterization. But if incommensurable concepts are involved no ecumenical, theory-neutral description is feasible. There is no way to coherently construe the claim that the two incommensurable descriptions refer to the same experiment or observation. The conclusion of the argument is that theories couched in incommensurable concepts cannot be compared empirically.

Kuhn accepts the empirical comparability of such theories on the ground that incommensurability is always restricted to just a few concepts. The large majority of the concepts involved have equal meaning and thus provide a shared conceptual basis (Kuhn, 1983, pp. 670–671; Hoyningen-Huene, 1993, pp. 218–222; Irzik & Grünberg, 1998, p. 212). This means that large chunks of the content of the theories affected by incommensurability can be intertranslated homophonically, and empirical comparability is secured for these parts. This suggests that translation is supposed to fail and that empirical comparison is thought to be ruled out for the incommensurable parts of theories. Consequently, if incommensurability extended to the entirety of the relevant concepts no empirical comparison should be feasible. Kuhn's position entails that "local" incommensurability does no serious harm to the comparative empirical evaluation while comprehensive or "global" incommensurability is vicious.

But in fact, empirical comparability is not ruled out for the incommensurable parts of theories. The point is that incommensurable theories need to be comparable in some respect in order to generate a non-trivial translation problem in the first place. A host of theories is not translatable into one another without anything significant coming out of it. Darwin's theory of natural selection is not translatable into hydrodynamics; quantum mechanics cannot be rendered in the concepts of Zen. In order for non-translatability to become a significant issue at all, such cases need to be excluded. The obvious way to do this is to draw on one of the defining features of incommensurability, namely, inconsistency of the laws involved. Incommensurable concepts are not translatable since the relevant laws, as specified within each of the theories at hand, are incompatible with one another (see section 3). In this vein, Feyerabend distinguishes between competing and independent theories and restricts incommensurability to concepts from theories of the former kind (Feyerabend, 1972, p. 304). But no such inconsistency occurs in one of the just mentioned examples.

The same result emerges if the earlier analysis of the reasons underlying untranslatability is taken into account. The translation failure of incommensurable concepts is of a particular sort. Each potential conceptual analog of a given concept either respects the empirical constraints or the theoretical integration (see sections 6 and 8). In the earlier discussion I laid stress on the fact that these would be analogs failed to satisfy both demands and were unsuitable as adequate translations for this reason. But at this juncture the converse aspect deserves emphasis: incommensurable concepts exhibit a particular type of relationship to one another. Advocates of incommensurability are committed to acknowledging some such relationship. If there was none, untranslatability would be a mere triviality.

Consequently, the significance of incommensurability requires that there is some range of phenomena that the theories jointly address (Hoyningen-Huene, 1993, p. 219). The point is that this common ground is sufficient for enabling one to compare some of the empirical consequences of the theories involved. Consider a particular experiment for which both theories claim responsibility. Each of them captures the outcome by using its own observational vocabulary. For instance, one employs the term "phlogiston release" and the other one "combination with oxygen" to describe what was happening. Since there is incommensurability involved, the two concepts cannot be translated into one another.

Here is an example. According to the phlogiston theory, metalic calces are like ash in that their principle of combustibility had escaped (see section 5). In order to transform a calx into the corresponding metal some phlogiston source is needed. However, the so-called red calx of mercury (HgO) has the remarkable distinction of being reconvertible into mercury by means of intense heat alone and without any identifiable phlogiston supply. The oxygen theory is compatible with this finding while it constitutes a blatant anomaly of the phlogiston account.

There is worse to come for the phlogiston theorist. Lavoisier collected the gas released during the reduction of red calx. This gas is oxygen by Lavoisier's lights and it should be suitable for retransforming metallic mercury into red calx. This can be achieved, indeed, by using mild heat. The gas collected in the reduction process is consumed completely during this calcination, suggesting that the reduction of the calx is in fact the reversal of its production. Moreover, the substance generated in this experiment is indistinguishable from the red calx obtained by employing oxygen from a different source. Lavoisier's conclusion was that oxygen was set free from the calx during the initial reduction process and that this oxygen was suitable, and sufficient, for reconverting the metal into the calx. On the phlogistic account, by contrast, something should be taken up, rather than given off, during reduction, and something should be released, rather than consumed, during calcination. It is true, phlogiston theorists came up with auxiliary hypotheses to rescue the theory from this counterexample. But the point is that both parties took responsibility for accounting for the chemical reaction at hand. No adherent of the rivaling approaches hesitated to accept the phenomenon as being of relevance. The phenomenon is that mercury becomes red and brittle in mild heat and that the metallic luster reappears in intense heat. There was no quarrel among the factions about this fact. This undisputed fact was in agreement with the oxygen theory whereas it contradicted the phlogiston theory.

Consequently, one theory was successful on the very turf where the other was defeated. And at least one such range of common relevance has to exist so as to create non-trivial incommensurability in the first place. Consequently, all pairs of incommensurable theories necessarily possess at least one realm of phenomena which they jointly address and which provides the basis for their empirical comparison (see Papineau, 1979, pp. 137–138; see also Laudan, 1977, pp. 142–144).

10. EMPIRICAL COMPARABILITY, THEORY-LADENNESS AND TOKEN-TYPING

One of the characteristics of this procedure for comparing incommensurable theories empirically is that the success or failure of the examination is judged against the background of one's own commitments and standards. No need for translation arises. The reasoning from translation failure to empirical non-comparability proceeds by arguing that claims made using inexpressible concepts cannot be understood and, consequently, cannot be put to empirical scrutiny (see section 9). But it is in no way mandatory to have the empirical claims of one theory checked by the adherents of another theory. A theory may well be tested by its own followers, and to them the relevant claims are by no means obscure. The only thing necessary to proceed from empirical test to empirical comparison is a shared realm of phenomena. Advocates of each theory have to acknowledge responsibility for coping with these phenomena (which may be disparately understood in either theory). This much of a common ground is secured by the mere fact that we are dealing with incommensurable theories.

It might be objected that this procedure abandons one of the chief distinctions of the scientific method, namely, public or inter-subjective control of all the claims entertained. For the sketched approach to empirical comparison appears to restrict access to the claims at issue to the supporters of the corresponding theory – which would be a definitely unwholesome feature. However, Kuhn rightly distinguishes between translation and language acquisition (Kuhn, 1970, section 6; 1983, pp. 676–677; 1990, pp. 4–5; see section 3). It is entirely possible to learn a new theoretical language by studying the theory. It follows from the context account that concepts can be learned by familiarizing oneself with the pertinent theoretical context. One may become bilingual – and yet be at a loss to translate. Translation proceeds by adequately tying together concepts from different languages, and this is more demanding than being able to locate concepts within the network of the pertinent language. It follows that the empirical comparison of theories with incommensurable concepts can be performed by a single person. Although such concepts are confined to a single theory, bilingual speakers manage to switch back and forth between two of them. Polyglot scientists are in a position directly to compare empirically claims cast in incommensurable concepts.

The upshot is that empirical comparison does not demand a translation of the relevant theoretical principles into one another. The standards for appropriate translations are harder to satisfy than the requirements for empirical comparison. Empirical comparison demands that instantiations of observational consequences can be correlated, whereas translation requires the mapping of theoretical concepts under the joint preservation of their inferential relations and their conditions of application (see section 4). Empirical comparison needs the identification of an experiment or phenomenon as lying within the domains of application of both theories involved. Such an identification does not require an agreement on the concepts appropriate for capturing the observations at hand. Rather, joint identification is possible on the basis of the description of the equipment employed or of the circumstances under which the relevant effect is supposed to show up. And nothing beyond joint identification is needed for this purpose.

Consequently, no theoretically disinfected level of data description is required. The identification of relevant phenomena may use all theories shared by the two incommensurable approaches. This common pool of observation theories may vary in comparing different such approaches. Nothing like an observation basis for all theories is requisite; rather, what is accepted as observation basis may be different for each pair of theories. This shows that theory-laden evidence may well be brought to bear on the empirical comparison of incommensurable theories. This sketch provides some idea of how one might arrive at identifying phenomena that are taken as relevant by both approaches in question. Partial agreement on auxiliary theories is the mechanism that allows scientists to reach agreement on the relevance of phenomena across the abyss of incommensurable concepts. Whereas the general argument given in section 9 amounts to a sort of existence proof of some such shared domain, the more detailed consideration of shared observation theories circumscribes means by which this domain can be identified.

For this reason, empirical comparison need not be impaired by the shift in reference and the cross-classification that is characteristic of incommensurable concepts. It is not detrimental to an empirical comparison of the relevant theories that the phlogiston theory assumes abstract property bearers which are non-existent by the oxygen theory's lights; nor is this endeavor distorted by the fact that the phlogiston theory alleges connections among properties that are dissected by the oxygen theory and vice versa (see section 8). Empirical comparison demands a correlation among *observational tokens*; translation needs a connection among *theoretical types*. The former correlation is necessarily realized for pairs of incommensurable theories. If it were missing the theories at hand weren't incommensurable. Only if some common ground exists between theories can any conflict or inconsistency arise between their laws. Theories addressing completely disjoint sets of phenomena are compatible with one another and hence cannot be incommensurable. It is precisely the amount of shared features which makes translation failure non-trivial in the first place that secures the possibility of empirical comparison.

It is at this juncture that the sketched argument against the identification of pieces of evidence loses its force. Feyerabend intimates that this identification requires shared recourse to types of experiment or observation and advances apprehensions as to this assumption (see sec 9). But the token typing may well proceed differently in either theory. Two experiment tokens may be taken to belong to the same type in one theory and to different types in the other. Of course, completely singular events are out of the question. Rather, the relevant level of token-typing includes observational circumstances that are recognized within both approaches. The point is that the incommensurable concepts themselves need not be employed for the purpose of identifying the relevant phenomena.

It is true, joint identification of such phenomena is requisite and demands in turn some shared concepts. One might object, then, that two theories which exhibited entirely incommensurable conceptual frameworks were incomparable empirically. But such a scenario is incoherent. Without any phenomenon acknowledged as lying within the scope of both theories, we are not dealing with incommensurability at all, but rather with a trivial case of non-translatability. "Global incommensurability" which allegedly affects the entirety of relevant concepts is ruled out for conceptual reasons. Incommensurability cannot be truly comprehensive. Some concepts have to be shared in order to make the issue arise in the first place.

To be sure, the range of empirical overlap between incommensurable theories might be narrow and insufficient for an unambiguous comparative evaluation. But problems of that sort may arise for any pair of theories. There is no guarantee that in comparing two theories one clearly comes out first. But uncertainties of this kind haunt empirical comparison in general and have nothing specifically to do with incommensurability. The upshot is that theories containing incommensurable concepts can be evaluated as to their comparative empirical achievements – albeit subject to those constraints that restrict empirical comparison in general.

11. CONCLUSION

There are three conclusions to be drawn from these considerations. First, incommensurability is real. The notion can coherently be reconstructed and positive instances of theories with incommensurable concepts can be specified. Second, incommensurability does not pose a serious threat to the objectivity of science or its commitment to experience. Incommensurability is no obstacle to empirical examination and it requires some option of empirical comparison. Still, it is of lasting importance as a problem in the philosophy of language. Incommensurability is the translation problem associated with the context account of meaning. It constitutes the successor problem to Quine's indeterminacy of translation. Whereas Quine's claim was based on the verification theory of meaning, Kuhn's and Feyerabend's views emerge within the context account. And whereas Quine is prepared to accept a large number of distinct but equally appropriate translations, Kuhn and Feyerabend suggest that there may not be a single adequate rendering. Thus, the basis and substance of the two positions are fairly disparate. Incommensurability is the follow up paradox of translation which continues to haunt us after the demise of verificationism.

Third, in one respect incommensurability continues to be of epistemic significance. It contributes to undermining a cumulative view of scientific progress according to which science manages incessantly to pile up truths upon one another. The lesson incommensurability teaches is that losses occur as well. In the course of theory change, some scientific achievements are conceptually reframed beyond recognition. In particular, the occurrence of reference shifts poses a serious threat to the claim that scientific theories accomplish an ever deeper understanding of the same objects and processes. Actually, one of the targets Kuhn and Feyerabend had aimed at by introducing the argument from incommensurability was the overthrow of the cumulative view of scientific progress.[15] In this respect the incommensurability thesis retains some force after all.

Bielefeld University

ACKNOWLEDGEMENTS

I am grateful to the editors for their most valuable remarks on an earlier version of this paper.

NOTES

[1] This argumentative strategy allows me to leave other linguistic approaches out of consideration (although I briefly touch upon the causal theory of reference; see section 8). My claim is that possible problematic side effects of incommensurability can be defused on the basis of the very account that gave rise to the relevant worries. As far as incommensurability is concerned, there is no need to advance alternative linguistic accounts.

[2] Kuhn agrees with Jerry Fodor in basing kinds on lawful generalizations; kinds are established by laws (Fodor, 1974, pp. 101-102). A law entails that its instantiations are similar in certain respects so that its adoption involves a conceptual structuring of the pertinent realm of phenomena; in particular, it involves assumptions as to what is alike and what is not. For example, it's a law that all protons possess the negative value of the electron charge and a half-integer spin. Consequently, all protons are of the same kind in that they equally display these properties.

[3] In fact, Feyerabend roughly anticipated aspects of this Kuhnian approach. Feyerabend claimed that incommensurability obtains "if a new theory entails that all concepts of the preceding theory have zero extension or if it introduces rules ... which change the system of classes itself." (Feyerabend, 1965, p. 98)

[4] Note that the context theory is sufficient for coping with the relevant meaning differences so that there is no need to resort to some other semantic account so as to elucidate meaning. In particular, meaning ascription should not be based on possible world semantics – as the present author incoherently did in Carrier, 2000, section 2.

[5] This "broad" conception defended by Brandom does not invoke "wide content" in the sense of Putnam's thought experiment. Wide content includes reference, whereas the inferential role conceived "broadly" only comprises empirically accessible states such as perceptions and actions.

[6] Hoyningen-Huene gives precisely these two features (namely the relations between concepts and their conditions of application) as the key elements of Kuhn's notion of the meaning of empirical concepts (Hoyningen-Huene, 1997, pp. 234–235).

[7] If "phlogiston escape" is assumed to be equivalent to "oxygen bonding", the capacity to release large amounts of phlogiston is tantamount to the capability of combining with large quantities of oxygen.

[8] Admittedly, this involves a pinch of rational reconstruction. Lavoisier himself retained traces of principles reasoning. Oxygen, for instance, was assumed to confer the property of acidity on its compounds. One often finds that the founder of a theory is less coherent or stringent in his or her argumentation as one might wish. Consequently, what is later to become the standard formulation of a theory frequently goes back to the disciples rather than to the master himself. Newton's mechanical writings display an ongoing flirtation with concepts from the impetus theory, and Darwin adopts large chunks of Lamarckian reasoning.

[9] The alternatives presented in no way exhaust the range of possible translations. Another alternative for coordinating phlogistic terms with concepts from the oxygen theory was suggested by Kitcher and involves a context specific connection of these concepts. Drawing on another version of the phlogiston theory, Kitcher claims that "phlogiston" refers to hydrogen under certain circumstances and refers to nothing under others (Kitcher, 1978, pp. 530–532, 539–540). But splitting up the relevant contexts entails that the inferential relations are not preserved (as Kuhn was not slow to point out; Kuhn, 1983, pp. 674–678). And the occasional loss of reference shows that the conditions of applicability are not preserved either.

[10] There are more examples of the kind. "Length", "velocity", and "mass" in classical electrodynamics and special relativity are incommensurable in the same sense. See Carrier, 2001, section 4.

[11] This marks the chief conceptual discrepancy between impetus theory and Aristotelean physics. The latter features "natural motion" as an additional type of motion. Natural motion continues without external or internal force and is rather maintained by the body's striving toward the place appropriate to its nature. At its natural place the body comes to rest. Impetus theory abandons the concept of natural place. Put conversely (as Kepler did), each place is considered a natural place.

[12] See, however, Sankey, 1994, chapter 2, for the defense of an interesting approach to this effect.

[13] It may be gleaned from the discussion above that causal descriptivism also grants reference changes (see section 8). There is no quarrel between theoretical contextualism and causal descriptivism regarding this issue.

[14] Feyerabend, 1970, p. 219; Kuhn, 1993, p. 330. Kuhn's anti-cumulative approach is restricted to revolutionary periods (in which incommensurability is thought to occur). Normal science, by contrast, is considered a "cumulative enterprise" (Kuhn, 1962, p. 52).

REFERENCES

Brandom, R. (1994). *Making It Explicit. Reasoning, Representing and Discursive Commitment*. Cambridge Mass.: Cambridge University Press.

Carrier, M. (1993). "What is Right with the Miracle-Argument: Establishing a Taxonomy of Natural Kinds." *Studies in History and Philosophy of Science* **24**: 391–409.

Carrier, M. (1994). *The Completeness of Scientific Theories. On the Derivation of Empirical Indicators within a Theoretical Framework: The Case of Physical Geometry (Western Ontario Series in the Philosophy of Science 53)*. Dordrecht: Kluwer.

Carrier, M. (2000). "Incommensurability and Empirical Comparability: The Case of the Phlogiston Theory." In P. Gärdenfors, K. Kijania-Placek and J.Wolenski, eds., *Proceedings of the 11th International Congress of Logic, Methodology and Philosophy of Science*. Dordrecht: Kluwer.

Carrier, M. (2001). "Shifting Symbolic Structures and Changing Theories: On the Non-Translatability and Empirical Comparability of Incommensurable Theories." In M. Ferrari and I. Stamatescu, eds., *Symbol and Physical Knowledge*. Heidelberg: Springer.

Churchland, P. (1979). *Scientific Realism and the Plasticity of Mind*. Cambridge: Cambridge University Press.

Feyerabend, P. (1962). "Explanation, Reduction and Empiricism." In Feyerabend, 1981, pp. 44–96.

Feyerabend, P. (1965a). "Problems of Empiricism." In R. Colodny, ed., *Beyond the Edge of Certainty*, pp. 145–260, Englewood Cliffs: Prentice-Hall.

Feyerabend, P. (1965b). "On the 'Meaning' of Scientific Terms." In Feyerabend, 1981, pp. 97–103.

Feyerabend, P. (1970). "Consolations for the Specialist." In I. Lakatos and A. Musgrave, eds., pp. 197–230.

Feyerabend, P. (1972). "Die Wissenschaftstheorie – eine bisher unerforschte Form des Irrsinns?" In Feyerabend, 1978a, pp. 293–338.

Feyerabend, P. (1975). *Against Method. Outline of an Anarchistic Theory of Knowledge*. London: Verso, 7th edition. 1987.

Feyerabend, P. (1978). "Kuhns Struktur wissenschaftlicher Revolutionen: Ein Trostbüchlein für Spezialisten?" In Feyerabend, 1978a, pp. 153–204 (revised version of Feyerabend 1970).

Feyerabend, P. (1978a). *Der wissenschaftstheoretische Realismus und die Autorität der Wissenschaften (Ausgewählte Schriften I)*. Braunschweig: Vieweg.

Feyerabend, P. (1981). *Realism, Rationalism and Scientific Method, Philosophical Papers Volume 1*. Cambridge: Cambridge University Press.

Fodor, J. (1974). "Special Sciences (or: The Disunity of Science as a Working Hypothesis)." *Synthese* ***28***: 97–115.

Hanson, N. (1958). *Patterns of Discovery. An Inquiry into the Conceptual Foundations of Science*. Cambridge: Cambridge University Press, 1965.

Hoyningen-Huene, P. (1993). *Reconstructing Scientific Revolutions: Thomas S. Kuhn's Philosophy of Science*, trans. A. Levine. Chicago: University of Chicago Press.

Hoyningen-Huene, P. (1997). "Der Empiriebezug der Psychologie: Philosophische Bemerkungen zur Position Jochen Brandstädters." *Zeitschrift für Sozialpsychologie* **28**: 229–240.

Hoyningen-Huene, P. (1998). "Kuhn and the Chemical Revolution." In V. Abrusci et al., eds., *Prospettive della logica e della filosofia della scienza*, pp. 483–498, Pisa: Edizioni ETS.

Irzik, G. and T. Grünberg, (1995). "Carnap and Kuhn: Arch Enemies or Close Allies." *The British Journal for the Philosophy of Science* **46**: 285–307.

Irzik, G. and T. Grünberg, (1998). "Whorfian Variations on Kantian Themes: Kuhn's Linguistic Turn." *Studies in History and Philosophy of Science* **29**: 207–221.

Kitcher, P. (1978). "Theories, Theorists and Theoretical Change." *The Philosophical Review* **87**: 519–547.

Kuhn, T. (1962). *The Structure of Scientific Revolutions*. (1970, 2nd edition). Chicago: University of Chicago Press.

Kuhn, T. (1970). "Reflections on my Critics." In I. Lakatos and A. Musgrave, eds., pp. 232–278.

Kuhn, T. (1983). "Commensurability, Comparability, Communicability." In *PSA 1982, Volume 2*, pp. 669–688, East Lansing: Philosophy of Science Association.

Kuhn, T. (1987). "What are Scientific Revolutions?" In L. Krüger et al., eds., *The Probabilistic Revolution I: Ideas in History*, pp. 7–22, Cambridge Mass.: MIT Press.

Kuhn, T. (1989). "Possible Worlds in History of Science." In S. Allén, ed., *Possible Worlds in Humanities, Arts and Sciences*, pp.9–32, Berlin: de Gruyter.

Kuhn, T. (1990). "The Road Since Structure." In *PSA 1990, Volume 2*, pp. 3–13, East Lansing: Philosophy of Science Association.

Kuhn, T. (1993). "Afterwords." In P. Horwich, ed., *World Changes: Thomas Kuhn and the Nature of Science*, pp. 311–341, Cambridge Mass.: MIT Press.

Lakatos, I. and A. Musgrave, eds. (1970). *Criticism and the Growth of Knowledge*. Cambridge: Cambridge University Press.

Laudan, L. (1977). *Progress and its Problems. Toward a Theory of Scientific Growth*. Berkeley: University of California Press.

Marras, A. (1992). "Behavioristic Approaches." In M. Dascal et al., eds., *Sprachphilosophie, Philosophy of Language, La philosophie du language I*, pp. 704–717, Berlin: de Gruyter.

Papineau, D. (1979). *Theory and Meaning*. Oxford: Clarendon Press.

Putnam, H. (1973). "Explanation and Reference." In Putnam, 1975a, pp. 196–214.

Putnam, H. (1975). "The Meaning of Meaning." In Putnam, 1975a, pp. 215–271.
Putnam, H. (1975a). *Mind, Language and Reality, Philosophical Papers Volume 2*. Cambridge: Cambridge University Press.
Sankey, H. (1994). *The Incommensurability Thesis*. Aldershot: Ashgate.
Sankey, H. (1997). "Incommensurability: The Current State of the Play." *Theoria* **12**: 425–445.
Sellars, W. (1963). "Empiricism and the Philosophy of Mind." In Sellars, Science, Perception and Reality, pp. 127–196, London: Routledge.
Sellars, W. (1979). *Naturalism and Ontology*. Reseda: Ridgeview.
Ströker, E. (1982). *Theoriewandel in der Wissenschaftsgeschichte. Chemie im 18. Jahrhundert*. Frankfurt: Klostermann.
Wittgenstein, L. (1953). *Philosophical Investigations*. Oxford: Blackwell.

FRED KROON / ROBERT NOLA

RAMSIFICATION, REFERENCE FIXING AND INCOMMENSURABILITY

Abstract. Though Kuhn and Feyerabend introduced the idea of referential incommensurability, many have found their account problematic. Subsequent developments in the theory of reference which address some of these problems are reviewed here, from the Kripke and Putnam semantics to the Ramsey-Carnap-Lewis account of theoretical terms with recent modifications due to Papineau. However Stich argues that such theories of reference can do no work towards solving issues in the philosophy of science. To overcome this objection, and to remedy some remaining shortcomings in the Ramsey apparatus used, we provide some epistemic conditions for naming. The resulting account yields a less problematic understanding of incommensurability, one that is consistent with realism.

This essay is an excursus into the topic of how to understand theoretical terms and the way they refer, and the kind of pitfalls that they face – specifically, the danger of incommensurability as theories change. The first section of the paper traces some of the history of the problem beginning with Kuhn's and Feyerabend's introduction of the notion of incommensurability. In section 2 we briefly mention the theory of reference developed by Kripke and Putnam for natural kind terms which allows much theory change without referential variance. However, the causal theory of reference they develop does not provide us with an account of how non-observational terms get their reference fixed – the focus of this paper. In section 3 we move away from pure causal theories to those which re-introduce an element of descriptivism in the reference fixing definition, suggest a schema for defining some terms and discuss a number of cases which fit the schema or a modified version of it. But this is not intended as a general theory of reference. In subsequent sections we explore what such a theory might be like that arises from Ramsey's suggestions about defining theoretical terms. We give our reasons for by-passing Carnap's development of Ramsey's suggestions (section 4), and focus on Lewis' account (section 5) which is more directly concerned with how theoretical terms can be defined. In section 6 we discuss Papineau's further tripartite partition of the sentences of a theory into those which are, are not, or might be, involved in reference fixing. But as useful as this distinction is in providing notions of referential determinacy and indeterminacy, the distinction itself proves to be undermotivated.

One aspect of incommensurability that Papineau addresses is how, with change in theory, we draw a distinction between (a) referring to the same things but changing our beliefs about those things, and (b) referring to quite different things. According to Papineau we may have reached a point where we have run out of further semantic facts to help us decide, and that we ought to look to the sociology of the use of terms within a given scientific community to tell us how a community has decided between (a) and (b). This Stich finds congenial because he finds the distinction between (a) and (b)

P. Hoyningen-Huene and H. Sankey (eds.). Incommensurability and Related Matters, 91–121.

impossible to make in principle. But from Stich's point of view (discussed in section 7) it is not just that semantics cannot help us at this point. Rather there are no viable theories of reference at all, including that of Lewis. On Stich's deflationism about theories of reference, the task of trying to get any mileage out of a theory of reference concerning issues in the philosophy of science is a misguided one. To counter Stich we make a suggestion about the point of having names in our language (both natural and scientific) that we think overcomes both Stich's deflationism and the lack of motivation of Papineau's distinction.

Why bother with names at all? In section 8, we argue that there is an epistemic point to naming that is not captured by purely semantic considerations about names. A referential practice of using a term, say t, in our discourse must have a certain kind of warrant. For the practice to have an *objective warrant* not only must there be something that can serve as referent, but the following further conditions must hold. First, there are interests that the community of users of t have in the item t refers to that leads them to continue their discourse about this item. Whatever the interest, without it the referential practice would lose much of its point. There are many items in our midst for which we do not have names; but for those that do, we suppose that there is some continuing interest to be indulged in continued use of t. Second, we have some epistemic access to the item referred to by t which enables us to track the item in its space-time wanderings, or enables us to learn more about it either from our mere acquaintance as we track it, or from a properly conducted scientific inquiry. Here we talk of 'The Fact Finding Condition'; our epistemic access must be such that we get to know more of what we are talking about as we pursue our interest in t's referent. Without some epistemic access, whatever this may be, our interests could not be satisfied. And without a belief that there is something to which we have this kind of access, answering to such interests – without a kind of *subjective warrant* for the referential practice – the practice should strike us as pointless. The notions of subjective and objective warrant provide epistemic conditions, and a rudimentary pragmatics, for the point of the game of naming, and thus a setting for semantic theories for terms.

In section 9 we put some flesh on the bare bones of the theory of subjective and objective warrant by discussing a number of cases of how reference might be fixed in a manner satisfying these conditions. In the course of the discussion we give our response to Stich's deflationism and an account of how Papineau's distinction can be motivated (section 10). Section 11 shows why our theory does not allow the kind of rampant Kuhnian incommensurability that tolerates talk of 'world changes', and why realism is quite consistent with the limited 'theory incommensurability' we allow.

1. INCOMMENSURABILITAS FLORET

The *Oxford English Dictionary* tells us that two items are commensurable when they can be compared, or measured, by the same standard, a paradigm case being the expression of a ratio in terms of whole numbers. Ratios are then incommensurable when they are not expressible as a ratio of integers. More generally two items are incommensurable when they are not comparable by the same standard or in the same respect. Extending the metaphor to the historical evolution of scientific theories, any pair of theories could be commensurable or incommensurable in a number of ways. Thus a pair might be commensurable in the kinds of test data they admit, or incommensurable in the kinds of problems they solve or in the kinds of epistemic or cognitive virtues they display, and so on. Here we will focus on whether the terms in a pair of successive theories can be compared in respect of what they are about, i.e., their reference, denotation or extension.

The idea of referential and meaning incommensurability is well expressed by Feyerabend in a 1962 paper when he tells us what can happen in the transition from a theory T' to a wider theory T:

> What does happen is, rather, a *complete replacement* of the ontology (and perhaps even of the formalism) of T' by the ontology (and the formalism) of T and a corresponding change of the meanings of the descriptive elements of the formalism of T' (provided these elements and this formalism are still used). This replacement affects not only the theoretical terms of T' but also at least some of the observational terms which occurred in its test statements. (Feyerabend, 1962, p. 29)

The same idea is variously expressed in Kuhn's *The Structure of Scientific Revolutions* first published about the same time. Kuhn tells us that strictly speaking we cannot derive Newton's laws of motion from Einstein's Special Theory of Relativity, even on the standard limit assumptions, because referential incommensurability holds: "the physical referents of these Einsteinian concepts [spatial position, time, mass, etc] are by no means identical with those of the Newtonian concepts that bear the same name" (Kuhn, 1970, p. 102). Meaning incommensurability is mentioned in a number of places, one of the more notorious being the following remarkable claim about those who defended pre-Copernican astronomy against Copernican "new-speak" about the Earth:

> Part of what they meant by 'earth' was fixed position. Their earth, at least, could not be moved. Correspondingly, Copernicus' innovation was not simply to move the earth. Rather, it was a whole new way of regarding the problems of physics and astronomy, one that necessarily changed the meaning of both 'earth' and 'motion'. Without those changes the concept of a moving earth was mad. (Kuhn, 1970, pp. 149–150)

On an extreme reading, the semantic theory which lies behind these remarks says the following. Given a theory or conceptual framework T, for all (non-logical) terms t in T

(1) the reference of t (in T) = k iff k uniquely realizes the open sentence T(x), or most of it,

(2) the sense of t (in T) = the sense expressed by the term (¶x)T(x) (where '¶' is the definite description operator).

The holism implicit in this radical form of Kuhn's proposal has obvious and well-known difficulties. Most significantly, if all non-logical terms in T are to be defined this way, nothing at all seems to get defined since the very predicates that go into the *definiens* can themselves only be understood in terms of the *definiendum*. If, on the other hand, we take (1) and (2) as defining the reference and sense of all the terms in theory T together, we can surely never expect to get unique realization; so terms will at best always be empty (or, if we drop uniqueness, multiply ambiguous). It is difficult to see what sense to make of claims of incommensurability on this understanding of the underlying semantic theory.

Kuhn and Feyerabend clearly meant something weaker. To say that the terms 'motion' and 'Earth' changed their meanings in the transition from pre-Copernican to

Copernican astronomy requires there to be some common measure of change of meaning: a way of saying "the term meant such-and-such [e.g., has a fixed position] in the old theory, but means thus-and-so [e.g., doesn't have a fixed position] on the new theory", where 'such-and-such' and 'thus-and-so' are formulated using vocabulary available to both theories, thereby allowing clear display of the difference in meaning. So let us assume that some basic terms already have a determinate common meaning and reference (of course, it is hard to see how this can be possible on the Kuhn-Feyerabend framework, but without some such concession the game is lost). Then the following question arises for a term t which occurs in two sufficiently different successive theories T and T* whose other terms already have determinate meaning and reference: does t refer to the same item, and express the same sense (meaning), in both T and T*? Assuming the Frege principle that sense determines reference (extension), we cannot be assured that there is sameness of reference. In fact, where t occurs in sentences of T and T* that ascribe inconsistent predicates, the open sentences T(x) and T*(x) cannot be realized by the same item, and so the reference of t will not be the same in each theoretical context. That, we take it, is an important part of what Kuhn and Feyerabend meant by saying that the term t as it occurs in T is incommensurable with term t as it occurs in T*.

On the basis of their historical work, but with some such tacit background semantics, Feyerabend and Kuhn gave us an excess of, indeed rampant, incommensurability. More radical claims of Kuhn even spoke of different worlds (Kuhn, 1970, chapter X). In fact Kuhn made "world changes" a central, if somewhat ineffable, aspect of the notion of incommensurability. After the last cited remark Kuhn goes on to say of the 'earth' and 'motion' examples: "These examples point to the third and most fundamental aspect of incommensurability of competing paradigms. In a sense that I am unable to explicate further, the proponents of competing paradigms practice their trades in different worlds" (*loc. cit.*). Kuhn and Feyerabend readily embraced incommensurability as a new exciting discovery which exposed the dullness of the cumulative view of science commonly advocated by those with non-historical positivist inclinations. Exciting as it may be, however, the semantic background of Kuhn's and Feyerabend's claims about incommensurability is untenable.

2. INCOMMENSURABILITAS NON FLORET

A new theory of reference advocated by Kripke (1980) and Putnam (1975, chapter 12) for proper names and kind terms showed us how to avoid counter-intuitive and excessive claims of incommensurability. The "causal theory of reference", as it is often called, showed us how to fix the reference of names without appeal to Fregean senses or Russellian descriptions. A name is introduced in a quasi-baptismal manner for a perceptually present object, or samples of a kind, and is then transmitted through a community with the result that members use the name to refer to the same object or kind.

A plus for the causal theory is that it shows how the referent of kind terms can remain invariant while quite central beliefs about the kind vary. For example, both we and the Ancient Greeks introduced kind terms to refer to water and fire while holding different theories about them. (We pass over the resolvable difficulty that we and the Ancient Greeks in speaking different languages introduce different kind terms such as 'fire' and 'pur', or more correctly 'πυρ', to refer to the same kinds.) Aristotle produced a theory of these items which claimed that water and fire were two of the four basic pure elements (impurities and admixtures in any sample aside). We now think that water is a non-elemental compound; and fire is not even substance-like but radiation in the heat and light wave-band. Whatever deep theoretical difference there may be between ourselves and Aristotle, it is the very same thing, water or fire, that we and the Ancient Greeks talk about with our respective terms and that we both use to quench our thirst or to warm ourselves.

Such a theory also suggests what is wrong with Kuhn's claim about the reference of the proper name 'Earth'. The reference can be fixed either by pointing to the Earth, or by picking it out by means of contingently applicable descriptions such as 'the whole body upon which we are now standing', or somesuch. That the Earth is alleged to have a fixed position, or does not move (i.e., rotate on an axis, or orbit the Sun) is at best a contingent property within some theory about the motion of the Earth and not a definition of 'Earth', or an analytic truth about Earth, as Kuhn contends. If it were a definition then Aristotle, who investigates arguments for and against the Earth's motion in *On the Heavens* (Book II, chapters 13 and 14) would have found his task much easier. That the Earth moved was, for Aristotle, a contingent claim about the Earth, and one which had to be shown false through empirically based arguments; it was not to be settled by any investigation into meaning. It is this important feature about proper names such as 'Earth' that the causal theory preserves while the Kuhnian theory obscures it.

Even though it helps cut back on the incommensurability of some scientific theories, there are some problems the causal theory faces for kind terms. The first is sometimes called 'the *qua* problem'.[1] Consider the introduction of the term 'tiger' to refer to a kind of which some samples are perceptually present. For successful introduction one presupposition needs to be satisfied, viz., that there is some kind which the samples instantiate and not none. Further, tigers instantiate a number of kinds such as animal, mammal, vertebrate, and so on; so we need a means of specifying which of the several kinds the samples instantiate to fix a reference for the term to just one of those kinds. But it is hard to see how this can be done without supplying some description of what is being referred to. This suggests a retreat from a pure causal theory of reference to a causal-cum-descriptivist one into which true descriptive elements must enter. The second problem is that the causal theory cannot be applied to names of kinds that are not observable. We turn to this problem in the next section through a discussion of some examples of reference fixing for unobservables; we leave to later sections an account of what theory might back such examples.

3. SOME EXAMPLES OF REFERENCE FIXING FOR TERMS WHICH REFER TO UNOBSERVABLES

A natural extension of the causal theory of reference just outlined would introduce descriptive elements back into reference fixing, in something like the following way. Unobservable kinds often stand in correlatory, causal, part/whole, or other relations to observable objects and kinds. Given that observable objects and kinds are named (we use 'O' as an arbitrary name below), names for unobservables can, for a large number but not all cases, be introduced according to the following schema (where 't' is a term which is being introduced to refer to unobservable item k, P is a property and R is a relation):

(3) the reference of t = k iff k is an object which has property P and stands in relation R to O, or a kind whose members have these features.

To illustrate, consider the following medical example based on Semmelweis' discovery of the cause of childbed fever. The large Viennese maternity hospital at which Semmelweis worked was divided into two clinics called 'First' and 'Second'. In the Second Clinic pregnant women were attended by midwives while student doctors attended in the First Clinic where there was a much larger, disturbing incidence of childbed fever. Each morning the student doctors first had their practical pathology classes during which they examined corpses; next they examined pregnant women in the First Clinic. Semmelweis hypothesized that the difference in the incidence of childbed fever between the two clinics was due to 'something' on the hands of the student doctors, acquired from their early morning practical pathology classes, which was not removed by ordinary soap-and-water washing and which was passed on to the women. This 'something' caused the women to contract fatal childbed fever, the symptoms and pathology of which were well known at the time. He called this material 'cadaverous particles' because he thought that they were distinctive minute bits of matter obtained from dead bodies (Semmelweis, 1963, pp. 88–89). Though Semmelweis knew nothing about germs, bacteria and the like he was able, given his epistemic situation, to hook into a hitherto unrecognized and unobserved kind of thing which existed in the world and to introduce a name for it. Given the above story, an appropriate reference fixer would be:

(4) the reference of 'cadaverous particles' = k if k is a kind of particle, or bit of matter, instances of which are only found in decaying matter in corpses and, by being transferred from the hands of student doctors to women just before childbirth, are able to cause childbed fever in these women.

At this stage two comments on (4) need to be made. The first is that there is a sense in which part of the reference fixer reflects a false belief on Semmelweis' part. This introduces complexity into the actual historical example which is addressed in further discussion of the example at the end of section 6. The second concerns the use of the word 'kind'. This is not intended in the strong sense of a natural kind. Depending on the circumstances there is now known to be a cocktail of agents producing childbed fever, the most common of which are several species of the genus streptococcus (*streptococcus pyogenes*) and on some occasions staphylococcus, *E. coli* and even tetanus, all of which can bring about the requisite blood poisoning. In his introduction of the term 'cadaverous particles', Semmelweis did latch onto a 'something' that caused observed symptoms of childbed fever; the term is not empty. But it is not required that he introduce a name for a natural kind. As we will emphasize later, his term successfully refers but is ambiguous concerning the species of bacteria to which it refers.

Unless we are cautious about such reference fixers new problems can arise. One new problem is that quite austere reference fixers along the lines of (3), which leave property P very broad and effectively invoke only a causal relation, can hardly fail to fix a reference (unless the world is far more indeterministic than we think!). Thus phlogiston theorists, if they introduced their special theoretical term in the following way:

(5) The reference of 'phlogiston' = k if k is a substance which (somehow) causes the combustion of heated tin,

would have referred to oxygen and not phlogiston. To allow (5) would be to introduce a degree of scientific prescience for phlogiston theorists that goes contrary to both what the 18th century theory of phlogiston was assumed to be about, and the historical story told about how phlogiston theory of combustion was replaced by the rival oxygen theory of combustion.

Problems with (5) can be overcome by adding more descriptive elements from the theoretical beliefs that were held about phlogiston into the reference fixing definition. It was part of the theory of phlogiston that it be given off in the process of combustion thereby bringing about the reduction of metals like tin to a powder (a process called 'calcination'). This is in contrast to oxygen which chemically combines with the burning tin to form an oxide. Thus a new reference fixing definition could be:

(6) the reference of 'phlogiston' = k if k is a substance which causes the calcination of tin in a combustion process by being given off in that process.

This succeeds in fixing a reference by appealing to broad features of the causal mechanism whereby phlogiston is supposed to bring about its affects.[2] Such a theory is sometimes called 'causal descriptivism' (Sankey, 1994, 1997). But while the theory is successful for a large range of terms, caution is necessary. This approach might not succeed with all reference fixers for the following reasons. To use an example suggested by Niiniluoto (1997, p. 549), when the AIDS syndrome was identified the hunt for its cause, the HI-Virus, was on in many laboratories. Some had initially assumed that the HI-Virus acted by a particular causal mechanism. However it soon turned out that there were different sorts of virus which could not have acted by this particular causal mechanism and whose actual causal mechanism remained largely unknown. This suggests that for the case of 'HI-Virus' an austere reference-fixer modeled on (5) rather than (6) is the more appropriate reference fixer:

(7) the reference of 'HI-Virus' = family of viruses that cause AIDS (in diverse ways).[3]

Unlike the phlogiston case, a reference is fixed by whatever virus stands in a causal relation to the symptoms of AIDS no matter how variable the causal mechanism whereby the kind brings about AIDS. Any general theory of reference fixing should be able to account for both the phlogiston and HI-Virus cases.

Schema (3) can also be used to fix the reference of proper names, successfully for 'Neptune' before the planet was observed but unsuccessfully for 'Vulcan', a planet alleged to orbit the Sun inside Mercury. However terms can also be introduced on the basis of theory alone without the item named standing in any relation R to observables O. Such is the case for the following: Mendelejev's introduction of names for some elements on the basis of the principles underlying the construction of the periodic Table well before their detection; Pauli's introduction of the term 'neutrino' to stand for a massless electrically neutral particle that conserves energy and angular momentum in beta decay; 'spin' as the name of the property of intrinsic angular momentum of sub-atomic particles; Wheeler's introduction of the term 'black hole' to stand for the already well known state of a collapsed star anticipated by Schwarzschild; and so on.

The above examples of reference fixing for theoretical terms are merely illustrative of the ways in which descriptive elements can enter into the right-hand side *definiens*, thereby taking us well away from the pure causal theory of reference. Since only some, and not all, descriptive elements from theory are employed in the definitions, they do not take us all the way back to the problems of incommensurability that arise for Feyerabend and Kuhn who hold that the total theory in which a term occurs is to be invoked to fix its reference. However nothing is suggested by these examples concerning a *general* theory about the reference of theoretical terms. What we seem to have are a patchwork of cases, some to be understood along more or less pure causalist lines (ordinary proper names, perhaps certain natural kind terms like 'water'), others along causal-descriptivist lines, still others in ways that make no appeal to the way things causally impact on the world we can observe.

If we are to arrive at a general theory then a different approach needs to be taken. In the next two sections we continue our focus on the case of reference fixing for theoretical terms, but consider the distinctive approach to such terms that began with Frank Ramsey's work early last century and was continued by Carnap, and later, Lewis. In defending and articulating this approach, our discussion will draw some important connections between the problem of reference fixing for theoretical terms and for terms more amenable, as it seems, to a causal treatment, but it will do this in a way that steers clear of the problems attending Kuhn's and Feyerabend's assimilation of observational terms to theoretical terms. If we are right, what we say will, in the end, provide some of the generality and underlying motivation that was so conspicuously lacking from our earlier discussion of the patchwork of cases.

4. RAMSEY, CARNAP AND THE RAMSEY SENTENCE

In order to discuss Ramsey's account, in particular the idea of a Ramsey Sentence, we need to employ a distinction between the theoretical terms, 't_1', 't_2', ... 't_n', and observational terms, 'O_1', 'O_2', ... 'O_m', of a theory $T(t_1, t_2, \ldots t_n, O_1, O_2, \ldots O_m)$. Ramsey says little about how the distinction is to be drawn, although it is likely that to some degree he followed the path of verificationists of the time. Here we will follow David Lewis (1970, p. 428) for whom the distinction need not always be epistemologically based on what is immediate and what is inferred in perception. The difference could, for example, be that between *old* or *original* or *other* terms (O-terms) which have their reference and meaning antecedently fixed by whatever means, and the 'theoretical' or T-terms which are new terms introduced into the language of some theory at particular stages in the historical development of a science and for which reference or meaning has yet to be fixed. (This is already at odds, of course, with Kuhn's and Feyerabend's radical idea that the meaning and reference of all terms depend on the theories embedding the terms.)

Given theory $T(t_1, t_2, \ldots t_n, O_1, O_2, \ldots O_m)$, its Ramsey Sentence[4] T^R is well known:

$$T^R = (\exists x_1)(\exists x_2) \ldots (\exists x_n) [T(x_1, x_2, \ldots x_n, O_1, O_2, \ldots O_m)]$$

But the import of such 'Ramsification' is not always clear. One issue concerns the eliminability of theoretical terms. In T^R the theoretical terms of T have been syntactically eliminated; for some this is the main advantage of Ramsification in that it eliminates any need to deal with theoretical talk at all. However the existential quantifiers still range over theoretical items of a domain. Thus the term elimination evident in T^R does not imply entity elimination; in fact T^R through its quantifiers requires theoretical entities (or concepts) on pain of falsity.[5] One of Ramsey's original concerns in his posthumously published paper "Theories" (written in 1929) was whether theoretical terms could be given an explicit definition and thereby eliminated. How such definition is possible becomes clearer in David Lewis' treatment of theoretical terms. But first let us briefly look at Carnap's development of Ramsey's idea.

Using the idea of a Ramsey Sentence, Carnap[6] was able to divide a theory T into its purely analytic component and its synthetic component. To specify the analytic postulates of theory T we define the 'Carnap Sentence' for theory T, (T^C): $T^C =_{df} T^R \supset T$. The following relations hold between T, T^R, and T^C:[7]

(i) T and T^R are empirically equivalent (share the same observational consequences);

(ii) T entails T^R (this follows by existential generalization)

(iii) The conjunction of T^C and T^R entails the original theory T.

In the light of (i) the Ramsey Sentence T^R gives us all the empirical content of the theory. But suppose we want to know what fixes the reference of the n theoretical terms in T? Here the Carnap Sentence T^C might be expected to help, since it at least contains all the theoretical terms. But now we strike trouble. Given T^C, there are, in fact, three cases to consider concerning the way the open sentence $T(x_1, x_2, \ldots x_n, O_1, O_2, \ldots O_m)$ might be realized (namely, by exactly one n-tuple, by none, and by more than one), and these should result in different answers to the question of the reference of the theoretical terms. But T^C seems utterly unable to give us these answers.

First of all, the open sentence might be uniquely realized by precisely one n-tuple of entities. In this case, the Carnap Sentence T^C in effect tells us that the theoretical terms of T denote, in order, the elements of the n-tuple. But the truth of T^C doesn't, of course, guarantee that there is exactly one n-tuple. T^C is also true if there is no n-tuple which realizes T (since T^C then has a false antecedent, thus leaving T^C is true). In this second case, however, we surely think that the theoretical terms are empty terms, even though T^C by itself doesn't allow us to reach any such conclusion. Thirdly, in the case of realization by two or more n-tuples T^C can tell us only that the theoretical terms in T name some n-tuple or other but it can't tell us which – a curious situation since multiple realization of our theories is a defect which we might hope to have remedied.[8]

Because the Carnap Sentence can't differentiate these cases, it has nothing to say about reference fixing. We suspect that Carnap's silence on this matter simply reflects his conviction that theoretical terms were strikingly different from observational terms in their semantic function. There is a sharper focus on the reference of terms in Lewis' account, to which we now turn. Though it is common to refer to the Ramsey-Carnap-Lewis view of theoretical terms in the light of their historical connections, there are important differences between the members of this trinity.[9]

5. LEWIS ON THE DEFINITION OF THEORETICAL TERMS

For the sake of simplicity consider a theory T with one (new) theoretical term t highlighted, $T(t, O_1, O_2, \ldots O_m)$, with the O-terms understood in Lewis' sense of *original* or *other* terms. We can also include amongst the O-terms special terms such as 'cause', 'part', 'correlated', and so on, if they are needed to play a role in defining t. Finally, assume that the theoretical vocabulary consists of names rather than predicates, occupying positions open to first-order quantification. In line with the concerns expressed above, Lewis in effect requires that realization be replaced by unique realization. This allows him to define (or fix the reference) of t as follows:

(8) the reference of t = $(\P x)[T(x, O_1, O_2, \ldots O_m)]$

(or, more simply, t = $(\P x)(T(x)$, where '$T(x)$' abbreviates '$[T(x, O_1, O_2, \ldots O_m)]$'.[10]

(8) specifies that t refers to whatever uniquely realizes the open sentence '$T(x)$'; if there is no realizer, or if there are two or more realizers, it does not refer. A more relaxed view, advocated in later writings of Lewis,[11] allows that where there are two or more realizers there is indeterminacy of reference rather than lack of reference. The later modification is a small concession to incommensurability in the limited sense that terms may be introduced which turn out to have multiple referents; with the advance of science, such multiplicity is then removed. How such multiplicity arises and how it can be resolved will be discussed in the next section. However in admitting such indeterminacy of reference, there is no commitment to the sort of incommensurability found in Kuhn and Feyerabend. First, there is no wholesale holism in which (nearly) all (or most) of a theory is invoked to fix reference; this is ensured by the preservation of the O-term and T-term distinction, which can no longer always be understood epistemically. Lewis' definition defines T-terms in a stepwise fashion using previously defined O-terms. In contrast the Kuhn/Feyerabend approach tends to define all terms at once in a manner that smacks of circularity. Nor is there a proliferation of ontologies relative to a theory, or talk of differing worlds so commonly found in Kuhn's and Feyerabend's account of incommensurability. There is a big difference between indeterminacy of reference which can occur in the realist framework supposed here, and talk of multiple ontologies which undercuts realism.

Lewis makes some useful modifications to the account of reference fixing that concern referential continuity under correction of a theory (see Lewis, 1970, p. 432, pp. 445–447; 1994, pp. 416–417). Correction can arise in several ways such as through the acquisition of improved quantitative data at the observational level, through the improvement of the value of some physical constant, through the improvement of a law (such as the addition of more variables, etc) or through the recognition that some hitherto uncontrolled or unknown factors interfered with the way in which the theoretical item for which a term is being introduced interacted with other items referred to in T, and so on. Let T* be the corrected version of T. Strictly there is no unique realizer of T(x) since T has undergone correction; but let us suppose that there is some *near realizer* of T(x) and that there is a *best unique* near realizer of T(x). Also suppose there is a unique realizer of T*(x) and t names that unique realizer. Now t is a term in both T and T*. And t will refer to the same item in uncorrected T and corrected T* if and only if the unique realizer of T*(x) is the same as the best unique near realizer of T(x). In this way t refers to the same item in the change of theory from T to T*.[12]

Granted the above, it is easy to see how change (through correction) in the observational sentences of a theory leads to evolution in the theory without referential change; and the evolution of theory without referential change can even be extended to corrections and additions to the theoretical sentences of the theory. The notion of a near realizer[13] allows for improvements in a theory without referential variance; for example there are quite radical changes in theory from Bohr's various early, and later, theories of the electron without referential variance of 'electron'. It also makes it more difficult to mount the Putnam pessimistic meta-induction about the ontology of successive theories (Putnam, 1978, Lecture II, section 2). Earlier theories can be right about what exists even though wrong about what to say of them; later improvements to the theory retain the same items as referents but get less wrong about what is said of them. Still, there are remaining unresolved problems. Do we need to invoke all of theory T in which t occurs in order to fix its reference, or only most, or about half, or very little of T? Some useful suggestions have been made on this by David Papineau (1996).

6. PAPINEAU ON THE IMPRECISION OF DEFINITIONS OF THEORETICAL TERMS.

Papineau makes the point that we should not invoke so little of theory T in the Ramsey Sentence that 'T(-)' is multiply realized. And we should not invoke so much that it is realized by nothing. Rather what is wanted is enough theory so that 'T(-)' is uniquely realized (or has a unique near realizer, this qualification being understood from now on). To this end Papineau proposes a division of T into three parts: (a) those parts T_y ('y' for 'yes') which do contribute to the definition of t; (b) those parts T_n ('n' for 'no') which do not contribute to the definition of t; (c) those parts T_p ('p' for 'perhaps') which might contribute to the definition of t. The proposed tripartite division is not hard and fast; Papineau proposes that there can be more refined divisions such as overlaps between the different parts. We might even envisage a probability function from parts of T to values ranging from 1, the most definite 'yes' with maximum involvement in definitions, down to 0, the most definite 'no' in which there is minimal or no involve ment in any definition.

We will discuss shortly what might underpin the tripartite division. But given such a division, it is now possible to see how imprecision can arise in the definition of theoretical terms. Let us suppose that T_y is not so weakly specified that it has many realizations; rather it is strong enough to have a unique realization, but not so strong that it has none. And let us also suppose that there is a part of T, T_n, which is definitely not used in defining t. Now consider T_y-plus-T_p. Suppose that adding elements from T_p in a definition of t that uses T_y makes no difference to the entity picked out. Then there is imprecision in T concerning which of its parts are to be used to define t, since in moving from T_y to T_y-plus-T_p we move from minimum to maximum imprecision in the definition of t. However the reference of t remains invariant, so there is no imprecision at the level of reference.

Hence on Papineau's tripartite model, and even more elaborate variations on it, imprecision in definition does not always affect uniqueness and determinateness of reference. But sometimes it might do so. Two cases where there is indeterminateness of reference might be of relevance to issues of incommensurability.

(1) Suppose T_y has several realizers (so there is no unique reference for t fixed by T_y and t might be said to have indeterminate reference); but when most of T_p is added there is a unique realizer. Then there is indeterminacy of reference for t fixed by T_y that is not always removed by addition of parts of T_p but is removed when most of T_p is added. This kind of indeterminacy commonly arises when a term is introduced for what turns out to be a broad genus of a kind of thing, but there are several species of the genus that are sufficiently different because, say, they operate by means of different causal mechanisms. This might be another way of understanding the referential mechanism of the term 'HI-Virus' which may well indeterminately refer to a variety of viruses which operate by means of different causal mechanisms as the virus evolves in the body (see footnote 4). Papineau (1996, p. 18) suggests two illustrative examples, the first from current gene theory. Various kinds of lengths of DNA satisfy the undisputed criteria T_y for being genes; however, further claims from gene theory are needed to narrow down the kinds so picked out. The second example is from relativity theory. Both rest mass and relativistic mass satisfy the original Newtonian definition of mass as proportional to amount of matter; and they both satisfy a general form of Newton's force law. But what does the term 'mass' in Newton's theory refer to? Further criteria from the theory, if available, are needed to render the reference of the term 'mass' unique.[14] A final example is that of the term 'heavier' in pre-Newton theories of motion compared with Newton's theory which distinguishes between 'having greater mass than' and 'having greater weight than'. We do not want to say that the term 'heavier' lacked reference; nor that 'heavier' determinately referred to just one of *being more massive* or *being more weighty*. In the absence of a third thing, 'heavier' could be said to indeterminately stand for either; on such a view it remains unclear what needs to be added from Pre-Newtonian theory to render 'heavier' determinate in reference.[15]

(2) Suppose T_y-plus-T_p determines no referent at all but that if some of the assumptions of T_p are dropped then there is unique realization. Examples of this arise in science where there sometimes seems to be no way of distinguishing between a change of topic and saying different things about the same topic. Thus caloric was thought to be a substance which flows from a hot to a cold body. But if there is no such flowing substance does this mean that there is no caloric, or that there is caloric but it is not a flowing substance? Similarly electricity was thought to be a flow of a substance from a body with one charge to a body with the opposite charge. But again if there is no flowing substance does that mean that there is no electricity, or that there is electricity but that it is not the flow of a substance? Note that historically we took different decisions over caloric and electricity.

Papineau's response to these difficulties is to look to the micro-sociology of how thinkers actually use the terms that they are responsible for inventing and using in the first place. There are no further semantic facts to appeal to in resolving the problem of whether caloric or electricity fail to exist or whether we have simply changed our minds about their properties. Conservatives will cling to their old terms while the more radical will come up with new terms. This might, in part, explain why we have continued to talk of electricity but not caloric. This, he thinks, should not upset the realist (and it does not, on some conceptions of what a realist is). It does, however, introduce forms of indeterminacy that might please some advocates of incommensurability since it does acknowledge that we have run out of semantic 'hard facts' to decide matters of reference, and thus ontology. We will suggest later that something more can be said about the distinction between change in topic versus saying different things about the same topic.

Our answer is also tied up with the following problem. Though the tripartite distinction has important uses, nothing has been said so far about how it is to be drawn, beyond the claim that "some core assumptions involving t unquestionably *do* contribute to t's definition and other assumptions involving t unquestionably do *not* contribute, but beyond that it is indeterminate" (Papineau, 1996, p. 11, but with our t in place of his 'F'). Papineau does provide an account (*ibid.*, footnote 12) of how we can detect whether the tripartite model might apply to a given linguistic community, given the model in the first place. Thus to test a proposal concerning a candidate T_y - T_p - T_n distinction, simply ask members of the relevant linguistic community what they would say if, contrary to their opinion, there were no unique realizers of (a) T_y, (b) T_y & T_p, and (c) T_y & T_p & T_n. But even if we get agreement on the answers (something which strikes us as far from obvious), what would this show? The answers might simply reflect a division that is in principle variable, depending, in somewhat the way Quine defends, on the rest of one's theory of the world and on the concrete circumstances the parties find themselves in (let alone philosophical bias). So given the way Papineau wants to use the tripartite model, it cannot simply be based on patterns of community assent and dissent. We will later suggest a way in which something like that model can be motivated.

To complete this section, we will link the Lewis-Papineau account of reference fixing to one of the examples set out in section 3. There we considered a schematic form of a reference fixer:

(3) the reference of t = k iff k is an object which has property P and stands in relation R to O, or a kind whose members have these features.

Such a reference fixer accords with the tripartite distinction advocated by Papineau. To illustrate how this is so, consider one way of imposing schema (3) on an example, that of (4), the reference fixer for Semmelweis' term 'cadaverous particles':

(4) the reference of 'cadaverous particles' = k if (a) k is a kind of particle, or bit of matter, (b) instances of which are only to be found in decaying matter in corpses and, (c) by being transferred from the hands of student doctors to women just before childbirth, (d) are able to cause childbed fever in these women.

Using Papineau's tripartite model, we could say that (a), (c) and (d) are part of T_y and not part of either T_p or T_n. In fact all of these are part of the robust relational, but highly contingent, properties that obtained in the situation in which Semmelweis introduced his term. Once fixed the reference of the term remained invariant through further discoveries about the particles. Later Semmelweis discovered that instances of the kind did not only arise from dead bodies but also from live bodies in which there was decaying animal-organic matter. The idea that the particles only come from dead bodies is contained in the very connotation of the name that Semmelweis introduced, 'cadaverous particles' and it is underlined by the presence of (b) in the reference fixer. But in the light of Semmelweis' subsequent discovery, the property specified in (b), being from decaying matter in corpses, does not always hold of all instances of the kind. As a result of this discovery we do not want to say that there are no cadaverous particles; there are such particles but it is wrong to attribute property (b) to them.

That a reference is fixed in (4) ought not to be compromised by the fact that Semmelweis made a correction to the beliefs he employed by altering (b) to (b′), viz., being found in decaying animal-organic matter. There are two ways of proceeding at this point. One way would be to invoke Lewis' notion of a near realizer for the right-hand side of (4), treated as an open sentence. The particles are realizers of (a), (c) and (d); they are at best only near realizers of (b) but better realizers of (b′). Here we need to invoke the notion of best near realizer, or consider ways in which what is said in the reference fixer for (4) is only approximately true of the particles picked out. The other way of proceeding is to adopt Papineau's tripartite model in which clause (b) is not in T_y but in T_p instead. This is more natural in that it seems to invoke the idea of "definitional properties" such as (a), (c) and (d) while other properties such as (b) are not strictly "definitional". But only *seems*, because we will not take the tripartite model to necessarily reflect any analytic/synthetic distinction underpinning definitions.[16] In subsequent sections we will provide a motivation for taking the Papineau "tripartite" approach rather than the Lewis "near realizer" approach. However this is not to say that the notion of near realizers is not important for fixing the definition of some terms.[17]

We can place in T_n further discoveries about what kind of "particle" caused childbed fever, viz., something in the broad kind *bacteria*. None of these later discoveries either undercut the successful reference of Semmelweis' term, or introduced incommensurability in the items referred to in this portion of the history of medicine and the growth of our understanding of infection. However, what the example shows is that we need some criterion for saying why certain features do or do not fall into one or other of T_y or T_p or T_n. As we will argue, it is features of the situation in which Semmelweis introduced his term that provide a criterion. And the criterion has little to do with traditional theories of analyticity. If 'analytic' means both *a priori* and logically necessary, then there is nothing analytic about reference fixer (4). It turns on highly contingent relational properties of the particles such as that given in (c). In fact it was Semmelweis' prophylactic procedures for getting the student doctors to wash their hands in chlorinated lime rather than merely soap and water, as was common practice, that reduced the incidence of childbed fever in the First Clinic. Though important in achieving identifying reference, this can hardly be an analytic feature of the term that was introduced.

7. STICH'S OBJECTIONS

But there is a looming worry, one that mirrors our earlier worry about the patchwork-of-cases approach to the theory of reference. The worry is well described by Stephen Stich, who challenges not only Papineau's position but also the presupposition of all the approaches considered so far that there is a viable theory of reference at all. His objections can be found in *The Deconstruction of the Mind* (Stich, 1996). Stich is concerned to show why he now thinks his earlier eliminativism about mental states was wrong. His new view is reminiscent of Papineau's but far more radical: there really is no fact of the matter in the case of any such ontological decision. When we are confronted by a seriously mistaken theory, there is nothing principled that allows us to settle what we should say – no theory of reference, even one allowing for indeterminacy, no principles of ontological inference, etc. Rather, such matters are typically settled through a process of social negotiation in which politics, personalities, and social factors can all play a role: a kind of social constructivism.

To see how different this view is from Papineau's, consider his complaints against Papineau. First, he complains that Papineau's choice of a theory of reference is arbitrary and controversial. Papineau relies on Lewis' account of theoretical terms, but that account is a type of descriptivism, and descriptivism is no longer the only game in town. To the response that semantic intuitions will tell us which is the right kind of theory and for which terms, he counters that semantic intuitions are likely to be thoroughly compromised by the actual course of the history of our words; had things been a little different, he thinks, we might even have ended up saying that phlogiston is identical to oxygen. (Some causal theorists have taken this line with respect to certain terms often thought to be broadly theory-dependent such as 'belief' and 'desire'.) What reinforces such doubts is that Papineau's version of descriptivism continues to maintain a problematic distinction of sorts between meaning-constituting assumptions and others (never mind that it is a vague distinction), for what facts could possibly sustain such a distinction? As Stich reminds us, Quine's doubt about the analytic/synthetic distinction concern the basis of the distinction, not the idea that it is a precise distinction.

Stich's second main objection is even more radical. There may be no correct theory of reference if such a theory is supposed to be an account of some privileged word-world relation that we should care about, a theory that will provide us with the kind of ontological conclusions we also care about (for example, that atoms exist but not phlogiston). In fact, the only time we do or should care is when reference is treated as deflationary, but in that case the caring really concerns our ontology directly, not the fate of words under some preferred word-world relation. The fact that 'phlogiston' doesn't refer while 'atom' does, is just the fact that atoms exist, while phlogiston does not. Semantic ascent provides no insight into the meaning of the latter existential claim. More recently, Stich has complained that the appeal to a non-deflationary notion of reference to settle the substantive issue of whether or not the world contains atoms, phlogiston, and so on, constitutes a "flight to reference" that has held up progress in the philosophy of science (Bishop and Stich, 1998).

Of these criticisms, the second is the more radical and in many ways the more difficult to answer. Our reaction to its radical dismissal of theorizing about reference is to draw a distinction encouraged by Devitt's writings (see for example, Devitt, 1981). It is one thing to think that questions about what there is or isn't to be found in the world are to be construed as questions about language, to be answered by an appeal to a theory of reference. They are not. They are questions of ontology and the answers to such questions may be very secure indeed (Moorean facts, in some cases), much more secure than we have any right to be about our favorite theory of reference. It is quite another, however, to think that such responses can be elucidated by inquiring into the reference-conditions of the terms in question, with the initial ontological judgements seen as rooted in the implicit determination that there are things in the world that meet the associated reference-conditions. That thought seems correct to us. But we also think that such theorizing need not give us anything like an *a priori* theory, and, in particular, we think that we should not expect findings in this area to overturn secure ontological judgements such as the claim that atoms exist, while phlogiston does not. Nonetheless, theorizing of this kind is important, if only because it is sometimes used to establish contrary ontological conclusions, as has happened in some of the debates about realism and incommensurability. When this happens we need to understand what has gone wrong.

As for Stich, his refusal to countenance any use for the theory of reference seems self-defeating. It is not enough to look at the results of social negotiation to tell us whether phlogiston, atoms, and the like, exist because we need to tell a story about the focus of our negotiation.[18] If some scientific community decides to find a use for the term 'phlogiston' by dubbing some new chemical 'phlogiston' even though it has nothing to do with the causes of combustion (perhaps it is called 'phlogiston' because its presence in the brain causes bursts of anger, a metaphorical kind of combustion), that is not a way of establishing that phlogiston exists after all. Much more is needed, and we doubt that Stich has any way of saying what else is needed without engaging in his own speculations in the theory of reference.

But put all this to the side. Consider instead the first criticism, both its worries about how to determine which theory of reference is correct and its worries about Papineau's tripartite T_y -T_p - T_n distinction. Here Stich is surely onto something. First of all, it is often hard to adjudicate between the varying intuitions that motivate philosophers to go sometimes for descriptivism, sometimes for causal-historical theories; hard, too, to see why one theory should seem right for some terms and causal theories for others. Call this mix of problems *The Choice of Reference-Theory Problem*. (We encountered a version of this problem in section 3 in our concern about the inadequacy of the patchwork approach to the theory of reference.) Secondly, it is hard to see why we can be confident that there is a way of drawing Papineau's tripartite division at all, let alone in a way that clearly reflects something important. Call this the *Choice of Meaning-Partition Problem*.

Both the Choice of Reference-Theory Problem and the Choice of Meaning-Partition Problem focus on the apparent arbitrariness of certain choices one needs to make to settle the issue of the reference fixing mechanism of theoretical terms, and hence also the issue of whether theories that use the same term can rightfully be said to use it commensurably. They are problems that remain once we move from the extremely problematic Kuhn-Feyerabend approach (with its talk of world-making) to the Ramsey-Carnap-Lewis approach. On the surface they look just as threatening to the hopes for a viable, comprehensive theory of how our terms secure reference to the world. We think, in fact, that these two problems are connected, and that they should be solved together. But it should by now be clear that they cannot be solved by simply giving a more fully articulated theory of reference. Giving more theory simply tends to put the problem back another step. So in the next section we will begin with something different; this is not another schema of reference fixing but an attempt to understand why terms (observational or theoretical) are introduced in the first place. This yields a way of resolving the first problem. In the following sections we indicate how this response can be parlayed into a response to the second problem.

These responses bear an interesting relation to the concerns about reference, meaning an incommensurability with which we began this paper. Kuhn and Feyerabend did not, in fact, begin with a theory of reference or meaning, but with a daring account of the purposes of theorizing and the difficulties of transtheoretical debate. It was the radical way in which they put their point that then allowed others to abstract theses (1) and (2) of section 1. But now the boot is on the other foot. With the large number of often unmotivated variations one finds among theories of reference, it is time to ask for an overarching framework that might explain their various successes and failures – time to ask for an account that explains the nature of our referential practices in terms of the larger purposes secured by the need to talk and theorize about the world. Unfashionably, perhaps, we therefore propose a return of sorts to the kind of semantics-transcending orientation that we find in Kuhn's and Feyerabend's work. But unlike them we find it in something quite weak indeed, namely in the weak epistemic idea that our attempts to name and talk about things in the world must be appropriately warranted.

8. WARRANT AND THE CHOICE OF REFERENCE-THEORY PROBLEM

Consider again the idea of a causal-historical theory of reference of the sort promoted in the 70s and 80s as an alternative to descriptivist accounts of the meaning and reference of proper names and natural kind terms. The label 'causal' in the phrase 'causal (-historical) theory of reference' is in some ways strikingly misleading. By stressing the involvement of causal notions in familiar examples, it encourages the thought that the essence of a relation of name-reference is its causal character. The label thereby hides the *cognitive* nature of the sort of relationship that underlies and justifies our referring practices. We will argue that where the semantics provided by the Lewis/Papineau approach runs out, we need to appeal to cognitive and epistemic features of the game of naming to see if they can help us with the Choice of Reference-Theory Problem and the Meaning-Partition Problem. We will make a suggestion which deals with both at once.

To see this, consider ordinary proper names ('Edmund Hillary') and familiar natural kind terms ('water', 'tiger'). Such terms are deemed useful only because they are terms for things that are deemed to be cognitively available to us: we take there to be relatively rich epistemic modes of access to the things in question that allow us to track them as we gather non-trivial information about them. Thus Edmund Hillary was not named 'Edmund Hillary' simply on the basis of a particular pattern of causal ties between a group of early users of the name and a child, but on the basis of a perceptual connection linking this group to a recognizable human child – a resilient connection, furthermore, of a type available at other times and to others, that allows those connected to track the child through the course of its spatio-temporal meanderings. And the term 'water' does not denote the scattered substance water simply in virtue of a bare causal relationship which users of the term, or many of them, stand in to samples of water in the actual world, but in virtue of their resilient perceptual relationship to samples of water in the actual world: a connection of a type that allows them to judge that they are confronted by stuff that is liquid, transparent, and found roundabout here in lakes and rivers. Far from being just a causal relation, perception is, crucially, a *cognitive* causal relation, and it is the cognitive, epistemic character of the relation of perception, not its extensional causal character, that we think is of overriding importance as background setting for any theory of reference.

Some will think that, suitably elaborated, such a view harbors a kind of descriptivism – a causal descriptivism for proper names that assigns an important role to such properties as being an individual the speaker is acquainted with in such-and-such a way (Lewis, 1984; Kroon, 1987; Jackson, 1997). Others will insist that it is the patterns of acquaintance themselves that ground reference, so that the view remains a version of causalism. That debate is not important for the purposes of the present paper.[19] What is important is the need *for* a thick cognitive account of the sort of relationship between language-users and the world that underlies term-introduction and term-use. And it is this cognitive feature, we now argue, that also helps us understand the role of theory in the meaning – and reference fixing of theoretical terms.

Here is an account that abstracts from the standard cases by acknowledging this epistemic point. Suppose that t is a name in person (or community) P's idiolect for object, or kind of object, k. Let us say that *the referential practice of using t for k has* objective warrant *for P* iff:

(a) P has an interest in k of a kind that, all things being equal (including the satisfaction of (b)), makes it reasonable to have a referential practice of referring to k by name; and

(b) P has the relevant kind of epistemic access to k, that is, has the ability to pursue, more or less successfully, an enquiry that is appropriate to his interest in k.

And let us say that *the referential practice of using t has* subjective warrant *for P* iff:

> P *believes* that there is an x such that t is a name for x in P's idiolect, and such that the referential practice of using t for x has objective warrant for P.

A number of features of this notion of warrant need clarification.

(1) The idea of subjective warrant is in some ways more fundamental. After all, even the best informed of speakers may be wrong in thinking that the referent of t really exists. (This is what drove Russell to his notion of logically basic terms in which we cannot be wrong about their successful reference.) But those using or introducing the term t at least *believe* that it does. (In fact, they believe that the practice of using t is objectively warranted.) Let us call the belief that the referent of t exists 'The Existence Condition'.

(2) What explains clause (b) is the following. If, following a term-introduction, there is nothing we can usefully say about the object (or kind) denoted, if we cannot follow the term-introduction with a serious attempt to find out more about the object in line with the kind of interest we have in the object, then there is no point to the existence of such a referential practice. Names are not simply linguistic devices that serve the semantic purpose of making something the subject of discourse; they are, just as fundamentally, devices that serve the epistemic *point* of discourse – the acquisition (and, equally, the sharing) of the relevant kind of information. There is an important aspect to requirement (b), viz., that in introducing a name a person is able to pursue

their interest in the thing or kind named, supposing the Existence Condition to be fulfilled, and successfully come up with facts about it. Call this 'The Fact Finding Condition'. Note that this presupposes that, as we decide whether to introduce a term for k, we already have access of sorts to k. But there is no circularity here. The ability to refer to, and think about, k (which for descriptivists will again be explained in broadly descriptivist terms) may exist without there being the opportunity of a viable pattern of referring to k by name. From the present point of view, a crucial question is whether the access we have to k is access of a kind that provides relevant opportunities for fact finding.

(3) The idea of interest in (a) plays an important role in the Fact Finding Condition. It is bound to harbor a degree of relativity. Thus *qua* scientist P may have an interest in a certain chemical kind, its important intrinsic and extrinsic properties, how to test for its presence, and so on, and a scientist's consequent interest in the existence of a referential practice of referring to this kind with a special name in order to pursue this kind of enquiry. The focus of this interest is clearly quite different from that of the interest people have in other persons and their impact on medium-sized physical objects. Thus the focus of the associated enquiry into the item being named will also be different. In the case of a scientist's interest in a certain substance or kind, there is a concern with such questions as: How do we recognize its presence? What phenomena can it explain? What kind of causal impact can it be expected to have? What are its important intrinsic properties? And so on. There are also mixed cases. Our interest in the stuff called 'water' is primarily, but not exclusively, based on such salient properties and its more or less immediate impact on people. It is an interest in the wet potable stuff found in rivers and lakes, and our interest in the existence of a name for this stuff primarily reflects the need to indulge this interest. Not exclusively, however, for other questions naturally arise: for example, what are the underlying physical properties that explain these salient features? With such questions to the fore, it becomes easy to view water as a natural kind, although in a way that should allow the two uses of the term 'water', and the two kinds of interest they serve, to live alongside each other.[20]

(4) In all these cases a common dimension remains: unless we at least have the incipient ability to indulge our interest in an object or kind postulated by our beliefs, the existence of a referential practice of referring to that object or kind with a special name must lack epistemic point, and must count as a gratuitous inflation of language.

It is reasonable to hypothesize that referential practices must have at least subjective warrant if we are not to attribute such a gratuitous inflation of language to term-introducers; such practices must be seen as pointful to their practitioners, even if they are in error and there is nothing in the world their words latch onto. This gives the following perspective on the question of what determines the reference of a term. Speakers must implicitly think of the reference of their terms in such a way that the referential practice involving these terms is seen as appropriately pointful and hence as having subjective warrant. Hence to understand the reference fixing mechanism of the terms we must understand what it is about the entities or kinds posited that, according to those engaged in the practice, gives them reason to believe that the Existence Condition and the Fact-Finding Condition are satisfied. (We stress again, however, that since we are only talking of subjective warrant, a referential practice may satisfy these conditions without there actually being a referent and without our discovering any actual facts. This is surely as it should be. Our terms may be empty terms.)

To sum up, the above proposal says that what determines the reference of a term as it might be used by someone in a referential practice should be something – possession of (enough) associated properties, if one is a descriptivist; the thought that there is a causal network of the right kind if one is a causalist – that shows why this way of attempting to secure reference to something is part of a warranted and hence epistemically pointful referential practice. We should not, of course, presume that speakers can self-consciously rehearse this motivational blue-print. It is enough that the proposal describes constraints that, through our practice, we at least implicitly acknowledge.

When talking of Stich's reservations about Papineau's proposals, we listed two problems: the Choice of Reference-Theory problem and the Choice of Meaning-Partition problem. We have argued for a certain way of answering the first problem. On the proposal under consideration, it is wrong to think that there are two quite different theories of reference that – depending on how the clash of intuitions is to be resolved – may or may not apply to different kinds of terms. Or rather, while we can perhaps see these theories as strategically different for certain purposes they are not fundamentally different if the above proposal is right. The Choice of Reference-Theory problem is answered by stressing that the warrant that underlies term-introductions can have different sources: a *theory* (typically a causal-explanatory theory) in the case of theoretical terms, the availability of *perceptual access* in the case of non-theoretical terms.[21] In both cases, the presumption of a causal connection between term-introducers and the world is given a warrant-entailing cognitive construal. Failure of this presumption – as when a person begins talking about a new comet X on the basis of an illusion – means a failure of reference for the proper name, just as failure of a substantial part of a theory embedding a new theoretical term can mean failure of reference for the theoretical term.

The second problem, the Choice of Meaning Partition Problem, has a solution which immediately drops out of the above considerations. This is best seen by considering some examples of term introduction.

9. SOME APPLICATIONS AND ILLUSTRATIONS

9.1. Proper Names

We begin by applying the proposal to the case of ordinary non-theoretical terms such as proper names, as a way of testing it and seeing how its various components are to be construed in that familiar kind of case. Suppose you introduce a name for someone on the basis of seeing the person in front of you (given their wide-ranging impact, we can assume there is a pattern of interest in such middle-sized objects). What, given your beliefs, confers warrant on this introduction? Note that you believe that there is a person in front of you because you take yourself to *see* this person, thereby showing that the Existence Condition is satisfied (remember that this condition can be satisfied without the referent really existing; we are only talking of subjective warrant). What remains to be satisfied is the Fact-Finding Condition. But this condition is satisfied because you believe, once again, that you have perceptual access to the person. Because you believe you have perceptual access, you believe not only that there is something you can single out in thought, but also that you are in a position to determine facts about this person in various familiar ways (including ways that indirectly build on perception, such as induction or inference to the best explanation). Fact finding need not be an individualistic enterprise, of course; it is difficult to see how we could do without relying on others, especially if, as often happens, we are not in an especially fortunate epistemic situation with respect to the objects we wish to find out about. Causal-historical transmission chains of the kind emphasized by Kripke (1980) acquire much of their importance from this feature of our epistemic situation.

Putting all this together, it is your belief that you and your community have a degree of perceptual access to some particular object that gives warrant to the referential practice. If your belief is wrong (if, for example, you are hallucinating), then the term-introduction failed to secure a referent for the term – and precisely because the belief was wrong, not because the practice wasn't subjectively warranted. How wrong before the term lacks a referent? That is vague, but tolerably so. So long as we can correct many of our perceptual mistakes, charity rules. If there is chronic unreliability over a large and important set of claims, as when we are hallucinating, we would deny that our relationship has the required cognitive features.

9.2. Theoretical Terms

Turn now to the case of theoretical terms, in particular the case of terms introduced on the basis of causal-explanatory theories designed to explain certain puzzling phenomena. What should be clear is that while perception may put us in causal touch with the determinate cause of these phenomena, using just about any theory of causality you care to mention, it need not put us in anything like *cognitive* touch with the cause.

9.2.1. The Case of Semmelweis' "particles"

Semmelweis' "cadaverous particles" were not observable, even though on some occasions there was a distinctive smell that remained on the hands of doctors that had carried out autopsies even after extensive washing in ordinary soap and water. Though Semmelweis and his fellow doctors were in direct causal contact with the kind of "particle" that caused childbed fever, that was even true before Semmelweis developed his theory. So what were the additional cognitive features, and what were the interests involved, that helped determine that the term 'cadaverous particles' had both subjective and objective warrant? First, the Existence Condition: they had good reasons to believe that they had latched onto some entity given that they were able to affect the incidence of childbed fever by asking the doctors to wash their hands not in soap and water but in a chlorine solution. Second, the Fact-Finding Condition: the above enabled researchers to focus upon what was on the hands of the doctors that arose from dead bodies and conduct further research, in particular noting the effect that washing in chlorine solution had on the incidence of childbed fever. Note that the Fact-Finding Condition can be satisfied despite ignorance of any plausible causal mechanism where by the particles bought about their effects. Had Semmelweis merely said: "I give up. I imagine that it is a certain sort of bit of matter that is causing the fever, but I have no idea how. Let's call them 'childbed fever particles'", it is hard to imagine the practice taking hold. Why suppose it was a particle? How would one begin to investigate the nature of the particle and the prevention of its effects?

Now focus on Papineau's tripartite distinction. What is of prime significance are the two relational facts concerning the item that Semmelweis had discovered. The first is the causal effect of bringing about the usual symptoms of childbed fever. The second is the highly contingent relational facts of transmission of some material from dead bodies to women, this transmission being lessened by washing in chlorine solution. These relational facts constitute the core of T_y, viz., k is a kind instances of which are too small to observe (i.e., are particle-like), and can cause childbed fever if transmitted from dead bodies to women about to give birth. What is of less importance is what the too-small-to-see particles are like, and what the various sources of the particles might be. On this Semmelweis had little or nothing to say. Or he had some false beliefs about their origin which he later corrected; the particles did not only come from corpses but from living bodies that contained some decaying organic matter (though even this is not correct as some infections arose without any contact with such bodies). As we have argued at the end of section 6 such widening of the source of the particles should not affect the reference fixing. As to the nature of the particles, further examination was left to later researchers such as Lister and Pasteur. Semmelweis fixed a reference to the 'particles' without either knowing anything of their nature or how they causally acted, or even forming an analytic definition of the term he used to refer to them.

The involvement of a detailed causal mechanism is even thinner in other cases. Thus consider some of the other cases mentioned above, for instance Mendelejev's introduction of names for some elements on the basis of the principles underlying the construction of the periodic Table well before their detection. Sometimes little more than abstract mathematics drives the thought that there is some physical entity to be denoted. Again when Murray Gell-Mann used symmetry considerations suggested by the group-theoretic 'eightfold way' to argue for the existence of a missing baryon which he called the 'omega-minus baryon', he instituted a warranted referential practice whose satisfaction of the Existence Conditions was rooted in abstract group theory. (The Fact-Finding Condition was an easier matter, of course, since there was a great deal of existing particle physics to rely on; significantly, it was not long before the omega-minus baryon was observed in bubble-chamber photographs at the Brookhaven National Observatory.)

9.2.2. The Case of Witches

The term 'witch' is a relatively well-understood term of folk-sociology, at least as it came to be used in 17–18th century western societies. What gave (subjective) warrant to this use was the possibility of broadly theological explanations of a range of more or less mysterious evils by appeal to the existence of human agents possessed of demonic, magical powers. With a well-understood interest in the type of individual who might exhibit this kind of agency, users of the term were able to exploit this interest in appropriate ways, in particular by using ways of identifying such individuals and their activities that were suggested by the theory in conjunction with various background beliefs. But it didn't take long for the theory to suffer a cognitive collapse: the arbitrariness of the background assumptions and the lack of consensus even among adherents of witch-discourse made the resulting discourse widely suspect. Not surprisingly, witch-theory was relatively short-lived in western society, destroyed by the very thing – the desire to learn and talk about witches and their craft, apparently needed to explain evils of various kinds – that warranted the introduction of witch-talk in the first place.

However the case might be more complicated if one takes into account more naturalistic views. In societies with no scientific medicine to speak of, witches might be picked out as those who prepare herbal concoctions to cure various illnesses, perhaps using incantations ("sorcery") to secure their efficacy in the absence of scientific explanations. Now our 17–18th century forebears might have attributed, wrongly, demonic powers to such people, and that might have been the source of the theological conception. (In fact, this probably happened relatively early, with the rise, in the late Middle Ages, of the view that all magic and miracles that did not clearly come from God must have its origins in the Devil). If so, this would explain an ambiguity at the heart of our concept of witch. Thus consider an earlier time at which there is an agreed T_y for a naturalistic view of witches, and suppose the naturalistic theory has, at a later time, a supernaturalist view grafted on to it. We surmise that this became the T^*_p part of an emerging new theory of witches T^*, with T^*_y broadly the same as T_y. In this case different users of the term 'witch' precisify it in different ways. On one way, given by the original T_y, say, there are witches. On another, given by T^*_y-plus-T^*_p (or some subset), there are no witches. The competing core theories relate to different interests – a naturalistic interest in the activities of certain healers, say, and a theological interest in explaining the origin of certain evils in terms of the activities of people capable of such healing – that are both substantial enough to generate subjective warrant for the practice of using the term 'witch'. Such, we conjecture, may well be a useful way of understanding the history of the term 'witch'.

9.2.3. The Case of Phlogiston

The case of 'phlogiston' is more straightforward. There is no use of 'phlogiston' on which "phlogiston exists" is true; in particular, phlogiston is not identical to oxygen. The reason we provide is this. A person can be in causal contact with the substance responsible for combustion and calcination simply by watching flames and yet not stand in a cognitive relation to this substance, for the mere fact that they rightly believe, or even know, that there is a substance or other causally responsible for certain perceived effects scarcely suffices for the acquisition of relevant information about the substance. Much more is needed.

It was, we claim, the framework of phlogiston theory that provided the early phlogiston theorists with their own mode of cognitive access to the substance they postulated as the cause of combustion/calcination, in line with their attendant interest in this substance which in turn derived from their interest in understanding the puzzling phenomena of combustion and calcination. It was phlogiston theory's account of how combustion and calcination were caused, supplemented by needed background assumptions as well as the availability of perceptual access to objects and processes implicated in phlogistonian processes, that provided the wherewithal for pursuing phlogistonian discourse. By suggesting that combustion and calcination were caused or constituted by the escape of a certain substance from bodies undergoing these chemical changes, the theory gave a unifying explanation of combustion and calcination, and thereby gave its supporters reason to think that such a substance did indeed exist (the Existence Condition).

But the theory also suggested relatively clear ways of enlarging our knowledge of this substance, thereby showing that the Fact-Finding Condition was satisfied. In conjunction with widely accepted background theory and beliefs, the theory yielded (1) techniques for locating, measuring, and even isolating samples of the substance, as well as (2) strategies for determining the part played by the substance in the formation of other substances like sulphur (Stahl, 1730). In addition, the theory provided (3) a theoretical matrix to help in the construction of phlogiston-based explanations of other phenomena (for instance, color).[22] Without phlogiston theory's vivid picture of the causal powers of phlogiston its proponents would not have had the advantage of (1), (2) and (3) for extending the field of their more or less reasonable beliefs about this postulated cause of combustion and calcination.[23]

Like classical witch-theory, phlogiston theory too began to suffer a species of cognitive collapse, not only on the basis of the *ad hoc* nature of some of its explanations (Lavoisier, 1785), but especially after Lavoisier's experimental proof that something was absorbed from air during combustion and calcination. Phlogiston theorists' attempts to reconcile this fact with their account of chemical reactions was soon confronted by a range of puzzles and inconsistencies (resulting, in the main, from quantitative studies by Lavoisier and his followers), with every attempt at resolution seeming to generate new puzzles (Kitcher, 1993, pp. 272ff.). In our view, there is little systematic difference between such a case and the case of language rooted in perceptual illusion; both kinds involve pervasive incoherence among experiences that resist plausible explanation, and eventually requires the judgement that what was thought to exist does not in fact exist. This way of seeing matters once again encourages the thought, notably missing from Ramsey and Carnap, that theoretical terms are continuous with other kinds of terms, subject to some of the same epistemic vagaries.

9.2.4. The Case of Electrons

Consider finally (and very briefly) a term like 'electron'. Thomson took electrons to be a particular kind of negatively charged particle or corpuscle that makes up cathode rays, is a constituent of all matter, and is far less massive than atoms and molecules. This was a new kind of constituent of matter, eminently deserving of denotation by a special term.[24] There clearly was warrant for introducing talk of such a particle. In brief: The Existence Condition was satisfied because Thomson took himself to have shown that cathode rays were deflected in the right direction by both magnetic and electrostatic fields, yielding the same small mass to charge ratio (close to 10^7) for each kind of deflection, no matter what the source, and thereby suggesting that whatever was deflected was a uniform kind of particle smaller than known states of matter. The Fact-Finding Condition was satisfied for related reasons; the very circumstances that show the Existence Condition to be satisfied also suggested that participants in electron-talk could learn more about electrons by engaging in the right kind of experiments, for example identifying matter carrying these point charges and determining the mass/charge ratio in experiments conducted by Thomson and later more famously by Millikan. Astoundingly, these values were robust under many different methods of determination. Less than two decades after Thomson's work, Bohr's quantum model of the atom was able to account for atomic spectra by assigning quantized energy levels to the electron. Later developments established possession of an intrinsic, quantized spin. Far from the cognitive collapse that attended both phlogiston theory and witch theory, what we seem to have in the case of electron-talk is a vigorously growing body of theory in which the evidential successes of earlier theories are for the most part preserved and our understanding is deepened. Unlike the case of 'phlogiston' or 'witch' (on its theological use), the subjective warrant for talk of electrons became objective warrant. It is noteworthy that all this can happen in the case of electron without knowing its nature or essence and without providing an analytic definition of 'electron'.

10. WARRANT AND THE CHOICE OF MEANING-PARTITION PROBLEM

Turn now to the Choice of Meaning-Partition Problem, the problem of how we can be confident that there is a way of drawing Papineau's tripartite division at all, let alone in a way that clearly reflects something important. But it should now be clear why the broadly pragmatic answer given to the first problem also helps us with answering this second problem. For the stress on matters epistemic means that any support for the Ramsey-Carnap-Lewis way of seeing things will be conditional. This is because a theory that embeds a theoretical term will normally be a complex, discursive instrument; yet our answer to the Choice of Reference-Theory problem emphasized the role of theory in providing for the satisfaction of the Existence and Fact-Finding Conditions – and entire theories, given their complex nature, will often be far more elaborate than this emphasis entitles them to be. It is the excess that, on the present picture, constitutes what is theoretically distinctive in theories about some common object or kind singled out by the warrant-providing core theory.

Here is an abstract presentation of the situation. Consider a Lewis-type term-introduction of the form: let (new) term t apply to the unique entity satisfying all (or even most) of the predicates $T_1, \ldots T_n$. Semantically this seems in order. Epistemically, however, it may be quite pointless, since we may have no reason whatsoever to believe that $T_1, \ldots T_n$ are jointly and uniquely satisfied (the Existence Condition in the account of 'warrant'); and we may know of no fact finding techniques appropriate to the term even when we do know that $T_1, \ldots T_n$ are jointly and uniquely satisfied (the Fact-Finding Condition). Now consider terms introduced in the context of explanatory theories. Given enough background theory, a sufficiently detailed description of the way in which a postulated entity accounts for certain phenomena is able to satisfy both conditions at once. On any term-introducers' own reckoning, they have reason to believe that this description is satisfied, since they believe that they have good reason for thinking that something like this account provides a reasonable explanation of the phenomena to be explained; and the description of the explanatory mechanism, if sufficiently determinate and sufficiently enriched by background theory, will suggest the fact finding techniques needed to satisfy the Fact-Finding Condition. Theorists thus armed will then be in a position to suggest other bits of theory (in particular bits of theory that make good on the promise of satisfaction of the Fact-Finding Condition), claims that incorporate experimental results, suggest explanations of other phenomena, and so on.

One final point. So far we have argued for a certain way of understanding the privileged role of the core of a theory, seeing this as importantly similar to the role of acquaintance in our understanding of the mechanism of reference of proper names, say.

That provided an epistemic motivation for something Stich rejected in Papineau's account, the idea of a split between the reference fixing and non-reference fixing part of a theory. But Papineau, of course, also argues that there is no precise core/non-core split. And that too seems right to us, although we offer a differ account of such vagueness. Suppose, for example, that a theory introducing a certain term is used to generate an explanation of some new set of facts which then acquires overwhelming importance because of the way it is used to increase our 'knowledge' of this entity (imagine that we began to get much of our 'knowledge' about electrons by exploiting Thomson's 'plum pudding' model of the atom). When this happens, it is possible to view the expanded theory as the core that really warrants the practice of using the term, even though it is equally possible to view the explanation as something that falls out of the fact finding practices encouraged by the theory (and background beliefs), so that the failure of the explanation doesn't compromise the existence of the entity or kind posited.[25]

More generally, there may surely be no determinate way to identify a unique core theory that must be satisfied for the term to have a referent; there may be various competing cores, all close enough to the original theory and large enough to generate (subjective) warrant for the practice of using the term in question. The characterization of warrant – specifically, its appeal to a vaguely delimited ability to acquire warranted belief and to a shifty notion of interest – makes this inevitable. If a theory is somewhat wrong in its explanations, or wrong in some of what it sets out to explain, that scarcely matters, so long as this doesn't greatly affect its fact finding potential. How much error is too much error? That is vague, but why should this matter? We know that there is vagueness too in how close to paradigm perception our access to objects has to be before we declare our attempts at demonstrative reference a success or a failure.

11. INCOMMENSURABILITAS RESURREXIT?

If referential commensurability is the issue of how there can be referential invariance with theory change, then the above account shows how this is possible in a wide range of cases. In many cases we can identify warrant-conferring core theories that license other claims, using fact finding techniques recommended by such cores and with the aid of background theory and empirical observation. But because there need be no unanimity about either background theory, the results of empirical observation, or the way in which the fact finding techniques are applied (especially if we include inference to the best explanation), the application of such techniques is bound to result in competing theories built around the same core. In short, same reference fixing conditions, different theories.

However, the account also leaves open the possibility of widespread semantic indeterminacy, which we think important to highlight but which does not have the radical implications of Kuhnian incommensurability. Using the radical sense in which incommensurability entails different subject matter, it is easy enough to identify uninteresting cases of incommensurable pairs of theories (Keynesian economics and Mendelian genetics are incommensurable in this sense). The interesting cases would be those that exhibit such a radical difference in ontology yet are, in some sense, about the same domain. Kuhn's classic case is the alleged radical incommensurability of Newtonian and Einsteinian mechanics in which terms like 'mass' have different referents in the different theories. We considered this case briefly in section 6 and argued that, like the term 'heavier', there is semantic indeterminacy for the term 'mass' used by Newtonians which is uncovered by, and then resolved within, a subsequent theory.

In many cases, that is our preferred way of understanding how forms of incommensurability might arise. We saw in the previous section how there may be vagueness in the choice of relevant warrant-conferring conditions. Different choices in core-theory may be possible, and even likely (think again of the case of 'witch'). Something like this may well underlie some of the charges of incommensurability in the literature, but like Papineau we are inclined to see such cases as largely benign examples of the way in which vagueness can be resolved in different ways.

Forms of incommensurability may also arise in other, related ways. Consider the situation in which a term t is introduced for a theory T with its associated tripartite distinction T_y - T_p - T_n. Suppose the introduction of t has objective warrant, so that (some subset of) $T_y + T_p$ yields a referent for t, while also providing techniques and heuristics for advancing our "knowledge" about the referent of t. Now suppose that the investigation based upon T proceeds apace, supplemented, perhaps, by new theories; and suppose that this investigation provides new explanations of phenomena involving the referent of t, new experimental techniques for investigating the referent, new lines of research, a new understanding of the "nature" of the referent, and so on. Take the term 'cadaverous particles'. Semmelweis was able to discover only relatively little about the nature of his particles (although a lot about preventive measures). But by the time of Lister, Pasteur, Koch and others, there was a new understanding of what these 'particles' might be, and new techniques for investigating them using microscopes. Not only did they have a new theory of, and new techniques for investigating, what Semmelweis had discovered; they even introduced a new name in the light of their theory for the same old thing. New theories can allow for new ways of conceptualizing old things to which reference had already been made, and so also new reference fixing conditions for terms for these old things.

Here we are assuming no change of reference. But we should also allow for the removal of indeterminacy from the reference fixing conditions of some old term. Thus in the case of Semmelweis' term there is indeterminacy of reference (and not lack of reference) since, unbeknownst to Semmelweis, several different species of bacteria were picked out using his reference fixer (4). More generally there might be a number of equally good deservers of the name t in theory T, where the inclusion of T_p does not remove the indeterminacy. However such indeterminacy might well get removed when a new T* comes on the scene, as before, and the new T^*_y - T^*_p - T^*_n distinction determines a single object or kind. Such is the case when, with the advance of atomic physics, isotopes of chemical elements have to be recognized as well as the chemical elements themselves. More complicated is the case in which biological kinds have been established on the basis of phenotypic characteristics using some theory of classification. But then the discovery of new tools for determining the DNA contained within cells may lead to new biological classifications. Old terms might then be recognized to be indeterminate and new terms are introduced to name the various sub-species; or new boundaries have to be drawn in which earlier classifications are overturned and new re-classifications made, each new kind being furnished with a new name. As before, there is no reason to suppose that this process is accompanied by a kind of radical incommensurability that licenses talk of "different worlds".[26]

12. CONCLUSION

This paper has looked at various ways of understanding the way theoretical terms refer, focusing on the Ramsey approach as exemplified by Lewis, and then Papineau's suggestions as to how that approach might be further developed. It also canvassed some problems – the Choice of Reference-Theory and Meaning-Partition Problems – that arise naturally from Papineau's suggestions, especially those raised by Stich. It finally develops some linked answers to these problems in terms of certain weak epistemic constraints concerning the point of using names in the first place. Along the way we gave reasons for our intuitive tendency to let the reference fixing parts of a theory include some things but not others; reasons too for thinking that there is bound to be a degree of indeterminacy in what gets included, and why this doesn't (greatly) matter.

None of this casts any doubt on the techniques of Ramsification discussed in this paper, but what it does show is that we need more than just Ramsification if we are to understand how theoretical terms work, and if we are to be confident about the kind of semantic distinctions that Papineau in particular wants to use. We need – not surprisingly, in our view – to understand how epistemology matters to semantics.

University of Auckland

ACKNOWLEDGEMENTS

Thanks to the Center for Philosophy of Science at the University of Pittsburgh for supporting some of this work while the first author was a Fellow at the Center. Special thanks to Hans Rott for some very incisive criticisms of an earlier version of this paper as a commentator at the conference where it was first presented.

NOTES

[1] This has been dubbed the '*qua*-problem' by Devitt and Sterelny who discuss it in their 1987 chapters 4.4 and 5.3. The *qua*-problem was earlier raised in Papineau (1979) chapter 5.7.

[2] The issues raised here were discussed in Nola (1980) and Kroon (1985).

[3] Later, of course, it became possible to differentiate different kinds of HI-Viruses, in particular HIV-1, the virus responsible for most cases of AIDS to date, and HIV-2, a less aggressive virus more common among West Africans. Also HIV-1 'was found to have at least ten distinct genetic subtypes that might vary in transmissibility' (Thagard, 1999, p. 126).

[4] As far as well can tell the Ramsey Sentence was first so called by Hempel in section 9 of his "The Theoretician's Dilemma" first published in 1958; see the reprint in Hempel (1965). Ramsey's idea first occurs in his "Theories" of 1929 (see Ramsey, 1990), though it is not always clear to what purpose he wished to put his idea. We will not explore Ramsey's original paper here.

[5] This matter is put even more strongly in Hintikka (1998) by presenting the issues in terms of model theory. Theory '$T(t_1, t_2, \ldots t_n, O_1, O_2, \ldots O_m)$' will have a full model $M(t_1, t_2, \ldots t_n, O_1, O_2, \ldots O_m)$ of which there is a substructure, an observational model $M(O_1, O_2, \ldots O_m)$. What Ramsification reveals is that theory T restricts the class of observable models by saying that they must be imbeddable in the full model for T but without saying anything about the theoretical entities themselves beyond this structural requirement.

[6] See Carnap (1966) chapter 26 on the Ramsey Sentence, and chapter 28 for his account of analyticity in a theoretical language and what has come to be called the 'Carnap Sentence'.

[7] See Tuomela (1973) chapter III, section 2 for proofs.

[8] The argument given in this paragraph comes from Lewis (1970, pp. 431–434). We need not rehearse here the considerations Lewis raises against accepting that all our best theories are multiply realized and for the view that "good" theories will be uniquely realized. See Lewis (1970, pp. 432–434). What we do allow is that

theories might at some stage have multiple realizations but that with the advance of science they become uniquely realized, unique realization being one of the virtues we wish our theories to exemplify.

[9] For a useful account of the differences between Ramsey and Carnap, the employment of the "Ramsey Sentence", and what seems to be Carnap's independent discovery and use of it, see Psillos (1999, chapter 3 and 2000).

[10] There are three points to be made about (8). The first is that it can be expressed more generally. If there are n theoretical terms in T to be defined, then we can fix the reference of each as follows where i = 1, 2, ... n: the reference of $t_i = (\P x_i)[(\exists!x_1) \ldots (\exists!x_{i-1})(\exists!x_{i+1}) \ldots (\exists!x_n)\, T(x_1, x_2, \ldots x_n, O_1, O_2, \ldots O_m)]$. Here we have again used first order variables x_j in the Ramsey Sentence rather than second and higher order quantifiers. Lewis shows how names can be introduced to stand for any entities such as properties, relations, classes, functions, or whatever. For example, in the case of 'a is an electron' we can rewrite this as 'a instances electronhood'. The theoretical predicate 'is an electron' is transformed into the name 'electronhood'. The relation '– instances ...' will be part of the non-descriptive vocabulary of T and is such that both gaps are filled by names, the first a name for an object, the second a name for a universal or property. See Lewis (1970, p. 429).

The second point is that Lewis intends the above to provide a definition of theoretical terms while we have used it only as a reference fixer for theoretical terms. The difference between Lewis' more broad account and our more narrow account affects nothing in what follows. Finally like Lewis we shall assume an appropriate free logic for definite descriptions.

[11] In his 1970 paper "How to Define Theoretical Terms" Lewis adopts the view that where there are two or more realizers there is no reference; but in "Reduction of Mind" (1994, p. 417) the position is more relaxed so that sometimes, or even always, we should allow for ambiguity in reference rather than lack of reference.

[12] It is possible to drop the assumption that T*(x) has a perfect unique realizer and say that it too is open to correction by a later theory T**. In this case let us suppose that there is also a best unique near realizer of T*(x). Then t in T and T* refers to the same item if and only if the best unique near realizer of T is the same as the best unique near realizer of T*, or the item t refers to is a better (unique) near realizer of T*(x) than it is of T(x).

[13] Talk of near realization in Lewis (1994, p. 417) is even more relaxed; strict satisfaction can be replaced by satisficing across a number of dimensions. O'Leary-Hawthorne (1994) also get us to imagine trade-offs between more perfect realizers which arise out of gerrymandered sets of entities and much less perfect realizers which are not gerrymandered but are natural kinds. Here there is a trade off between the degree of perfection of the realization and features of the kinds of entities that are invoked as realizers (such as whether they are natural kinds or not). See also Niiniluoto (1997, section 3) in which the reference of t in a theory T to object a is defined in a number of related ways to accommodate the fact that a is not a perfect unique realizer of T; a might realize *most* of T which contain t; or the reference is a where a maximizes the degree of approximate truth of T relative to a (i.e., how close T is to being true about a), or maximizes the degree of truthlikeness of T relative to a; and so on. Niiniluoto also indicates the importance of such considerations against Putnam's pessimistic meta-induction.

[14] Field's position that the Newtonian term 'mass' is indeterminate in reference between rest mass and relativistic mass (Field, 1973) is challenged in Earman (1977) who says that Newtonian mass refers to rest (proper) mass. Following Earman's paper, Fine (1977) cites the authority of Einstein who though the same. Citing Einstein's authority is akin to the sociological influences mentioned by Papineau and Stich as a way of resolving problems that arise for referential continuity with theory change.

[15] The example of 'heavier' was discussed by Hartry Field, a participant at the conference at which this paper was originally given.

[16] Here we follow the position set out in Kripke (1980) in which reference fixers like (4) are known to be *a priori* true by those who introduce the term. But the reference fixers do not express analytic truths; nor are they necessary truths. However nothing precludes the use of necessary properties of a kind, if there are such, in a reference fixer that introduces a name for the kind.

[17] In fact the notion of realizers and near realizers is more fundamental. The notion of a realizer is required by the Papineau account as the entity which (perfectly) realizes, say, T_y. However it might be the case that even in Papineau's T_y that there is no unique perfect realizer but only a unique best near realizer to be the referent of a term. There are a number of complexities here that we only touch on and do not explore in a more nuanced account.

[18] The role for sociology adumbrated in Papineau (1996, p. 19) is greatly expanded upon in Stich (1996, pp. 74–82).

[19] Kroon (1987), Lewis (1984), and Jackson (1997) argue for such a version of causal descriptivism, which they take to be a kind of causal theory made self-conscious. On this kind of theory, causal descriptions rather

than causal relationships (or even causal relationships supplemented with descriptive conditions, as in the weaker account of causal descriptivism defended in Sankey (1994)) determine reference.

[20] It is the vagueness in what is relevantly important and of interest that, in our view, generates the different intuitions in Putnam's well-known Twin-Earth thought-experiment about the reference of 'water' (Putnam, 1975, chapter 12). Once we hear Putnam's manner of presenting the thought-experiment, with its stress on the radical difference in underlying structure between what is called 'water' on Earth and Twin-Earth respectively, the 'natural kind', rigid designator, way of resolving the vagueness becomes particularly salient. But there is another resolution of vagueness on which 'water' stands for a more inclusive, non-rigidly designated kind. (As David Lewis has suggested, given the way in which Putnam's presentation raises the notion of a natural kind to salience it is no wonder that so many think he has won the day on this issue; Lewis, 1994, pp. 423–424.)

[21] Causal-explanatory theories explain some phenomenon in terms of the causal efficacy of a (kind of) entity, as in the case of phlogiston theory. But our account is generous enough to allow that in some cases there may be nothing approaching perceptual access or a causal-explanatory theory. This is so in the case of Mendelejev's table which allows the introduction of names for elements in the periodic table, for example, or Murray Gell-Man's introduction of 'omega-baryon-minus' on the basis of the group-theoretic eight-fold way.

[22] The explanation of color proceeded via the view held by Macquer and others that phlogiston is the matter of light 'fixed' in combustible bodies. For details of ways in which phlogiston theory satisfied (1), (2) and (3), see Partington (1962). Not surprisingly, there was considerable interaction among (1), (2) and (3). Having isolated samples of the substance based on an implementation of (1), and having arrived at reasons based on observation and background theory for thinking that the substance had such-and-such features, theorists were able to utilize these features to help with (2) and (3). Conversely, armed with implementations of (2) and (3), theorists were able to offer alternative suggestions for implementing (1).

[23] Note that the justification of these beliefs were sometimes based on shared heuristics (Zyglow and Simon, 1986), but often relied on hidden assumptions that themselves were in dispute between different theorists, and on explanatory moves that seemed simply arbitrary (Beretta, 1993).

[24] As Thomson (1897) notes, the direction of deflection was enough to establish the electric nature of cathode rays, but was not enough to show that a new kind of very small particle was involved (p. 302).

[25] The following example may be relevant. Consider the relation between late phlogiston theory and the kind of theory Richter and Gren proposed in the late 1700s, which claimed a role for a substance they called 'phlogiston' in explaining luminary phenomena but not combustion or calcination (see Partington, 1962). We suspect that the frosty reception on the part of chemists like Suckow to this use of 'phlogiston' had much to do with the fact that for some earlier phlogiston theorists the explanation of luminary phenomena was one of their theory's successes. In contrast Richter and Gren sided with oxygen theorists where combustion and calcination were concerned and tended to see luminary phenomena as requiring independent explanation in terms of the emission of a principle or substance in the course of chemical combustion, which they called 'phlogiston'. The condition 'the principle emitted from bodies during combustion, thereby causing the appearance of light [and heat]' was therefore regarded as a reference-fixing condition for 'phlogiston' by Richter and Gren but only a plausible theoretical by-product of phlogiston theory by many earlier theorists.

[26] The above story has been told using Papineau's tripartite distinction. However a similar story can be told using Lewis' notion of near realizers. If T* is a 'corrected' version of T, then T and T* cannot have the same unique near realizer, but whatever is the (unique) realizer of T* is also a best near unique realizer of T. To take into account the example from biology cited above, something like O'Leary-Hawthorn's modification (see footnote 14) needs to be made. Realization is now something which takes place across a number of dimensions; not only is there degree of near realization to take into account but also the extent to which the kinds which do the realizing are gerrymandered or approach natural kinds.

REFERENCES

Beretta, M. (1993). *The Enlightenment of Matter: the Definition of Chemistry from Agricola to Lavoisier.* Canton: Watson Publishing International.
Bishop, M. and S. Stich. (1998). "The Flight to Reference, or How *Not* to Make Progress in the Philosophy of Science." *Philosophy of Science* **65**: 33–49.
Carnap, R. (1966). *Philosophical Foundations of Physics*. New York: Basic Books.
Devitt, M. (1981). *Designation.* New York: Columbia University Press.
Devitt, M. and K. Sterelny. (1987). *Language and Reality*. Cambridge: MIT Press.
Earman, J. (1977). "Against Indeterminacy." *Journal of Philosophy* **74**: 535–538.
Field, H. (1973). "Theory Change and Indeterminacy." *Journal of Philosophy* **70**: 462–481.
Fine, A. (1977). "Appendix." *Journal of Philosophy* **74**: 538.
Feyerabend, P. (1962). "Explanation, Reduction and Empiricism." In H. Feigl, G. Maxwell, and M. Scriven, eds., *Minnesota Studies in the Philosophy of Science, Volume 3*, pp. 231–272, Minneapolis: University of Minnesota Press.
Hempel, C. (1965). *Aspects of Scientific Explanation.* New York: The Free Press.
Hintikka, J. (1998). "Ramsey Sentences and the Meaning of Quantifiers." *Philosophy of Science* **65**: 289–305
Jackson, F. (1997). "Reference and Description Revisited." *Philosophical Perspectives* **12**: 201–218.
Kitcher, P. (1993). *The Advancement of Science.* Oxford: Oxford University Press.
Kripke, S. (1980). *Naming and Necessity*. Cambridge Mass.: Harvard University Press.
Kroon, F. (1985). "Theoretical Terms and the Causal View of Reference." *Australasian Journal of Philosophy* **63**: 143–166.
Kroon, F. (1987). "Causal Descriptivism." *Australasian Journal of Philosophy* **65**: 1–17.
Kuhn, T. (1970). *The Structure of Scientific Revolutions.* 2nd edition. Chicago: University of Chicago Press.
Lavoisier, A. (1785). "Reflexions sur le phlogistique." In *Oeuvres, Volume 2* (Paris, 1862–1893), pp. 623–655.
Lewis, D. (1970). "How to Define Theoretical Terms." *Journal of Philosophy* **67**: 427–446.
Lewis, D. (1984). "Putnam's Paradox." *Australasian Journal of Philosophy* **62**: 221–237.
Lewis, D. (1994). "David Lewis: Reduction of Mind." In S. Guttenplan, ed., *A Companion to the Philosophy of Mind*, pp. 412–431, Cambridge Mass.: Blackwell.
Niiniluoto, I. (1997). "Reference Invariance and Truthlikeness." *Philosophy of Science* **64**: 546–554.
Nola, R. (1980). "Fixing the Reference of Theoretical Terms." *Philosophy of Science* **47**: 505–531.
O'Leary-Hawthorne, J. (1994). "A Corrective to the Ramsey-Lewis Account of Theoretical Terms." *Analysis* **54**: 105–110.
Papineau, D. (1979). *Theory and Meaning*. Oxford: Clarendon Press.
Papineau, D. (1996). "Theory-Dependent Terms." *Philosophy of Science* **63**: 1–20.
Partington, J. (1962). A *History of Chemistry.* London: Macmillan & Co. Ltd.
Psillos, S. (1999). *Scientific Realism: How Science Tracks the Truth.* Routledge: London.
Psillos, S. (2000). "Rudolf Carnap's 'Theoretical Concepts in Science'." *Studies in History and Philosophy of Science* **31:** 151–172.
Putnam, H. (1975). *Mind, Language and Reality. Philosophical Papers, Volume 2.* Cambridge: Cambridge University Press.
Putnam, H. (1978). *Meaning and the Moral Sciences.* London: Routledge and Kegan Paul.
Ramsey, F. (1990). *Philosophical Papers*. Cambridge: Cambridge University Press; edited by D. H. Mellor.
Sankey, H. (1994). *The Incommensurability Thesis.* Aldershot: Avebury.
Sankey, H. (1997). "Incommensurability: The Current State of Play." *Theoria* **12**: 425–445.
Semmelweis, I. (1963). *The Etiology, Concept and Prophylaxis of Childbed Fever*, trans. by K. Carter. Madison: University of Wisconsin Press (first published in German in 1860).
Stahl, G. (1730). *Philosophical Principles of Universal Chemistry.* London: Osborne and Longman.
Stich, S. (1996). *The Deconstruction of the Mind.* New York: Oxford University Press.
Thagard, P. (1999). *How Scientists Explain Disease*. Princeton: Princeton University Press.
Thomson, J. (1897). "Cathode Rays." *Philosophical Magazine* **44**: 293–316.
Tuomela, R. (1973). *Theoretical Concepts.* Vienna and New York: Springer-Verlag.
Zytkow, J. and H. Simon. (1986). "A Theory of Historical Discovery: The Construction of Componental Models." *Machine Learning* **1**: 107–137.

HAROLD I. BROWN

INCOMMENSURABILITY AND REALITY

Abstract. Realists hold that science seeks to discover entities and process that exist in nature independently of whether these are accessible to our senses. This pursuit requires the development of new concepts as part of the research process, and thus requires learning new modes of thought, which is the basic source of incommensurability. I sketch a theory of conceptual content that explains how new concepts are introduced as modifications of existing concepts, and how the resulting continuities allow innovators to promote new ideas in a coherent manner. An account of evidence is proposed that explains how items under study constrain the choice of concepts, thereby limiting the scope of incommensurability and promoting pursuit of the realist goal.

My aim in this paper is to argue for a pair of theses about the role of incommensurability in the development of science. I will, however, put off formulation of these theses until the end of the paper and devote most of my space to discussing four key notions: incommensurability, concepts, scientific realism, and external evidence. The main purpose of these discussions will be to defend some substantive positions on debated issues. Thus while I will discuss various positions that have been held on these topics, I will not be concerned with detailed explication of existing views. By the time I am ready to formulate my main theses, most of the arguments on their behalf will already be in place.

1. INCOMMENSURABILITY

I want to begin with an initial sketch of incommensurability. These remarks will be preliminary since they will make use of notions that I will discuss below, so I will further clarify my account of incommensurability as we proceed.

Incommensurability can appear within a scientific field when a new approach requires ways of thinking that are significantly different from established ways of thinking. Typically the new approach has features that are incompatible with aspects of the older approach. A new approach may involve elimination of older ways of thinking, introduction of new ways of thinking, or a combination of the two. There are at least two kinds of changes in scientific thinking that generate problems of incommensurability: conceptual change and changes in scientific methodology, by which I mean changes in the standards for evaluating the relevance and acceptability of scientific claims.[1] In this paper I will focus on cases of incommensurability that derive from conceptual change.

Incommensurability occurs only when two approaches to a single subject matter are in competition. Distinct frameworks – such as the rules of basketball and the standard

P. Hoyningen-Huene and H. Sankey (eds.). Incommensurability and Related Matters, 123–142.

model in physics – regularly exist together in a culture and in an individual mind without generating the problems addressed under the incommensurability rubric because there is no reason to choose between them. The means by which we recognize a genuine dispute are highly contextual, but in practice there is little difficulty – as the examples that follow will indicate.

Notice that I am approaching incommensurability in cognitive terms: The issues in question arise when people must learn to think about a subject matter in new ways. The need for new modes of thought is a pervasive feature of the development of human knowledge. As Descartes emphasized, it is an unhappy fact about us that we are not omniscient; if we were, we would have no need for research or for a methodology for evaluating knowledge claims. Moreover, as advocates of naturalistic epistemology regularly point out, the appropriate methodology for developing human knowledge depends on the kind of cognitive equipment we have available. For example, sense perception plays a central role in epistemology because we get our information about the items around us by means of our senses.[2] If we were beings of a different sort – say, beings that get information via a form of radiation that directly modifies neural tissue, or beings that already possess all truths in a dim form and need only meditate properly to draw them out – a significant portion of our epistemology would be addressing different issues and need different approaches. By the same token, questions about the nature and consequences of conceptual change arise because reflection on individual human development and human cognitive history show pretty clearly that conceptual change occurs. The approach I am taking to the development of science is a natural extension of the traditional empiricist view that we have no direct insight into the nature of reality, and includes in this view the thesis that we have no direct insight into the appropriate concepts for thinking about any subject. The development of appropriate concepts is one of the tasks that science pursues.[3]

Since I am going to focus on conceptual change, it is important that we keep in mind how pervasive such change is in the development of science. Most discussions in the literature turn on a few important examples, but there is a much larger set of concepts in contemporary science and mathematics that was developed over the history of these subjects, and that have to be learned in the course of an individual life. Consider just a few examples of scientific and mathematical concepts whose history we can trace: ultraviolet radiation, boson, isotope, isomer, gene, telomere, logarithm, gamma function, partial derivative, and on and on. Understanding each of these concepts requires mastery of many other concepts that have been introduced in the course of our history. Nor is the development of science the only source of new concepts. Examples based on technological developments include carburetor and surrogate mother, while other fields have provided such concepts as split infinitive, separable prefix, futures option, feudalism, home run, and secretary of state.[4] I will, however, stick to scientific examples in this paper.

Conceptual change also includes cases in which concepts are dropped. It is more difficult to list concepts that no longer occur in science than to list those that do occur, so I will mention some stock examples – natural place, celestial spheres, phlogiston, and caloric – to introduce the point. However, historical research on specific fields shows that there are many other cases in which researchers introduced new concepts and then dropped them before they became part of a scientific tradition. For example, from 1901–1903 Rutherford and Soddy were studying thorium in their attempt to understand radioactivity. At one stage of this study they thought that thorium was not itself radioactive. Rather, they believed that thorium underwent a change that produced thorium X, which is the source of the radioactive radiations they detected. They succeeded in separating thorium X from thorium, but the remaining thorium retained about 25% of the radioactivity no matter how much thorium X they removed. To account for this, they postulated a second radioactive component that could not be separated from thorium by chemical procedures, and that provided the source of the residual radioactivity. A little earlier Becquerel had found similar behavior in the case of uranium. Soddy redid Becquerel's investigations and extended the account to this element. But the concepts of separable and inseparable radioactive constituents were short lived. With regard to these constituents, one historian remarks:

> Historically, such constituents belong to the same category as phlogiston and caloric. This is another example of a conclusion drawn on the basis of available evidence and within a theoretical framework ultimately found to be erroneous. But in April 1902 Rutherford and Soddy, having overcome their initial scepticism, fully embraced active constituents, both separable and inseparable, produced and made active in the process of transformation, as the general explanation of radioactivity. (Trenn, 1977, pp. 75–76)

This example is far from unique. Consider another phenomenon that was noted in early studies of radioactivity: A material placed near thorium or radium would itself become radioactive. The Curies explained this by postulating a form of radioactive induction on the model of magnetic induction. Rutherford and Soddy also noted that there were some substances that appeared only briefly in the course of radioactive change. They initially took these to be a new form of matter that they called 'metabolons'. *Radioactive induction* and *metabolons* were concepts that had rather short life-spans.[5]

It may be useful to mention a bit more of the historical context here. X-rays had been discovered in 1895 and radioactivity in 1896. By 1902 alpha rays and beta rays had been distinguished, and the latter identified as streams of electrons. It had not yet been established that alpha rays are material particles or that they carry an electrical charge – they were generally believed to be a kind of secondary X-ray caused by the beta rays.[6] Gamma radiation had been identified, but its significance was not clear and it played no role in the brief period I am discussing. Einstein's account of the relation between matter and energy, Rutherford's discovery of the nuclear structure of the atom, the concept of an isotope, Bohr's theory of the atom, and the discovery of protons and neutrons were all in the future. We are in the midst of a period of intensive research at the interface between chemistry and physics during which new concepts were being formulated, tried out, modified, and dropped. Similar examples can be found in the development of other fields.

My general point, then, is that conceptual development is a central feature of the ongoing human attempt to describe and understand the world we live in – and that we pursue this attempt without any *a priori* guidance. A brief consideration of some of the literature on incommensurability will help clarify why I want to invoke this concept here. Recall, first, that when Feyerabend and Kuhn introduced this notion there was a prevailing view of the nature of concepts that had been derived, with modifications, from the empiricist tradition. On this view, our concepts divide into two classes: basic concepts which get their content by direct association with observables, and auxiliary concepts that, somehow, get their content from the basic concepts. On the oldest version of this view, auxiliary concepts are constructs out of basic concepts, and are completely replaceable by combinations of these concepts. Thus auxiliary concepts are genuine 'auxiliaries': they are introduced for convenience, and are eliminable for purposes of epistemic analysis. However, the development of science led to the pervasive introduction of theoretical concepts, and these proved to be resistant to analysis in empiricist terms. Early in the twentieth century, philosophers attempted to resolve these concepts into basic concepts on the traditional model. This approach ran into serious difficulties, and was followed by a series of alternative accounts of the relation between theoretical concepts and observation concepts (cf. Brown, 1979, chapter 3). By the late 1950s there was a prevailing view which held that the empirical content of theoretical concepts derived from a combination of implicational relations among axioms plus correspondence rules that related these axioms to a set of observation concepts. The key point for present purposes is the central role that observation concepts played in empiricist philosophy of science. In particular, since this view held that the content of any concept is wholly determined by a combination of observation concepts and logical relations, the introduction of a new concept into science required only the reorganization of already available materials. In addition, logic and observation provided a smooth path by which one could learn new concepts.

Feyerabend and Kuhn challenged this entire approach when they challenged the existence of a special set of observation concepts. (See Khalidi, 2000 for a recent account.) Without the common medium provided by observation concepts, the psychological process of learning new concepts becomes rather more difficult than it seemed according to the prevailing view. Kuhn emphasized the central role that this lack of a common medium plays in his thought when he explained that he had taken the term 'incommensurability' from mathematics where it is used to express the lack of a common unit for measuring, for example, the side and diagonal of a square:

> Applied to the conceptual vocabulary deployed in and around a scientific theory, the term 'incommensurability' functions metaphorically. The phrase 'no common measure' becomes 'no common language'. The claim that two theories are incommensurable is then the claim that there is no language, neutral or otherwise, into which both theories, conceived as sets of sentences, can be translated without residue or loss. (Kuhn, 1983, p. 670; cf. Kuhn, 1977b, p. 301)

Kuhn went on to emphasize that this lack of a common measure does not imply that no comparison is possible, but that comparisons are of a different sort than comparing two different items expressed in a common language. The cognitive form of incommensurability that concerns me in this paper comes to the fore when we consider learning the new concepts. However this is done, it requires a different – and presumably more difficult – transition than would be involved if the new concept could be completely formulated in terms of already available concepts.

A linguistic example may help underline the psychological point. The possessive adjectives in English have an odd characteristic: In the third person singular (his, her, its) – and only in the third person singular – these adjectives carry information about the gender of the possessor. French (to take but one example) is more systematic; none of the possessive adjectives carry information about the gender of the possessor. To be sure, all French adjectives come in both masculine and feminine forms, but the gender of an adjective is determined by the gender of the noun being modified. The choice of a possessive adjective has nothing to do with the gender of the person whose possessions are being discussed. Everyone, whether male or female, says '*ma mère*' and '*mon père*', and there are no French words that play the same linguistic roles that 'his' and 'her' (in this use of this ambiguous word) play in English. Thus an English speaker cannot learn the French forms by simply translating between individual words in the two languages. In French it would be fairly easy to write a story that introduces a character by describing objects that the character possesses without revealing the character's gender. Translation of the story into English would not be straightforward. (See Kuhn, 1983, pp. 679–681 for further examples.) In psychological terms, this situation requires an English speaker learning French to learn some new patterns of thought without making a smooth transition via a simple mapping of one term onto another, or mapping both onto some third language. It is this psychological gap that I am focusing on in my use of the term 'incommensurability' and I suggest that this psychological gap is the fundamental phenomenon. Other issues that are commonly discussed under the incommensurability rubric, such as change of meaning and change of reference, are important but derivative from this more basic source. For example, one important form of reference change in the development of science involves reorganization of reference classes – such as when Copernicans included the earth in the same class as the planets, and the sun in the same class as the stars. But these reclassifications result from changes in the concepts being used to think about a particular domain.[7] I will have more to say about such cases shortly, but first I must explain how I am using the notion of a concept.

2. CONCEPTS

I want to approach the subject by recalling that when Putnam argues that meanings are not in the head, he distinguishes meanings from concepts which are in the head (Putnam, 1975, pp. 217–218, 226, 245, 248). I am concerned with concepts in just this sense: Concepts are mental entities that represent some items of interest. This use of the term 'concept' is common in cognitive science.[8] There is no settled account of how concepts are implemented in the brain, and conceptual systems are regularly studied as abstract structures with the understanding that an adequate account must eventually be cashed out in terms of neural structures (cf. Thagard, 1992, pp. 28–29). This is the approach that I will follow here.

Once we focus our attention on the development of knowledge, the need for conceptual change as science proceeds leaps to the mind. Consider one common mode of scientific research: scientists pick out items that they consider to be natural kinds and attempt to develop concepts that will describe their properties. As research continues, drastic changes may occur both in our understanding of which items are members of a single kind, and in the ways in which we think about these kinds. For example, at an early stage in the development of chemistry, long before modern notions of chemical structure appeared, some theorists picked out earth, water, air, and fire as basic kinds into which the physical world is divided, and introduced two pairs of contrary properties – hot/cold and wet/dry – as a basis for understanding the characteristic features of these elements. Water, for example, was described as the cold, wet element, and earth as the cold, dry element.[9] From a contemporary perspective these researchers made mistakes both in their selection of the natural kinds and in the concepts they used to describe them. When we select a set of items as members of a single kind we fix the referents of the concepts we seek to develop, but to select items as constituting a single kind is to propose a revisable hypothesis. The reorganization of reference classes is a common outcome of research on classes we have selected. Still, once such classes have been selected, researchers can formulate hypotheses as to their distinguishing characteristics. When they become confident that these hypotheses are correct, they build them into the concepts they use to describe the items in these classes. For the moment I want to emphasize that the early chemists' concept of, say, water as the cold, wet element was a genuine concept, even though we no longer consider the concept to be an adequate description of its subject matter. The same point holds for the concept of phlogiston, which was introduced at a later stage in the development of chemistry. Those educated in the appropriate traditions could have provided analyses of these concepts, and historians can now recapitulate those analyses. I am going to concentrate on concepts for most of this paper, but referents will eventually force their way into the discussion.[10]

In order to understand conceptual change and the way in which incommensurability is generated, we need an account of how conceptual content is determined. I have already sketched the prevailing view in philosophy of science when Kuhn and Feyerabend began talking about incommensurability, and I noted how logic and an observation language were thought to provide a medium in which the content of all concepts could be formulated. As a result, on this view incommensurability is always avoidable in principle.

However, with hindsight we can see that there is considerable scope for conceptual change even on the most traditional empiricist approach. These changes will occur on the level of the auxiliary concepts, but most of the concepts we use in everyday life and in science are auxiliary concepts. Human beings create these concepts in response to experience, and we transmit them to our children as part of the process of teaching them how to deal with their physical and social environments. On the usual empiricist epistemologies, auxiliary concepts embody beliefs about phenomena that tend to occur together in some regular fashion, and research can lead us to introduce new auxiliary concepts and eliminate others from our active conceptual repertoire. We can, for example, conclude that Hesperus and Phosphorous are two manifestations of a single entity, and replace the two distinct concepts with a single concept. We can find that jade or earth is not a single natural kind after all, and introduce new concepts to capture the relevant distinctions. Learning new auxiliary concepts will often involve learning new ways of thinking about and responding to experience, and may be quite as challenging psychologically as it is on alternative theories of concepts since people do not learn new concepts by building them up from observation concepts. In other words, even if the empiricist account of concepts were correct, the point remains that human thinking takes place mainly in terms of auxiliary concepts, where the challenges of adapting to new modes of thinking do arise.

Kuhn and Feyerabend rejected this view of scientific concepts, opting instead for a view that had long been in the literature, although of only limited popularity in the empiricist tradition. On their view, concepts occur only as members of systems of concepts, and conceptual content is determined by relations among these concepts. No concepts are intrinsically basic, and experience plays no role in determining conceptual content. Rather, when concepts are related to experience the concepts provide an interpretation of that experience.[11] One consequence of this approach is that changes in relations among concepts generate changes in all the concepts in a set. Since there is no universal set of basic concepts to which all other concepts are reducible, when a new set of concepts is introduced we often end up with two conceptual systems that are not mutually translatable. On this approach, then, it is no longer possible to avoid incommensurability even in principle.

It is important to be clear that the conceptual holism involved in the Kuhn-Feyerabend account is local: a system of concepts is peculiar to a specific scientific subject. We are not talking here about a single massive system that includes all of an individual's concepts, or all of the concepts prevalent in a society, or even all the concepts of physics. A full account of how local holism works is too complex to be developed here, but I want to note some key points. Research fields overlap in a variety of ways, but there is also a considerable range in which fields and sub-fields are conceptually independent. Evolutionary biologists need not attend to the relativistic concepts of space and time; considerable change can take place in high-energy physics without a direct impact on some areas of physics, or on biology, geology, or psychology – although as consequences are recognized, a change in any one of these fields may have impacts on any of the others. Changes at the most fundamental levels of a hierarchical structure will have immediate impacts on many fields. So, a change in the conceptual structure of special relativity or basic quantum mechanics will directly impact much of physics and many related fields. Still, the point remains that there is a significant degree of conceptual isolation between various disciplines and sub-disciplines. That Kuhn was aware of this point is indicated by various remarks about the importance of mini-revolutions that affect only a small number of practitioners in a sub-discipline (cf. Kuhn, 1970, pp. 6–7, 180–181; see also Kuhn, 1983, pp. 670–671). Indeed, I think that both Kuhn and Feyerabend were clear that they were discussing only concepts in specific areas of science, not global conceptual systems.[12] People typically wield several independent conceptual systems. It is not unusual to find a single individual who has mastered concepts involved in relativity physics, gardening, antique collecting, and chess. In addition, a single individual may have two incompatible conceptual systems available, and shift between them as the need arises – a point that should be familiar to anyone who has studied both relativity and classical mechanics (cf. Kuhn, 1983).

The holistic approach to concepts opens up possibilities for conceptual innovation that are not available in standard empiricist theories of concepts exactly because holism does not recognize a set of intrinsically basic concepts. Still, I think that the Kuhn-Feyerabend version of holism is not sufficiently rich to account for the varieties of concepts that we use or for the fine structure of the process of conceptual change. I want, then, to sketch a more complex form of local holism which I will draw on in the remainder of this paper.[13]

The content of scientific concepts is determined, I propose, along three dimensions. Kuhn and Feyerabend emphasized the *first* of these dimensions: implications among the concepts in a system. The need for a *second* dimension derives from the point that central scientific concepts are introduced to describe extra-conceptual items – that is, items that exist (if they exist at all) independently of the concepts we use to represent them.[14] It is, however, a familiar point that a set of concepts specified only by mutual relations can be given many different extra-systemic interpretations. A full-blown system of descriptive concepts must include some additional element that relates it to a specific subject matter. This additional element plays a role in determining the content of those concepts and requires some further discussion.[15]

I suggest that this second feature is a set of criteria for assessing whether a concept is instantiated. In many cases this role is played by criteria for picking out instances of the concept in question; this is typically the case for items that are easily observable or have easily observable properties. But there are cases in which this approach does not apply. Consider, for example, the top quark. The evidence for the existence of these entities consists of a body of recorded interactions plus an argument for the conclusion that it is extremely improbable that none of these interactions involved a top quark. In this case physicists never pick out a particular interaction as one in which the particle occurred, nor a specific phenomenon that involved a top quark. Still, they have a well-developed concept of a top quark, and they had this concept before they designed experiments to determine whether these particles exist. My proposal is that their criteria for what counts as evidence for or against the existence of top quarks is part of the content of that concept.

I have avoided talk of 'reference' in discussing this second dimension because in the course of research we sometimes discover that a concept does not refer – that nothing in the relevant domain instantiates that concept. Now the point I want to stress is that we can have a genuine descriptive concept that lacks instances but that includes, among its constitutive features, an account of the kind of evidence that is relevant for deciding if it has instances. We can, for example, describe what Aristotelian physicists would count as evidence for the occurrence of natural and violent motion even though we now hold that no such phenomena – as understood in Aristotelian physics – exist in the physical world. In general, we must distinguish between formulating the content of a descriptive concept and asking whether that concept is instantiated. We must understand the content of a concept before we can undertake to determine if it has instances. I will argue in section 3 that the putative referents of our concepts do play a vital role in research, but my present point is that conceptual content is determined independently of the actual existence of referents and independently of any decision as to whether such referents exist.

The *third* dimension is the function that a concept plays in our thinking.[16] The guiding idea here is that we introduce concepts into a descriptive system in order to achieve a specific cognitive end, such as describing a newly encountered phenomenon, or making a distinction that was not previously made. For example, when Newton introduced the concept *mass* as a physical property that is distinct from, but proportional to, *weight* he invented a conceptual function that did not exist in previous frameworks. The new conceptual role was motivated by several features of his new dynamics. These include his treatment of the common phenomenon of weight as a force between two bodies that depends on the distance between them and an intrinsic property of each. It is this new intrinsic property that Newton sought to capture in the concept of mass.[17] In a similar way, the Aristotelian concepts of natural and violent motion served to distinguish two fundamentally distinct classes of terrestrial motions. One result of the seventeenth-century revolution in physics was the elimination of those concepts from the conceptual framework of physics.

Given these three dimensions, a complete account of a descriptive concept will include accounts of its implicational relations to other concepts in a system, criteria for determining whether it is instantiated, and an account of the function that this concept serves. These dimensions are not totally independent of each other; changes in one dimension can force changes in the others. Thus Newton's introduction of the mass concept involved new implications and required new techniques for determining mass. Still, in specific cases conceptual change can be more drastic on one dimension than on others, and each dimension can provide a significant degree of continuity through that change. This approach thus allows us to understand how relatively small changes at key points in a conceptual system can yield major innovations in the way scientists think about some aspect of nature, while maintaining continuity with an older view. For example, introduction of the concept of an isotope involved changes in chemical thinking on each of our three dimensions. *a*) It introduced a new conceptual function – distinguishing varieties of a single chemical element with different atomic weights. Such a function was incompatible with the prevailing view which held that each element is characterized by a unique atomic weight. *b*) It eliminated the implications between elements and weights that was central to the older conceptual framework. *c*) It eliminated the key criterion that had been used to assess the purity of a sample of an element. In the older framework a non-integral atomic weight was sufficient evidence that the sample was not pure. After the discovery of isotopes such evidence became irrelevant to chemical purity because the researcher might be dealing with a sample consisting of a mixture of isotopes.

These changes had important consequences for the practice of chemistry. One immediate consequence was the elimination as pointless of a central experimental research project of nineteenth-century chemistry: determination of the precise weight associated with each element. In addition, a new project was brought into existence: the search for a new view of the defining characteristics of a chemical element. Still, these changes left much chemical knowledge and practice intact. The periodic table was not changed, although its basis had to be completely rethought. Standard chemical and spectrographic criteria for identifying elements endured, but those that licensed inferences from a weight determination to an element had to be dropped. Many procedures of physical chemistry remained unchanged, although here too procedures that depend on weight relations had to be reworked to some degree.

Introduction of the concept of an isotope involves incommensurability because the older framework does not have the conceptual resources needed to express the new ideas. In the present example this is an understatement since if we were to attempt to introduce isotopes into the older framework without making any other changes, we would generate an inconsistency. Indeed, from the perspective of the older framework, the concept of an isotope is just a contradiction in terms. At the same time, the new framework *by itself* cannot be used to express accurately the concepts of the older framework. The new framework eliminates the implications between a specific weight and a specific element that were constitutive in the older framework. In the new framework, not only can instances of a given element have different weights, but instances of different elements can have the same weight. The transition from the older framework to the new one involves new ways of thinking and important changes in chemical practice. Nevertheless, we are not dealing with a case in which one conceptual system is dropped and another invented *ex nihilo*. The change takes place by selective additions to and deletions from the older system, along with reorganization of some of the older links among concepts.

3. SCIENTIFIC REALISM AND EXTERNAL EVIDENCE

Eventually I am going to defend a pair of theses about the relation between incommensurability and scientific realism, so I must next explain what I understand by the latter doctrine. The form of scientific realism I want to defend amounts to two claims.

(SR1) One aim of scientific research is to learn about the nature of various domains; doing so amounts to seeking a conceptual system that correctly describes items in a domain.

I want to emphasize several points about this thesis. *a*) I do not take this to be the only aim of science. The aim of finding ways of predicting the unobserved is compatible with this realistic aim, and may even be the best that can be done in a given field at a particular stage of its development. *b*) "Knowing the nature of reality" is not a single all-or-nothing accomplishment. Scientific research takes place in a number of disciplines, each with its own domains, sub-domains, and associated conceptual systems. Many of these domains can be studied independently of each other, and we can learn a great deal about specific aspects of nature without coming close to knowing everything. *c*) Progress towards knowledge in a domain need not occur by means of successively closer approximations to the correct account. We may have a well developed but radically incorrect view of a domain, and this incorrect view may provide a basis for continuing research in that domain – even for research that will lead to its own rejection. *d*) Elsewhere (Brown, 1990) I considered the major attempts to show that the realistic aim is misconceived in principle, and argued that these attempts fail. For present purposes I want to highlight just one point. Many contemporary anti-realists hold that the view of concepts as human creations implies that the realist aim is an impossible dream.[18] I think this is a fundamental mistake. One function of concepts is to allow us to think about items that are not themselves concepts – much as language allows us to speak about items other than language, and photographs allow us to study items that are not themselves photographs. The thesis that our concepts are created, not found, is quite compatible with the possibility that a conceptual system may accurately describe its domain – and that we may have good reasons for believing this to be the case. This last remark brings me to the second thesis of the version of scientific realism I am proposing:

(SR2) Our ability to pursue the realist aim has been improving primarily because of the development of improved means of gathering evidence by which to evaluate the adequacy of specific conceptual systems.

I want to elaborate on this claim, which requires that we consider the nature of the evidence mentioned.

The key point about such evidence is that it is derived from attempted interactions with the part of nature we are seeking to describe. This process is 'theory-dependent' in many ways, but this theory-dependence does not determine the outcome when we test a theory. I have avoided the term 'theory' until now, and a full account of theories is not possible here. However, for present purposes a theory can be viewed as the assertion that a particular conceptual system, or cluster of conceptual systems, provides the correct description of its domain.[19] It is a basic working hypothesis of this approach that nature exists independently of our attempts to describe it, and that the outcome of an interaction with some aspect of nature need not match the expectations generated by our theories. Unexpected outcomes are commonplace in science. Without the thesis that nature is independent of human concepts and beliefs, Kuhn's discussions of the role of empirical anomalies in scientific research – including his crucial argument that anomalies are not always counter-instances – make no sense.[20]

I can now explain why I hold that our ability to pursue the realist aim has been improving. Through the development of scientific instrumentation we have learned to interact with a much wider portion of nature than we could when we were limited to the senses we evolved on the surface of this planet. We gather evidence throughout the electromagnetic spectrum, we study radioactive emissions, we produce new interactions among items in particle accelerators, and so on. These developments vastly extend the range of evidence by which we test our theories. But this is only part of the story. We have also learned to get much more precise results than our predecessors could, and – through the development of both mathematics and computing power – to subject these results to more sophisticated analysis. All of these developments have increased the constraints that an acceptable scientific theory must accommodate.

Before proceeding I want to respond to two common objections that will occur to many. *First*, anti-realists may object to my use of current scientific concepts, such as *electromagnetic spectrum*, to describe the items what we study; but such descriptions were used only for ease of communication. Even if such concepts are abandoned in the future, one point will remain: Each time we develop a new kind of instrument we generate a new kind of interaction with nature whose outcome must be dealt with by subsequent theories. *Second*, those who object that scientific instruments generate 'artificial' phenomena are, in a sense, correct but are missing the point. Use of these instruments provides information about how nature responds under various circumstances. The more varied these circumstances, the greater the body of evidential constraints on our theories.

Returning to the main line of argument, note that I have attributed no special role to our senses in this sketch. I have even avoided the term 'empirical' in my account of evidence because of the historical association of this term with sensory experience. Our senses do have a role to play: they are the conduit by which information makes its way into our brains. But this is a pragmatic constraint on the design of instruments, not the foundation of our epistemology or our semantics. The important feature of the theory-testing process is that these tests are attempts to interact with the items about which a theory makes claims – items that exist or fail to exist independently of the theory being tested. In other words, theory testing is a process of attempting to interact with the presumed referents of our concepts. It is at this point that these referents, if they exist, constrain our theories and thus constrain the appropriate concepts for describing them. Moreover, even if the concepts embodied in a theory do not refer, we are still interacting with a part of nature that the theory picks out as relevant. As a result, the testing process can provide the evidence that leads us to revise or reject these concepts.

Nothing in this account guarantees that we will eventually achieve a correct description in any given field – which is why I formulated the realistic thesis as an aim. But nothing precludes this outcome, and we may be more successful in some fields than in others. Given that the aim is pursuable, no more is needed to justify its pursuit than the fact that some people are curious about what nature is like.

4. INCOMMENSURABILITY AND REALITY

I am now ready to state the first major thesis of this paper:

(IR1) Pursuit of the realist goal requires incommensurability because this pursuit requires the introduction of new concepts and the elimination of old concepts.

The basis for this claim is a set of fairly mundane observations about human history and cognition; many of these observations have already been introduced. Concepts developed early in our cognitive history often turn out to be mistaken: sometimes they describe items that do not exist; sometimes they embody incorrect, or partially incorrect, accounts of items that do exist; and sometimes they group items on the basis of features that, by later lights, fail to capture their important properties. In addition, earlier stages in the development of knowledge typically lack concepts for describing items that have not yet been encountered. Consider again the old distinctions between natural and violent motions and between the terrestrial and celestial realms, as well as the relation between the identity of an element and its weight. The five elements that the ancient Greeks believed to constitute the entire universe provide an example well worth pondering. One of these, ether, presumably does not exist.[21] None of the remaining four 'elements' are elements according to the contemporary concept that is associated with that term (which does not require that elements be *elementary* in a particularly deep sense); only water is now considered a chemically distinct substance. For most scientific purposes earth, water, air, and fire would not now appear as members of the same class.

For obvious reasons, early science focused on easily observable properties that often turned out to be superficial in more ways than one. Studies of radioactive decay, for example, have taught us much more about the nature of physical objects than studies of their colors. As a result, key steps in the development of a science have required the invention of new concepts that embody ways of thinking which cannot be captured using the older concepts. Someone who is thinking in terms of the older concepts will often have to learn new ways of thinking in order to understand and evaluate a new theory. And once the new concepts have taken hold, it may require considerable effort to learn to think in ways that once were commonplace.

We have now reached a point at which questions of theory evaluation come to the fore. But it is important to distinguish two different questions about evaluating theories: whether the comparative evaluation of two theories is *rational*, and whether this evaluation gives us any reason for believing that the successful theory is a better candidate for a *correct description of its domain*. I want to develop this distinction and each of its sides.

One reason for making this distinction is that the question of the rationality of theory change applies to anti-realists as well as to realists. Kuhn and Feyerabend, whose work moved the question of rational scientific change to the center of current philosophy of science, were both anti-realists (although not initially in Feyerabend's case). There are many varieties of anti-realism, but I will discuss only instrumentalism. For instrumentalists, incommensurability and the associated problems of comparing alternatives arise because we must learn which phenomena are reliable indicators of which other phenomena – and our views on this matter can undergo radical change. Consider what was involved in learning to predict the likely occurrence of disease from the presence of mosquitoes, or from the outcome of chemical tests on local drinking water, or from a failure of physicians to wash their hands. The problems of rationally justifying radical new proposals for predicting phenomena are quite as difficult from an instrumentalist perspective as on a realist view. Nor can instrumentalists simply fall back on predictive success as a universal test. Predictions made under real-life conditions are rarely perfect, and this provides wide scope for debating predictive adequacy. In addition, criteria for what counts as predictive success also change. A particularly striking example occurred when scientists concluded that only probabilistic predictions are available in some domains.

Two points provide the basis for an account of rational theory change in the face of incommensurability. These points are familiar, but they underline the value of a cognitive perspective on these issues. *First*, those who create new concepts in a field, such as Galileo and Einstein, are typically educated in terms of, and fully understand, the framework they would supersede. *Second*, it is the task of the innovators to provide their colleagues with a motivation for undertaking the work of learning the new concepts. I will highlight two ways in which this motivation can develop. Some may become interested in a new approach because they are aware of serious problems in an existing framework, and hope that the new framework will not have parallel problems. In this case, one may have to learn the new framework before even its promise can be evaluated. But another approach is sometimes available that does not require mastering a new framework before one can recognize that it has merits. This occurs when the innovators display new results that can be described in terms of concepts already available in the older framework, but that cannot be accounted for using the resources of that older framework. New results come in two varieties: results that directly contradict expectations generated by the older view, and results for which the older view generates no expectations at all. Galileo was pursuing the first tack when he predicted that a stone dropped from the mast of a moving ship would land at the foot of the mast. The prediction of a gravitational red shift from general relativity illustrates the second case.

Realists, however, face an additional problem because we have an additional goal: We want to describe nature as it is and adjust our beliefs accordingly. So we must ask what reasons we have for believing that a description may be on track. One consequence of the way that the problem of rational conceptual change arises for realist and anti-realist alike is that rational change, by itself, does nothing to assure that a newly adopted conceptual system is a better candidate for a correct description of its domain. I suggest that the only reason we have for thinking that a conceptual system may correctly describe items in its domain derives from the evidence we acquire as a result of attempts to interact with those items. But here too we must be careful. With the exception of extreme social constructivists, anti-realists also recognize that science must accommodate evidential constraints derived from interactions with nature. As Kuhn emphasized,[22] it is the attempt to solve problems that arise out of interactions with nature that distinguishes science from many other endeavors. To motivate realism I must, then, defend the thesis that accommodating external evidence provides positive grounds for thinking that a theory may embody an accurate description of its domain. I want to approach this question by exploring the nature of these constraints more closely, beginning with an argument due to Greenwood (1990).

Greenwood was concerned with the way in which Quine's sweeping holism has been used to argue that *any* scientific hypothesis can be protected against an evidential challenge by making adjustments elsewhere in the system. On the contrary, Greenwood argued, it is difficult to make these adjustments exactly because of the way the members of a system are linked together. Changes aimed at accommodating one set of troublesome observations will often have consequences that bring other parts of the system into conflict with other observations. "The 'web of belief' tends to reduce rather than increase the room for intellectual maneuver when faced with recalcitrant observations" (Greenwood, 1990, p. 566). The larger the system, the harder it is going to be to find a set of adjustments that protects the preferred members and still accommodates all the evidence. Consider, for example, the hypothesis that planets move in square orbits. It is not plausible that our current belief system can be adjusted to include this hypothesis and still be brought into conformity with all currently available evidence. Moreover, this example immediately suggests an infinite set of other orbital shapes that cannot be taken seriously.

Now I am urging that scientific conceptual systems have a considerably more restricted range than they do in Quine's account. In this respect Greenwood's point holds to a lesser degree, but its full force returns when we consider the growing mass and precision of the evidence that a serious scientific theory must accommodate. This has an effect that is quite similar to the sheer size of a Quinean system. The goal of protecting favored elements of a conceptual system – no matter what the evidence – becomes very difficult to pursue as the evidence becomes more varied and precise. Moreover, according to the account of evidence I have proposed, this evidence arises from interactions with nature. As a result, constraints provided by external evidence are *constraints provided by nature*. And so we arrive at my second main thesis:

(IR2) Evidence provided by interactions with nature significantly limits the range of incommensurable conceptual systems that can provide plausible descriptions of a domain.

Thus we have reasons for believing that well-tested theories describe items in their domain because our choice of theories is limited by those items. I now want to consider how this form of realism stands up against the two main types of anti-realist philosophy of science in the current literature that also accord a fundamental role to evidence – the Kantian and empiricist views.

Consider, first, the Kantian approach typified by Kuhn (cf. Brown, 1975; Hoyningen-Huene, 1993). The anti-realist consequences of this view are supposed to derive from the role that concepts play in all of our thinking about nature, along with the thesis that we create our concepts – which change as science develops. But this view is significantly different from Kant's approach which posits a single permanent set of fundamental concepts and forms of sense that are constitutive of both science and nature. Even in Kant there is a substantive question about why space, time, causality, existence, and so forth cannot characterize things themselves. Strictly speaking, Kant's thesis is that we cannot *justify* applying these beyond the range of possible experience – but that thesis turns on Kant's attempt to justify these as *a priori* features of experience. It is central to the Kuhnian approach that the concepts at issue are not *a priori* features of all experience. But then Kant's arguments against the application of these concepts to things themselves cease to be relevant. Leaving Kant's arguments aside, the fact that we formed a set of concepts to describe items we have never experienced does not entail that those concepts cannot correctly describe the items in question – it only entails that we initially lack positive reasons for believing that our concepts describe these items. This shifts the question to the kind of evidence that could overcome this initial situation, and I have provided an account of evidence that addresses this question. Given this account there are no reasons for thinking that nature ultimately consists of items that, in principle, elude our cognitive grasp.

At this point arguments from incommensurability – that is, arguments from alternative conceptual systems – enter the discussion. These arguments come in two quite different versions. One of these is the view I have already addressed, which holds that any thesis can be defended against any body of evidence. Notice that even the Quinean version of this approach does not claim that entire frameworks can be defended against any body of evidence – only that selected parts of a framework can be so defended – and I have provided a counter-argument to this limited thesis. But there is another, even more modest, anti-realist argument that we must consider. This argument holds that it is always possible to find *an alternative* incommensurable framework that will accommodate any body of evidence. I have not argued against this view, and I do not think there is any general argument that can address it. Rather, this is a question to be decided in various scientific domains on a case-by-case basis. It may turn out that there are some domains in which only one defensible view is available. It is not enough to claim that another view is always possible; one must make the case by constructing such an alternative. When this is done, one common outcome is to point to other evidence that will distinguish the two views. Still, the version of realism I am defending leaves it an open question whether there are domains in which we will encounter alternative frameworks among which we cannot decide. If this does occur we will have learned something about the limits of our ability to learn about nature – and we will have learned this from the development of science. From a naturalist perspective this is wholly appropriate: The nature of human knowledge is one of the domains about which we learn as science proceeds. A key feature of the form of realism I am defending is that it avoids any prejudgment about what science will teach us in any domain.

I turn now to empiricist forms of anti-realism. Hempel provided a classic statement of the empiricist view: "Scientific systematization is ultimately aimed at establishing explanatory and predictive order among the bewildering complex 'data' of our experience, the phenomena that can be 'directly observed' by us" (Hempel, 1965, p. 177). Hempel goes on to note the "remarkable fact" that success in this project has been greatly enhanced by the introduction of laws that refer to entities we cannot directly observe, and that this is a problem for empiricists. He resolves the problem by arguing that the introduction of non-observables would be unnecessary if the only aim of science were to establishes *deductive* relations among observables. "But if it is recognized that a satisfactory theory should provide possibilities for inductive explanatory and predictive use and that it should achieve systematic economy and heuristic fertility, then it is clear that theoretical formulations cannot be replaced by expressions in terms of observables only ..." (Hempel, 1965, p. 222). This view is anti-realist in the sense that it does not include the study of the substantial portion of nature that we cannot observe among the goals of science. The philosophical basis for this position lies in the combination of empiricist theories of concepts and of evidence that I discussed earlier in this paper. I have offered alternatives to both of these that undercut the anti-realist arguments. The accounts of concepts and of evidence I have proposed provide positive reasons for thinking that science teaches us about all of nature, not just about the portions we can detect with our unaided senses.[23]

5. CONCLUSION

The discovery of incommensurability raises methodological problems that were not recognized at earlier stages in the development of the philosophy of science, and indicates that the process of evaluating scientific theories is more complex and less certain than many would like it to be. But incommensurability is not just a source of problems, it is also a source of opportunities. The incommensurability thesis embodies the recognition that we can, and do, overcome limitations built into our ways of thinking at a given point in our history. The view of incommensurability as solely a source of problems arises, in part, from a failure to distinguish clearly between science and a philosophy of science. Incommensurability provides a serious challenge to the form of logical empiricism that was dominant in the 1950s. But to take this as a challenge to science is to make the dubious assumption that science must operate in accordance with logical empiricist methodological strictures if it is to be a genuine source of knowledge. Many of us who reject this assumption conclude that we are dealing with one more philosophical failure, and we are thus engaged in a process of philosophical reconstruction. I have been urging that such a reconstruction should include a genuine break with two of the foundation stones of traditional empiricism: its theory of concepts and its theory of evidence. Kuhn and Feyerabend took the first of these steps, but not the second. Together, these moves away from the empiricist tradition open the way for a view that Feyerabend, in some of his periods and some of his moods, might have found congenial: that the development of incommensurable conceptual systems is one important tool for the advancement of science. Or, as Sellars once put it: "After all, it is characteristic of modern science to produce deliberately mutant conceptual structures with which to challenge the world" (Sellars, 1953a, p. 337).

Northern Illinois University

NOTES

[1] See especially Doppelt (1978); for a recent discussion see Brown (1996).

[2] I use 'item' as a neutral term to cover entities, processes, events, or whatever else we may encounter or postulate in the course of our investigations.

[3] The same point holds for the development of methodology.

[4] I am not referring here to the U. S. Secretary of State, but to a position in Illinois, where the Secretary of State is an elected official in charge of the department that licenses drivers and automobiles.

[5] For details of this history see Pais (1986); Romer (1964; 1970); Trenn (1977). The concept of radioactive induction was abandoned when it became clear that radioactive products of previous decays were being deposited on the materials in question. Metabolons turned out to be radioactive isotopes with very short half-lives, but the concept of an isotope had not yet been formulated at the time in question.

[6] Recall that X-rays are produced when a stream of electrons hits a target.

[7] See Shapere (1974) for this use of 'domain'.

[8] See, for example, Fodor (1998); Margolis and Laurence (1999). Concepts should not be assimilated to Fregan 'senses' – at least as Frege used this term. Frege made a three-fold distinction between "a sign, its sense, and reference" (Frege, 1960, p. 58), where a sense is a sign's mode of presentation. Thus, for Frege, senses presuppose linguistic entities, and Frege is quite clear that senses are not mental entities – he reserves the term 'ideas' for this role (Frege, 1960, pp. 59–60). I want to emphasize that I am not treating concepts as linguistic entities. One reason for avoiding this identification is that it is an open question whether non-linguistic animals possess concepts. As Sellars notes (Sellars, 1975, pp. 303–304) if animals behave in ways that are sufficiently analogous to ways we behave, then the best explanation may involve attributing concepts to them. In any case, the issue should not be decided by fiat. Still, when dealing with humans it is appropriate to take linguistic evidence as evidence about underlying concepts.

[9] The ancient Chinese distinguished five basic kinds – water, earth, fire, metal, and wood – and thought about them in terms of a quite different conceptual machinery (see Brock, 1993, chapter 1).

[10] It is worth noting that the concept of a *natural kind* is itself a concept in a theory of how the world is structured and of how research should be pursued – a theory that is subject to reconsideration. One impetus for such reconsideration might come from the proliferation of isotopes. Elements are chemically identical, and may thus be considered members of the same kind for a wide variety of purposes. But all elements have isotopes, which have different atomic weights. As a result, different isotopes of an element will have different diffusion rates and may be separated in a centrifuge. In this context, the isotopes are not the same stuff. A more dramatic example occurs when we consider nuclear processes. Some isotopes of an element may be radioactive, while others are not, and this is a significant difference in kind for some purposes. Consider, as an example, thorium, which has more than twenty-five isotopes, all radioactive, with half-lives running from microseconds to billions of years. Further complications are introduced by the existence of allotropic forms of elements – especially the recent proliferation of allotropes of carbon – and of isomers of complex molecules. The usual accounts of natural kinds involve the view that there is one correct way of dividing up nature, and this may turn out to be uninteresting and unilluminating for many scientific purposes.

[11] The core issue is whether the content of scientific concepts is wholly or partially determined by experience, or whether the experience that is relevant to scientific research always already includes conceptual elements. The latter view requires an account of the source of conceptual content that is independent of experience. C. I. Lewis, (1956; however this is a reprint of a book originally published in 1929) provides a striking example of a philosopher whose overall epistemology has many empiricist elements, but who explicitly rejects the usual empiricist view of concepts, and holds that concepts are completely determined by implicational relations to other concepts. For Lewis there are no intrinsically basic concepts, and conceptual analysis requires mapping out relations between concepts; it does not involve breaking down complex concepts into elementary parts that are derived from experience. If Lewis were writing in the 1960s or later he would be viewed as holding a radical version of the 'concept-ladenness' of perception. Sellars, who seeks to synthesize elements from traditional empiricism and traditional rationalism, also holds that all perception involves conceptual elements and that "conceptual status" is determined by relations among concepts (e.g., Sellars, 1953a; 1953b; 1956). Examples of this approach from philosophers associated with the idealist tradition include Blanshard (1939) and Collingwood (1940). Harris (1970, chapters VII and VIII) discusses overlaps between the views of Blanshard, Collingwood, and Kuhn.

[12] One of the unfortunate results of treating conceptual systems as languages is that it tends to obscure this point. See Sankey (1994, chapter 4) for a good discussion.

[13] This view is built on the work of Wilfrid Sellars. See Brown (1986; 1991) for more details and further discussion. However, work done since I wrote those papers has altered my interpretation of Sellars and my views on the adequacy of Sellars' approach in the form in which he left it. For example, my account of Sellars' notion of an *entry transition* (1991, pp. 328–331) extends this notion beyond the limited set of examples that Sellars discussed, but the notion is still not adequate for dealing with cases such as the top quark; see the discussion below in the main text. In addition, the third dimension that I discuss is not an explicit part of Sellars' theory of concepts, although he makes many suggestions that point in this direction. I cannot provide an adequate defense of this theory of concepts here. Ultimately such a defense depends on showing how the theory provides a basis for studies of conceptual development, and also provides a deeper understanding of the nature and function of conceptual analysis than is now available. This work is in progress.

[14] I avoid the familiar practice of describing these items as 'mind-independent' since minds are a subject for scientific study, and thus one of the items that we seek to represent by means of concepts. The important point is that descriptive concepts serve to represent items other than themselves, and can fail to represent those items correctly. Not all concepts play a descriptive role; some have a purely formal role, such as the various components into which a vector may be resolved. Sellars distinguishes between formal, descriptive, and prescriptive concepts which play different roles in cognition and action, but I am limiting discussion here to descriptive concepts.

[15] In some of his writings Kuhn does include a role for ostension in learning some concepts. (See for example, Kuhn, 1977a, and the discussion in Hoyningen-Huene, 1993, pp. 77–81.) There are several issues that must be sorted out here. Kuhn seems to be discussing how a child may learn to make certain classifications. But, as Hoyningen-Huene points out, the account does not apply to a completely naive child since participation in the practice of learning by ostension requires a good deal of prior knowledge. Moreover, the strong Kantian element in Kuhn's epistemology militates against such a view: Concepts have to be in place before we can pick out objects. Kuhn also suggests that ostension, or something like ostension, can play a role when conceptually sophisticated scientists are attempting to teach their way of thinking to students or to other scientists. (See especially Kuhn, 1970, chapter 10, and the discussion in Brown, 1983, pp. 19–20.) It is, however, not clear what role, if any, ostension plays, for Kuhn, in determining the conceptual content. Indeed, it is not clear that Kuhn has a well-developed theory of concepts (or of linguistic meaning), but it is clear that he stresses the role of relations among concepts in determining their content, and the need to grasp an entire system in order to understand its constituent concepts. (See for example, Kuhn, 1970, pp. 101–103; 1983; 1990.) A similar emphasis pervades Feyerabend's writings (e.g., 1962; 1975, chapter 17).

[16] Whether a concept has a descriptive function is one instance of this, but at the moment I am after another point that applies among descriptive concepts.

[17] This is only part of the story. Newton's second law introduced the thesis that force is proportional to acceleration, not to velocity. This proportionality required a new constant that Newton also identified with mass. It soon became clear, however, that Newton had introduced two new conceptual roles which were synthesized into a single role in general relativity.

[18] I use the term 'anti-realist' to refer to those who hold, for whatever reasons, that the attempt to understand the nature of the world is not a legitimate aim of science.

[19] There are obvious affinities between this approach and the semantic view of theories. On that view a theory consists of a set of models, which is to be distinguished from the hypothesis that one of those models provides a description of some scientific domain.

[20] See Brown (1995) for elaboration of this point and for a more detailed account of the view of evidence that I am summarizing here.

[21] To be sure, some scientists still use the term and debate the existence of some universal material pervading spacetime. But this new ether is conceived of in terms that would be quite unintelligible to the ancients, and plays a role in physics that they did not imagine. Its existence is debated on the basis of evidence that they would not be able to understand without undergoing a great deal of conceptual reeducation.

[22] E.g., when he described normal science as an "attempt to force nature into the preformed and relatively inflexible box that the paradigm supplies" (Kuhn, 1970, p. 24). Similar remarks occur on pp. 112, 129, 150, *et passim.*

[23] An alternative form of empiricist anti-realism has been developed by van Fraassen. Adequate discussion of van Fraassen's views would require another paper, but two points are worth noting here. First, he shares the standard empiricist view that science is concerned only with relations among items that we can observe (van Fraassen, 1980, pp. 11–19). Second, he is much less clear than his predecessors on what counts as an observable. He holds that scientific theories determine what is observable by us (van Fraassen, 1980, pp. 57–58), but it is far from clear that this determination can avoid appeal to the non-empirical portions of those theories (cf. Fine, 1986, pp. 142–147). In addition, what is observable by us depends on who we include under the rubric "us," and van Fraassen holds that this may change as a result of "ideological or moral decisions" (van Fraassen, 1980, p. 18).

REFERENCES

Blanshard, B. (1939). *The Nature of Thought.* London: George Allen & Unwin.

Brock, W. (1993). *The Norton History of Chemistry.* New York: W. W. Norton and Company.

Brown, H. (1975). "Paradigmatic Propositions." *American Philosophical Quarterly* **12**: 85–90.

Brown, H. (1979). *Perception, Theory, and Commitment: The New Philosophy of Science.* Chicago: University of Chicago Press.

Brown, H. (1983). "Incommensurability." *Inquiry* **26**: 3–29.

Brown, H. (1986). "Sellars, Concepts and Conceptual Change." *Synthese* **68**: 275–307.

Brown, H. (1990). "Prospective Realism." *Studies in History and Philosophy of Science* **21**: 211–242.

Brown, H. (1991). "Epistemic Concepts." *Inquiry* **34**: 323–351.

Brown, H. (1995). "Empirical Testing." *Inquiry* **38**: 353–399.

Brown, H. (1996). "The Methodological Roles of Theory in Science." In B. Rhoads and C. Thorn, eds., *The Scientific Basis of Geomorphology*, pp. 3–20, Sussex: John Wiley and Sons.

Collingwood, R. (1940). *An Essay on Metaphysics.* Oxford: Clarendon Press.

Doppelt, G. (1978). "Kuhn's Epistemological Relativism: An Interpretation and Defense." *Inquiry* **27**: 33–86.

Feyerabend, P. (1962). "Explanation, Reduction and Empiricism." In H. Feigl and G. Maxwell, eds., *Minnesota Studies in the Philosophy of Science, Volume 3*, pp. 28–97, Minneapolis: University of Minnesota Press.

Feyerabend, P. (1975). *Against Method.* London: New Left Books.

Fine, A. (1986). "And not Antirealism Either." In *The Shaky Game*, pp. 136–150, Chicago: University of Chicago Press.

Fodor, J. (1998). *Concepts.* Oxford: Clarendon Press.

Frege, G. (1960). "On Sense and Reference." In P. Geach and M. Black, trans., *Translations from the Philosophical Writings of Gottlob Frege.* 2nd edition. pp. 56–78, Oxford: Basil Blackwell.

Greenwood, J. (1990). "Two Dogmas of Neo-Empiricism: The 'Theory-Informity' of Observation and the Quine-Duhem Thesis." *Philosophy of Science* **57**: 553–574.

Harris, E. (1970). *Hypothesis and Perception.* London: George Allen & Unwin.

Hempel, C. (1965). "The Theoretician's Dilemma." In C. Hempel, ed., *Aspects of Scientific Explanation*, pp. 173–226, New York: The Free Press.

Hoyningen-Huene, P. (1993). *Reconstructing Scientific Revolutions: Thomas S. Kuhn's Philosophy of Science*, trans. A. Levine. Chicago: University of Chicago Press.

Khalidi, A. (2000). "Incommensurability." In W. Newton-Smith, ed., *A Companion to the Philosophy of Science*, pp. 172–180, Oxford: Blackwell Publishers.

Kuhn, T. (1970). *The Structure of Scientific Revolutions*. 2nd edition. Chicago: University of Chicago Press.

Kuhn, T. (1977a). "Second Thoughts on Paradigms." In *The Essential Tension*, pp. 293–319, Chicago: University of Chicago Press.

Kuhn, T. (1977b). "Theory Change as Structure Change." In R. Butts and J. Hintikka, eds., *Historical and Philosophical Dimensions of Logic, Methodology and Philosophy of Science*, pp. 299–309, Dordrecht: D. Reidel.

Kuhn, T. (1983). "Commensurability, Comparability, Communicability." In P. Asquith and T. Nickles, eds., *PSA 1982, Volume 2*, pp. 669–688, East Lansing: The Philosophy of Science Association.

Kuhn, T. (1990). "Dubbing and Redubbing: The Vulnerability of Rigid Designation." In C. Savage, ed., *Minnesota Studies in the Philosophy of Science, Volume 14*, pp. 298–318, Minneapolis: University of Minnesota Press.

Lewis, C. (1956). *Mind and the World Order*. New York: Dover Books.

Margolis, E. and S. Laurence. (1999). "Concepts and Cognitive Science." In E. Margolis and S. Laurence, eds., *Concepts: Core Readings*, pp. 3–81, Cambridge, Mass.: MIT Press.

Pais, A. (1986). *Inward Bound*. New York: Oxford University Press.

Putnam, H. (1975). "The Meaning of 'Meaning'." In *Mind, Language and Reality: Philosophical Papers, Volume 2*, pp. 215–271, Cambridge: Cambridge University Press.

Romer, A., ed. (1964). *The Discovery of Radioactivity and Transmutation*. New York: Dover Publications.

Romer, A., ed. (1970). *Radiochemistry and the Discovery of Isotopes*. New York: Dover Publications.

Sankey, H., (1994). *The Incommensurability Thesis*. Aldershot: Ashgate.

Sellars, W. (1953a). "Inference and Meaning." *Mind* **62**: 313–338.

Sellars, W. (1953b). "Is There a Synthetic *A Priori*?" *Philosophy of Science* **20**: 121–138.

Sellars, W. (1956). "Empiricism and the Philosophy of Mind." In H. Feigl and M. Scriven, eds., *Minnesota Studies in the Philosophy of Science, Volume 1*, pp. 253–329, Minneapolis: University of Minnesota Press.

Sellars, W. (1975). "The Structure of Knowledge." In H. Castañeda, ed., *Action, Knowledge and Reality*, pp. 295–347, Indianapolis: Bobbs-Merrill.

Shapere, D. (1974). "Scientific Theories and Their Domains." In F. Suppe, ed., *The Structure of Scientific Theories*, pp. 518-565, Urbana: University of Illinois Press.

Thagard, P. (1992). *Conceptual Revolutions*. Princeton: Princeton University Press.

Trenn, T. (1977). *The Self-Splitting Atom: A History of the Rutherford-Soddy Collaboration*. London: Taylor and Francis.

van Fraassen, B. (1980). *The Scientific Image*. Oxford: Clarendon Press.

MICHAEL DEVITT

INCOMMENSURABILITY AND THE PRIORITY OF METAPHYSICS

Abstract. I aim to reject a semantic doctrine, "Incommensurability", commonly attributed to Kuhn and Feyerabend. They also subscribe to the neo-Kantian metaphysical doctrine of "Constructivism" which stands opposed to "Realism". I argue that the Incommensurability issue comes down to the Realism issue. On the Realism issue I reject four arguments for Constructivism. Two Kantian arguments make the mistakes of using an *a priori* methodology and of not "putting metaphysics first". Two arguments by Hoyningen-Huene and his co-authors support relativism but do nothing to support the Kantian core of Constructivism. I conclude by arguing against "meta-incommensurability".

I share the common view that Thomas Kuhn and Paul Feyerabend enormously improved our understanding of the history and epistemology of science. Their suggestions about the semantics of science are interesting but, in my view, dubious at best. Their anti-realist views on the metaphysics of science are a disaster. I shall be concerned with that disaster and with its connection to the incommensurability issue.

1. INCOMMENSURABILITY

I shall argue against an incommensurability thesis. My argument is partly a response to the views of Paul Hoyningen-Huene and his co-authors, particularly to his *Reconstructing Scientific Revolutions* (1993), a sympathetic and wonderfully scholarly account of Kuhn's views. I shall draw on my *Realism and Truth* (1997). The thesis I shall reject is:

> **Incommensurability**: Terms in rival comprehensive theories in a domain differ sufficiently in meaning, especially in reference, to make the theories incomparable.

What sorts of theory comparison is Incommensurability against? A straightforward sort would arise if the two theories share referents. As a result, some parts of one theory would be logically inconsistent with some parts of the other, and some parts would entail some parts of the other. But there could be more complicated sorts of comparison where referents were not shared; for example, Michael Martin's case where terms in the two theories have overlapping referents (1971); or Hartry Field's case where the theories share *partial* referents (1973). The intuitive idea of theory comparison is that what one theory says about x or about Fs is in agreement or disagreement with what the other says about x or about Fs. Sometimes the disagreement might be over the very existence of x or Fs. Agreement and disagreement at the observational level is, of

P. Hoyningen-Huene and H. Sankey (eds.). Incommensurability and Related Matters, 143–157.

course, particularly important to theory comparison. Incommensurability denies that any of these sorts of comparison of rival theories is possible.

Clearly Incommensurability, with a capital "I", has been stimulated by the writings of Thomas Kuhn (e.g., 1970) and Paul Feyerabend (e.g., 1981a; 1981b) on incommensurability. But do Kuhn and Feyerabend actually embrace Incommensurability? Their writings on the matter are so notoriously various and vague as to leave plenty of room for disagreement over interpretation. Hoyningen-Huene argues that the incommensurability that Kuhn has in mind does *not* imply incomparability (1993, pp. 218–222). I will address this interpretative issue briefly in a moment, but it is not really my concern.[1] Whatever the truth of it, the rejection of Incommensurability is worthwhile for two reasons. First, the writings of Kuhn and Feyerabend have been commonly taken to imply Incommensurability: Hoyningen-Huene gives a long list of philosophers "among many others" who have construed Kuhn and Feyerabend in this way (p. 218n). Second, rejecting Incommensurability is sufficient to remove the worries occasioned by those writings for a realist view of science. Indeed, if the claims in those writings about meaning change and translation failure do not threaten theory comparison then, whatever their purely semantic interest, they do not pose any special problems for the epistemology and metaphysics of science.

2. REALISM AND CONSTRUCTIVISM

My position on the realism issue is captured by the following doctrine:

> **Realism**: Tokens of most common-sense, and scientific, physical types objectively exist independently of the mental.

At the observable level, the tokens in question are of stones, trees, cats, and the like; at the unobservable level, they are of electrons, muons, curved space-time, and the like.[2]

Realism stands opposed to a variety of doctrines. The one that concerns us most is usually called "Constructivism". It starts from two Kantian ideas. The first of these is that the knowable world of "appearances" – stones, trees, cats, and the like – is partly constituted by the cognitive activities of the mind. Kant called this world "the phenomenal world". According to Kant himself, the mind constitutes the phenomenal world by imposing *a priori* concepts. In recent times the restriction to *a priori* concepts has been removed: the mind may impose any concept, or a theory, or a language. The second Kantian idea distinguishes objects as we know them from objects as they are independent of our knowledge. The latter objects make up "the noumenal world" of "things-in-themselves". Only the noumenal world, forever inaccessible and beyond our ken, has the objectivity and independence required by Realism. The phenomenal world of familiar objects does not, as it is partly our construction. So, where the Realist thinks that stones, trees and cats are both knowable and independent, Kant thinks that they are knowable but dependent, being partly constituted by us and partly by an unknowable independent world.

Constructivism adds a third idea to these two Kantian ones: relativism. Kant was no relativist: the concepts imposed to constitute the known world were common to all humankind. Constructivists drop this universality. Different languages, theories, and world views are imposed to create different known worlds. In sum:

> **Constructivism:** The only independent reality is beyond the reach of our knowledge and language. A known world is partly constructed by the imposition of concepts. These concepts differ from (linguistic, social, scientific, etc.) group to group and hence the worlds of groups differ. Each such world exists only relative to an imposition of concepts.

So Constructivists believe in what Hoyningen-Huene aptly calls "the plurality-of-phenomenal-worlds thesis" (1993, p. 36).

Examples of constructivist worlds include the stars made by a Goodman "version" (1978); the constructed worlds of Putnam's "internal realism" (1981); the worlds built by a Whorfian language (1956); the many worlds created by the "discourses" of structuralists and post-structuralists. Most important for our purposes, according to Kuhn and Feyerabend the ontologies of scientific theories are constructivist worlds.

It is common to interpret Kuhn and Feyerabend as subscribing to Constructivism, but the interpretation has not been without its problems. It is comforting, therefore, to find Hoyningen-Huene (1993) giving an authoritative argument for this interpretation of Kuhn.[3]

3. THE RELATION BETWEEN THE REALISM AND INCOMMENSURABILITY ISSUES

No position on Realism *entails* a position on Incommensurability because Incommensurability is a semantic doctrine whereas Realism is not. Still arguments can be mounted from a position on Realism to a position on Incommensurability with the help of semantic assumptions. Consider this argument first:

> Constructivism → Incommensurability.

Adopt the common assumption that meaning is to be explained at least partly in terms of reference. Let T1 and T2 be examples of rival comprehensive theories. What might the terms of T1 and T2 refer to? According to Constructivism, they cannot refer to the same entities, because with the move from T1 to T2 the world changes. Indeed, although some entities exist-relative-to-T1 and others exist-relative-to-T2, no sense can be made of any of these entities existing "absolutely". So the potential referents, or even potential partial referents, of T1 terms are different from the potential referents of T2 terms. So there is no way that T1 and T2 can be compared in the way that concerns Incommensurability: they cannot agree or disagree about x or Fs because they are not talking about the same x or Fs.

In light of this let us briefly consider the earlier-mentioned issue about the interpretation of Kuhn. Since Hoyningen-Huene thinks that Kuhn is a Constructivist, what could be his basis for claiming that Kuhn holds T1 and T2 to be nonetheless comparable? One basis (1993, pp. 219–220) comes from Kuhn's talk in his later writings of incommensurability being merely "local" (1983, pp. 670–671). This implies a more moderate relativism, hence more moderate Constructivism, than we have been discussing, a move back toward Kant. The local conceptual differences between T1 and T2 may leave the theories with a lot conceptually in common. To the extent of what is in common, T1 and T2 construct the same world. Reference to that common world could enable some theory comparison.[4] But, to repeat, Incommensurability is worth rejecting whether or not, and to whatever extent, Kuhn is committed to it.

Call the negation of Incommensurability, "Commensurability", and consider:

Realism → Commensurability.

This argument is not so easy. According to Realism, the world remains constant through theory change and a certain part of that world is the common domain of T1 and T2. It can then be argued that, by and large, these theories succeed in referring, or at least partially referring, to parts of this domain and this provides sufficient basis for theory comparison.[5]

Assume that I am right about this. So we have a plausible route from Realism to Commensurability as well as one from Constructivism to Incommensurability. There are surely other plausible routes to Commensurability, ones that do not involve Realism. However, I claim that there are *no* plausible routes to Incommensurability that do not involve establishing, or simply assuming, Constructivism or some other form of antirealism. So, Incommensurability depends on antirealism.

Hoyningen-Huene, and his co-authors Eric Oberheim and Hanne Andersen, may very well agree. In their review (1996) of Howard Sankey's book defending a commensurability thesis (1994), they criticize Sankey's argument on the ground that he "presupposes realism" (1996, p. 133). In another work, Oberheim and Hoyningen-Huene insist that "incommensurability was not introduced within a realist context" but rather in (what I am calling) a Constructivist one (1997, p. 450). They go on to talk of the "blatant inefficacy" of my own arguments against incommensurability "from the perspective of the non-realist proponent" (p. 452).

It fits my prejudices nicely that the Incommensurability issue should rest on the Realism issue. And I am rather delighted by the stereotypical nature of this particular debate: German Kantianism versus Australian Realism (1997, pp. vii, x).

In the light of this, it would, of course, beg the question against Incommensurability to presuppose Realism. Oberheim and Hoyningen-Huene allege (pp. 451–452) that I did beg this question, but they are wrong. I did not presuppose Realism, I *argued* for it and against Constructivism.

4. SUMMARY OF AN ARGUMENT FOR REALISM

Here is a summary of my argument for Realism.[6] I start by observing that Realism about the ordinary observable physical world is a compelling doctrine. It is almost universally held outside intellectual circles. From an early age we come to believe that such objects as stones, cats, and trees exist. Furthermore, we believe that these objects exist even when we are not perceiving them, and that they do not depend for their existence on our opinions nor on anything mental. This Realism about ordinary objects is confirmed day by day in our experience. It is central to our whole way of viewing the world, the very core of common sense. Given this strong case for Realism, we should give it up only in the face of powerful arguments against it and for an alternative. There are no such arguments. That concludes the case for Realism.

What about *scientific* Realism, Realism about unobservables? The argument for it rests upon Realism about observables. The argument consists mainly in a simple but powerful inference to the best explanation: by supposing that the unobservables of science exist, we can give good explanations of the behavior and characteristics of observed entities, behavior and characteristics which would otherwise remain completely inexplicable.[7]

It can be seen, then, that the case for Realism partly rests on rejecting alternatives. In the present context, the alternative that we particularly need to reject is Constructivism. I shall start with some criticisms of Constructivism and then consider the arguments for it.[8]

5. CRITICIZING CONSTRUCTIVISM

Constructivism is surely the metaphysics of the twentieth century. It has its origins in the work of a great philosopher and has been urged by some outstanding ones. Despite this popularity, and the respect due to our elders and betters, we should not close our eyes to the fact that Constructivism is *prima facie* absurd, a truly bizarre doctrine. This emperor has no clothes.[9]

To start with, the idea of noumenal things-in-themselves is explanatorily useless and probably incoherent. Constructivists are attracted to things-in-themselves to provide an external constraint on theorizing. The plausibility of the view that there is *some* external constraint is, of course, overwhelming: there must be something outside us determining that some theories are better than others. However, things-in-themselves provide the appearance of constraint without the reality. Since we can, *ex hypothesi*, know nothing about things-in-themselves, we can know nothing about the mechanisms by which they exercise their constraint, nor can we explain or predict any particular constraint. For Kant himself, the very idea of *causal* constraint by the noumenal world is incoherent because CAUSALITY is one of the concepts imposed by us. So causality is part of the phenomenal world and cannot hold between the noumenal and phenomenal worlds. If this is not the position of a Constructivist, it surely ought to be. Why should causality be exempt from the rule of creation by imposition?[10] If it is not exempt, the Constructivist faces the same problem that has baffled Kant scholars for years: the nature of the non-causal constraint exercised by things-in-themselves.

Frederick Jameson captures the mysteriousness of the noumenal world (in discussing structuralism): it is "a formless chaos of which one cannot even speak in the first place" (1972, pp. 109–110). It is hardly ever mentioned without the protection of scare quotes or capital letters. Yet mentioned it often is. And, given the role that Constructivists want the noumenal world to play, it is not surprising that they should try to tell us about it. Yet, *ex hypothesi*, this is to attempt the impossible. For example, consider the problem of specifying the common domain of rival theories. The Realist can do this in terms of shared referents or, at least, shared partial referents. How can a Constructivist like Kuhn do it? Hoyningen-Huene points out that Kuhn appears to his critics to have the view that rival incommensurable theories "bear the same relation to one another as, for example, theories of the unconscious bear to theories on the stability of globular star clusters ...

[they] address differently constituted regions of the world" (1993, pp. 218–219). Hoyningen-Huene rejects the analogy, claiming that whereas "the latter theories have totally different domains ... incommensurable theories target *roughly the same* object domain, *as far as the world-in-itself is concerned*" (p. 219). But this talk, which treats the noumenal world as if it were the Realist world, makes no sense:[11] we can know nothing about what is targeted in an unknowable world.[12]

The noumenal world adds only an invisible fig leaf to the naked idealism of Constructivism. If the leaf is dropped we are left with a modified Constructivism, which seems to be the preferred view of Putnam (1981, pp. 61–62, 83) and the later Kuhn (1979). On this view, no account of constraints on our theorizing can be given. The modified Constructivist might deny that there are any constraints: we can think anything we like. That is not plausible (to put it delicately). Alternatively, he might claim that there are constraints but we can, in principle, say nothing about them: it is just an inexplicable brute fact that we cannot think anything we like. This replaces the earlier incoherence with silence. It is hardly an appealing position.

Worse still, if that is possible, is the idea that we make the known world of stones, trees, cats, and the like with our concepts. It is common to convey this idea with the help of the cookie-cutter metaphor: the dough (the noumenal world) is independent of the cook (us); the cook imposes cookie-cutters (concepts) on the dough to create cookies (appearances). But how could cookie cutters in the head literally carve out cookies in dough that is outside the head? How could dinosaurs and stars be dependent on the activities of our minds? It would be crazy to claim that there were no dinosaurs or stars before there were people to think about them. Constructivists do not seem to claim this. But it is hardly any less crazy to claim that there *would not have been* dinosaurs or stars if there *had not been* people (or similar thinkers). And this claim seems essential to Constructivism: unless it were so, dinosaurs and stars could not be dependent on us and our minds.

Finally, there is an old problem for relativism: arbitrarily excluding from the scope of the theory something dear to the theorist's heart. In this case, why do the languages, concepts, cultures, and so forth, that do the worldmaking not themselves exist only relative to ... ? Relative to what? Themselves? The "texts" themselves start to shimmer and lose their reality.

Constructivists typically vacillate between talk of theories or experience and talk of the world.[13] This vacillation is important to the appeal of their message. For, although it is false that we construct *the world* by imposing concepts on *the world*, it is plausible to suppose that we construct *theories of the world* by imposing concepts on *experience of the world*. The vacillation helps to make the falsehood seem true.

6. REJECTING TWO KANTIAN ARGUMENTS FOR CONSTRUCTIVISM

What then is the case for Constructivism?[14] It arises out of alleged problems for Realism. I shall start with two Kantian arguments and then consider two arguments that Hoyningen-Huene proposes on behalf of Kuhn.

The main ingredients for one argument come straight from Kant. How can we save knowledge in the face of Cartesian doubt? The gap between the knowing mind and the Realist world of independent objects is alleged to make knowledge of those objects impossible. So the world we know about cannot be the Realist world. The only sort of world we could know about is one we partly constitute with our theories. Finally, we add relativism to this Kantian brew: when our theories change radically, so must our worlds.

A semantic variant of this argument can be abstracted from contemporary antirealist discussions. The gap between the knowing mind and the Realist world makes it impossible to refer to that world. So the world we refer to cannot be that world but must be a world we construct. With radical theory change goes reference to a different world.

I think that these sorts of argument are implicit in the discussions of Kuhn and Feyerabend and have frequently attributed a semantic one to them. Oberheim and Hoyningen-Huene dismiss the attribution (1997, pp. 448–449). I am not convinced by their dismissal but will not argue the matter. It is important to see what is wrong with these sorts of argument, whether or not they are to be found in Kuhn and Feyerabend.

The argument I have attributed to Kuhn and Feyerabend starts from a description theory of reference according to which the reference of a term in a scientific theory depends on the descriptions (other terms) the theory associates with it: it refers to whatever those descriptions (or most of them) pick out. Now with theory change, particularly radical theory change, is likely to go the view that those descriptions do not pick anything out. So, if we take the Realist view that a referent must exist independently of theory, we must conclude that, from the new perspective, the term in the old theory does not refer. This will be true even of an "observational" term; think, for example, of the change in descriptions associated with 'The Earth' that came with the Copernican revolution. However, if we abandon Realism we can take the old terms to refer to entities constituted by the old theory, entities that exist relative to that theory but do not exist "absolutely".

Such arguments should give us pause. Speculations about what and how we can know and refer have led to disaster: a bizarre metaphysics. But why should we have any confidence in these speculations? In particular, why should we have such confidence in them that they can undermine a view as commonsensical as Realism? A Moorean point is appropriate: Realism is much more firmly based than these speculations that are thought to undermine it.[15] We have started the argument in the wrong place: rather than using the speculations as evidence against Realism, we should use Realism as evidence against the speculations. We should, as I like to say, "put metaphysics first".

Indeed what support are these troubling speculations thought to have? Not the empirical support of the claims of science. This is most obvious with the epistemological speculations, but it is fairly obvious with the semantic ones. Thus, no attempt is ever made to establish empirically that a description theory of reference is appropriate for these scientific terms. In brief, the support for these speculations is thought to be *a priori*.[16] Reflecting from the comfort of armchairs, Constructivists decide what knowledge and reference *must* be like, and from this infer what the world *must* be like:

A priori epistemology/semantics → *a priori* metaphysics.

The Moorean point alone casts doubt on this procedure and the philosophical method it exemplifies, the *a priori* method of "First Philosophy". But we can do better: the doubt is confirmed by the sorts of considerations adduced by Quine (1952, pp.

xi–xvii; 1961, pp. 42–46). These considerations should lead us to reject *a priori* knowledge and embrace "naturalism," the view that there is only one way of knowing, the empirical way that is the basis of science.[17] From the naturalistic perspective, philosophy becomes continuous with science. And the troubling speculations have no special status: they are simply some among many empirical hypotheses about the world we live in. As such, they do not compare in evidential support with Realism. Experience has taught us a great deal about the world of stones, cats, and muons but rather little about how we know about and refer to this world. So epistemology and semantics are just the wrong places to start the argument. Instead, we should start with an empirically based metaphysics and use that as evidence in an empirical study of how we know and refer: epistemology and semantics themselves become part of science, they become "naturalized":

Empirical metaphysics → empirical epistemology/semantics.

And when we approach our metaphysics empirically, Realism is irresistible. Indeed, it faces no rival we should take seriously.

Quine is fond of a vivid image taken from Otto Neurath. He likens our knowledge – our "web of belief" – to a boat that we continually rebuild whilst staying afloat on it. We can rebuild any part of the boat but in so doing we must take a stand on the rest of the boat for the time being. So we cannot rebuild it all at once. Similarly, we can revise any part of our knowledge but in so doing we must accept the rest for the time being. So we cannot revise it all at once. And just as we should start rebuilding the boat by standing on the firmest parts, so also should we start rebuilding our web. Epistemology and semantics are among the weakest places to stand.

We start with metaphysics. We have already summarized our argument for a Realist one. Does the history of science, so nicely revealed by Kuhn and Feyerabend, demand any modification of this Realism? As theories have changed, have we abandoned our belief in entities that we previously thought existed? First, consider observables. Theoretical progress certainly results in the addition of new observables, terrestrial and celestial, to our catalogue. But there have been very few deletions. Cases like witches, Piltdown Man, and Vulcan are relatively rare. There have been some mistakes, but there is nothing in our intellectual history to shake our confidence that we have steadily *accumulated* knowledge of the make up of the observable world. We have been wrong often enough about the *nature* of those entities, but it is *their* nature we have been wrong about. We have not been wrong about the fact of their existence. In brief, theory change is no threat to Realism about observables.

Furthermore, we should be sufficiently confident of this metaphysics to reject any theory of language that fails to fit it. It is not that the historical facts of theory change, together with a description theory of reference for scientific terms, show Realism to be false. Rather, those facts, together with Realism, show the description theory for those terms to be false. Many ideas for other theories of reference compatible with Realism have emerged in recent times.[18]

It is less easy to rebut Kuhn and Feyerabend on unobservables. It is plausible to suppose that we have often been wrong in thinking that an unobservable exists. Even there, Kuhn and Feyerabend's commitment to the description theory leads them to exaggerate our degree of error. Without these exaggerations, scientific realism is not in much trouble: while our views of, say, the subatomic particles have changed and evolved, we still believe in the entities posited by Bohr and Rutherford. *At most*, the history of science should make us cautious in our commitment to unobservables. It should not lead us into Constructivism (1997, section 9.4).

The discussion in this section is intricate enough to warrant a summary. The background to the discussion is our earlier argument for the plausibility of Realism (section 4) and the implausibility of Constructivism (section 5). The Kantian arguments for Constructivism and against Realism rest on speculations in epistemology and semantics. Against the background – the plausibility of Realism and implausibility of Constructivism – the Moorean point is that we should prefer Realism to the Kantian speculations; we should put metaphysics first. This point is good on its own but when supported by naturalism it is formidable. From the naturalistic perspective these speculations cannot be supported *a priori* and they do not come close to having the empirical support enjoyed by Realism. The arguments for Constructivism use the wrong methodology and proceed in the wrong direction.

I turn now to the arguments that Hoyningen-Huene offers on behalf of Kuhnian Constructivism.

7. REJECTING TWO KUHNIAN ARGUMENTS FOR CONSTRUCTIVISM

Oberheim and Hoyningen-Huene dismiss the idea that Kuhn and Feyerabend were led to incommensurability and Constructivism by their semantics. Rather, these doctrines "were the result of attempting to achieve a historical understanding of the development of science" (1997, p. 449). How did they result? We need to be particularly concerned with how the history of science is supposed to support Constructivism, because without Constructivism the case for Incommensurability collapses (section 3).

In his book, Hoyningen-Huene emphasizes Kuhn's revolutionary historiography of science which was to be the basis for his view of science: "Kuhn's goal is *to propose a new picture of science and scientific development, in particular of scientific progress, grounded in this new historiography*" (1993, p. 13). According to this historiography, episodes in the history of science are best studied and explained in their own right and not from the perspective of contemporary science. We need to see the world as the scientists of those times did, which may be difficult because those scientists saw the world so differently. This practice "requires an exact reconstruction of the period's conceptual system" (p. 20). It leads to "a more alien, yet at the same time more reasonably alien, scientific past than the old historiography" (pp. 22–23). Hoyningen-Huene goes on to claim that the experience of the historian practicing this new historiography justifies the plurality-of-phenomenal-worlds thesis for the practice "*may* produce a different phenomenal world – different, that is, by comparison with the historian's own phenomenal world" (p. 38). But, of course, this experience could justify the *plurality* of phenomenal worlds only if it has already been established that *there is a phenomenal world at all* in the relevant Kantian sense, a world partly constructed by the imposition of concepts. Perhaps Hoyningen-Huene takes this to have been established already by Kant. Mooreans and naturalists think that nothing could be further from the truth (section 6).

Kuhn and Feyerabend's illuminating view of the history of science alone does not support Constructivism. No more does their illuminating view of the epistemology of science. They claim that all statements, even "observation" statements, are epistemically theory-laden.[19] This claim is not novel – it is central to the Duhem-Quine thesis – but Kuhn and Feyerabend did more than anyone to establish it, producing a marvellous array of scientific evidence in its favor. Hoyningen-Huene sees an argument here for the plurality-of-phenomenal-worlds thesis: that thesis helps explain theory-ladenness (pp. 36–37). And so, in a way, it does. But this is not the *best* explanation because it is based on an implausible metaphysics. The best explanation is based on Realism. Against a Realist background theory-ladenness is readily explained by a naturalized epistemology that appeals to the psychological facts of belief formation. Theory-ladenness would provide a reason for going beyond Kant's metaphysics by adding a plurality thesis if we had already established that metaphysics. But we have not.

Constructivism combines the Kantian idea of a phenomenal world, the Kantian idea of a noumenal world, and relativism. Kuhn and Feyerabend's views of history and epistemology would support the addition of relativism to the two Kantian ideas but they do nothing to support the ideas themselves.

Let us take stock. I have argued that Constructivism leads to Incommensurability and Realism leads to Commensurability. Furthermore, I claimed that without Constructivism, or some other form of antirealism, there is no plausible route to Incommensurability (section 3). So, we can refute Incommensurability by establishing Realism. The case for Realism is so strong that we should give Realism up only in the face of powerful arguments against it and for an alternative (section 4). I have argued against Constructivism, the alternative that concerns us, emphasizing the bizarre and mysterious nature of the doctrine (section 5). Adopting first a Moorean and then a naturalistic perspective, I have rejected two Kantian arguments for Constructivism (section 6). We have just seen that the arguments that Hoyningen-Huene proposes on Kuhn's behalf presuppose rather than argue for the Kantian core of Constructivism. Constructivism is not only implausible, it is unsupported. Realism still stands. If I am right about all this, the case against Incommensurability is made.

But one matter remains.

8. REJECTING META-INCOMMENSURABILITY

Hoyningen-Huene, Oberheim and Andersen, struck by the inconclusiveness of the realism debate, make the tentative proposal that the debate itself involves incommensurability, what they call "*meta-incommensurability*": because of meaning differences "effective means of rational *meta-theory* choice are not yet at hand" (1996, p. 138). If they are right about this, of course, the argument I have just summarized, purporting to give a rational basis for choosing Realism over Constructivism, must be a failure.

In support of meta-incommensurability, they claim:

> there are several terms that change meaning when one crosses the line from realism to non-realism: namely, 'reality', 'world', 'theory comparison', 'fact', and even 'reference' itself... they purportedly refer to different things, based on the different metaphysical assumptions each party brings to the debate. (pp. 138–139)

This simply *assumes* that because the realist and the non-realist make different assumptions involving a word, and hence associate different predicates with it, its meaning and reference differ for them. This is to assume a description theory for these words. Why should we do that? We already know that description theories cannot be true for all words: these theories explain the meaning and reference of a word in terms of the meaning and reference of other words, a process that cannot go on for ever. Some words must "stand on their own feet" being explained (at least partly) in terms of some sort of relation to reality that is not mediated by another word (Devitt, 1996, pp. 159–160). We have been given no reason to suppose that the words in question are not of that sort.

The initial proposal of meta-incommensurability was cautious but the caution of Oberheim and Hoyningen-Huene soon disappears: "'truth', 'world', 'fact', 'theory comparison', and 'reference' ... *clearly* have different meanings for the realist and the non-realist" (1997, p. 453; my emphasis). Furthermore, they reject the idea that this thesis rests on semantic considerations, in particular on a description theory (p. 460).[20] Rather, it is "based on a contemporary historical case study" (p. 453). They see meta-incommensurability as the best explanation of a range of phenomena which they observe in the realism literature: "communication difficulties"; accusations that arguments are "circular" or "question-begging"; the sense that arguments are "indecisive" (pp. 453–459). This is ingenious. The phenomena that they identify cannot be denied. And if there is meta-incommensurability we would certainly expect these phenomena. But is meta-incommensurability really the best explanation of them? I think the answer is, "No".

I start with a qualification. I would be the last to claim that there are *no* meaning differences in the realism debate, nor that such differences play no role in producing phenomena *of the sort* identified: I have often complained of the confusion over the word 'realism' itself. So I accept that incommensurability *sometimes* plays an explanatory role of the sort Oberheim and Hoyningen-Huene suggest. But this sort of incommensurability can easily be removed with a bit of terminological care. It is very different from their meta-incommensurability rooted in metaphysical difference, difficult if not impossible to remove.

Here is a reason for thinking that meta-incommensurability *may not* be the best explanation of the phenomena that they identify. Communication difficulties, accusations of circularity, and similar phenomena often occur in disagreements where it is very implausible that there are meaning differences. These may be humdrum disagreements of everyday life or "low-level" scientific disagreements within a paradigm. It is clear, then, that the occurrence of such phenomena does not *depend on* meaning differences between positions. Sometimes we need another explanation of these phenomena.

What other explanations are available? I shall briefly describe four features of the cognitive life that might contribute to such explanations. In light of these, it seems to me clear that meta-incommensurability *is not the best* explanation of the phenomena of the realism debate.

The first feature is the theory-ladenness of observation, a feature much emphasized by Kuhn and Feyerabend: what we make of our experience depends very much on what we already believe, on what we expect. This applies to our linguistic experience as well. We interpret sentences that mean *p*, in our language as much as the speaker's, to mean *q* because, given our beliefs, including those about the speaker, that is what we expect her to be saying in the circumstances. We have "communication difficulties". This is surely part of the explanation of a phenomenon experienced by all journal reviewers: the author of a critical paper seems unable to read what is plainly on the page criticized.

The second is the general difficulty of putting together good arguments, and the undeniable extra difficulty of doing so in philosophy. It is very hard to think straight and particularly hard to do so about such abstract topics as realism. This difficulty is surely the explanation of many arguments in ordinary life and science that are circular, question-begging, or indecisive. And it is plausible to think that it is part of the explanation of these failures in many philosophical arguments.

The third is a bit more speculative. It seems plausible to suppose that we humans suffer from a certain rigidity in our thinking that makes it difficult for us to contemplate alien views. And the more alien and the more global the view, the greater the difficulty. So we should expect great difficulty in the realism debate. Sometimes, no doubt, difficulties with alien views arise from meaning differences but there seems no good reason for supposing that they all must. Why should we not regard it as a brute fact about us that, even with meanings constant, we find it difficult to "get our heads around" alien views? Certainly we do not know enough about psychology and semantics to rule this out.

Finally, a rather obvious feature of many debates in philosophy and elsewhere is the ego involvement of the participants: the participants are wedded to their views and "want to win". This can blind them to the faults in their arguments leading to question begging, circularity and the like.

Given the general availability of explanations built out of these four features, it is appropriate to invoke meta-incommensurability as an explanation of a particular phenomenon of question begging, circularity, etc, only if we have some independent reason for thinking that meaning differences are involved. This independent reason must arise from a semantic theory, however primitive. So an historical case study that does not invoke semantic considerations will not do the job of justifying a meta-incommensurability explanation of the phenomena of the realism debate, contrary to what Oberheim and Hoyningen-Huene desire. And, so far as I can see, the only semantic consideration that they can invoke to do the job is a description theory of reference. As I have pointed out, we have been given no reason to suppose that this theory is appropriate here.

I conclude that the meta-incommensurability thesis is false. My earlier argument for Realism and hence against Incommensurability stands.

Finally, it is just as well that the meta-incommensurability thesis is false because it is dangerous. If the thesis were true both sides of the realism debate would be immune to rational criticism. Post-modernists would relish such a conclusion but it should dismay the rest of us.[21]

The Graduate Center, City University of New York

ACKNOWLEDGEMENTS

A draft of this paper was delivered at the conference, "Incommensurability (and related matters)", held in Hannover in June 1999. Eric Oberheim was my commentator. I am indebted to his commentary for several improvements in my presentation.

NOTES

[1] Eric Oberheim and Hoyningen-Huene (1997) raise another interpretative issue. They distinguish between "Feyerabend and Kuhn's conception of incommensurability, in the sense that it was a result arrived at through historical analysis and ... a semantic conception of incommensurability within a realist framework". Whereas "contemporary literature" has been concerned with the semantic conception, Feyerabend and Kuhn's conception was not "restricted to such semantic issues" (p. 447). I have two comments. First, here and elsewhere, Oberheim and Hoyningen-Huene confuse the question "What is *the argument* for Kuhn and Feyerabend's incommensurability?" with the question "What is *the nature* of that incommensurability?". Whether or not the argument for (or against) their incommensurability is from historical analysis, within a realist framework, or whatever, is one thing, what that incommensurability *is* is another. (My "Maxim 1" makes an analogous point about the realism issue: "In considering realism, distinguish the constitutive and evidential issues"; 1997, p. 3). In particular, whether or not Kuhn and Feyerabend's incommensurability *is* semantic is a distinct matter from whether or not their argument for it is from "semantic theory" (pp. 447–452). My second comment addresses what Oberheim and Hoyningen-Huene have to say on the former question, the one about the nature of incommensurability. They reject the common view, which I share, that Kuhn and Feyerabend's incommensurability thesis is semantic. They insist that the thesis "was intended to involve more than semantic issues": it was intended to involve a neo-Kantian antirealism (p. 450). I have no doubt that Kuhn and Feyerabend subscribe not only to a semantic thesis like my Incommensurability but *also* to an antirealist metaphysics. I do doubt that they conflated them under the one term 'incommensurability'. But whether or not they did, they shouldn't: nothing but harm comes from the conflation of semantics with metaphysics. Or so I have argued (1997). To insist on distinguishing the semantic thesis from the metaphysical one is not, of course, to deny that the theses may be related. I shall explore the relation in section 3.

[2] I explain and argue for this nonsemantic characterization of realism in my (1997).

[3] Howard Sankey talks of Hoyningen-Huene's "novel interpretation of Kuhn's philosophy of science, which presents the latter within a neo-Kantian anti-realist framework" (1997, p. 437). What is novel, of course, is not the neo-Kantian (= Constructivist) interpretation itself, which has been widespread for years, but the thoroughness of the case for it, and the clear and detailed presentation of Kuhn's philosophy of science from that interpretative perspective.

[4] Hoyningen-Huene offers two other bases (pp. 220–221). I find these very unconvincing but will not argue the matter.

[5] The argument for Commensurability in my (1979) is implicitly an argument from Realism along these lines, and that in my (1997, section 9.6), is explicitly so (also Devitt and Sterelny 1999, pp. 227–228). Sankey (1994) is an extended argument of this sort; see also Sankey (1998).

[6] Devitt (1997), particularly chapters 5 and 7; see also (1999).

[7] This simple argument should not be confused with a popular one captured by the slogan "Realism explains success". The popular argument uses Realism to explain the observational success of theories where the simple one uses Realism to explain the observed phenomena (1997, section 7.3).

[8] For more details, see my (1997), particularly chapter 13.

[9] Richard Rorty thinks it absurd to say "that we make objects by using words" (1979, p. 276). Nicholas Wolterstorff thinks that in saying this the Constructivist must be "speaking in metaphor. If we took him to be speaking literally, what he says would be wildly false – so much so that we would question his sanity" (1987, p. 233). David Stove has this to say in his chapter "Philosophy and Lunacy: Nelson Goodman and the Omnipotence of Words": "the statement that worlds can be made with words: a statement which, as Hume said of the doctrine of the real presence, 'is so absurd that it eludes the force of all argument'" (1991, p. 31).

[10] According to Hoyningen-Huene, "Kuhn stipulates [the noumenal world] to be *spatiotemporal*, not undifferentiated, and in some sense *causally efficacious*" (1993, p. 34). One wonders what Kuhn's justification could be for this departure from Kant.

[11] As, indeed, Hoyningen-Huene admits in a footnote (n. 119)! The admission totally undermines his rejection. See also sections 2.2.c–2.2.e where Hoyningen-Huene wrestles mightily with Kuhn's attempts to describe the constraining role of the noumenal world (in the guise here of what Hoyningen-Huene calls "object-sided stimuli"). This discussion brings out nicely the futility of such attempts to speak the unspeakable.

[12] Alternatively, if we can know what is targeted in the noumenal world, why can we not know other things about that world? Why then does the noumenal world not collapse into the knowable Realist world? For more along this line, see Sankey (1997, pp. 439–440).

[13] For example: Whorf (1956, pp. 55, 162, 213, 253); Kuhn (1970, pp. 114, 117); Feyerabend (1978, p. 70); Latour and Woolgar (1986, p. 183); Hawkes (1977, p. 28).

[14] I am here seeking *rational* explanations of Constructivism; and I have done so at greater length elsewhere (1997, sections 13.4–13.7). But the popularity of a doctrine that is so bizarre and mysterious cries out for a different sort of explanation. For some learned, and very entertaining, suggestions, see Stove (1991). Stove thinks that antirealism, like religion, stems from our need to have a *congenial* world. For some suggestions by Georges Rey along similar lines, see my (1997, p. 257n).

[15] Steven Hales drew my attention to the Moorean nature of this point. Note that the point is not that Realism is indubitable, to be held "come what may" in experience: that would be contrary to naturalism. The point is that, *prima facie*, there is a much stronger case for Realism than for the speculations. (Thanks to Paul Boghossian.)

[16] In this respect, Kuhn and Feyerabend are very much part of the Establishment, despite the radical nature of their philosophy of science. They are part of a semantic tradition, one that includes the positivists before them and is still dominant to this day, that proceeds as if semantics is, at bottom, *rather easy*. At the level of terms (or concepts) we can rely on *a priori* intuitions about which features of a term (or concept) constitute its meaning. So to determine the meaning all we have to do is describe how the term (concept) is learned and used and we can simply "see" what its meaning is. Feyerabend remarks that "conversations about meaning belong in the gossip columns" (1981a, p. 113). Since his own writings are riddled with such conversations, we must see his remark as characteristic waggishness. Of course, one might wonder how an empirical theory of meaning *should* proceed. I think that it is very difficult to say. My attempt is (1996, chapter 2).

[17] A particularly important consideration against the *a priori*, in my view (1996, section 2.2), is the lack of anything close to a satisfactory *explanation* of a nonempirical way of knowing. We are told what this way of knowing is *not* – it is *not* the empirical way of deriving knowledge from experience – but we are not told what it *is*. Rey (1998) and Field (1998) have a more tolerant view of the *a priori*. My (1998) is a response.

[18] See for example, Kripke (1980), Putnam (1975), Dretske (1981), and Millikan (1984).

[19] Not to be confused with another thesis that they have also sometimes seemed to hold: that "observation" terms are *semantically* theory-laden. This thesis amounts to a description theory of those terms. One could believe the epistemological thesis that one's judgment about the application of, say, 'rabbit' in a certain situation depends on all sorts of background assumptions about rabbits, whilst holding the semantic thesis that the meaning of 'rabbit' depends not on its relation to any other term but entirely on its direct causal relation to rabbits.

[20] This is not the same, of course, as rejecting that meta-incommensurability *is* semantic (as the characterization that opens this section shows it to be). However, they sometimes write as if the confusion of the argument and nature questions that I criticized in their discussion of incommensurability (note 1) may also be present in the their discussion of meta-incommensurability (pp. 453, 461).

REFERENCES

Devitt, M. (1979). "Against Incommensurability." *Australasian Journal of Philosophy* **57**: 29–50.

Devitt, M. (1996). *Coming to Our Senses: A Naturalistic Defense of Semantic Localism*. New York: Cambridge University Press.

Devitt, M. (1997). *Realism and Truth*. 2nd edition. With a new afterword. (1st edition 1984, 2nd edition 1991). Princeton: Princeton University Press.

Devitt, M. (1998). "Naturalism and the *A Priori*." *Philosophical Studies* **92**: 45–65.

Devitt, M. (1999). "A Naturalistic Defense of Realism." In S. Hales, ed., *Metaphysics: Contemporary Readings*, pp. 90–103, Belmont: Wadsworth Publishing Company.

Devitt, M., and K. Sterelny. (1999). *Language and Reality: An Introduction to the Philosophy of Language*. 2nd edition. (1st edition 1987). Oxford: Blackwell Publisher.
Dretske, F. (1981). *Knowledge and the Flow of Information*. Cambridge, Mass.: MIT Press.
Feyerabend, P. (1978). *Science in a Free Society*. London: New Left Books.
Feyerabend, P. (1981a). *Realism, Rationalism and Scientific Method, Philosophical Papers, Volume 1*. Cambridge: Cambridge University Press.
Feyerabend, P. (1981b). *Problems of Empiricism: Philosophical Papers, Volume 2*. Cambridge: Cambridge University Press.
Field, H. (1973). "Theory Change and the Indeterminacy of Reference." *Journal of Philosophy* **70**: 462–481.
Field, H. (1998). "Epistemological Nonfactualism and the *A Prioricity* of Logic." *Philosophical Studies* **92**: 1–21.
Goodman, N. (1978). *Ways of Worldmaking*. Indianapolis: Hackett Publishing Company.
Hawkes, T. (1977). *Structuralism and Semiotics*. London: Methuen & Co.
Hoyningen-Huene, P. (1993). *Reconstructing Scientific Revolutions: Thomas S. Kuhn's Philosophy of Science*, trans. A. Levine (German edition 1989). Chicago: University of Chicago Press.
Hoyningen-Huene, P., E. Oberheim and H. Andersen. (1996). "On Incommensurability." *Studies in the History and Philosophy of Science* **27**: 131–141.
Jameson, F. (1972). *The Prison-House of Language*. Princeton, N.J.: Princeton University Press.
Kripke, S. (1980). *Naming and Necessity*. Cambridge, Mass.: Harvard University Press.
Kuhn, T. (1970). *The Structure of Scientific Revolutions*. 2nd edition. (1st edition 1962). Chicago: Chicago University Press.
Kuhn, T. (1979). "Metaphor in Science." In A. Ortony, ed., *Metaphor and Thought*, pp. 409–419, Cambridge: Cambridge University Press.
Kuhn, T. (1983). "Commensurability, Comparability, Communicability." In P. Asquith and T. Nickles, eds., *Proceedings of the 1982 Biennial Meeting of the Philosophy of Science Association*, pp. 669–688, East Lancing: Philosophy of Science Association.
Latour, B. and S. Woolgar. (1986). *Laboratory Life: The Construction of Scientific Facts*. 2nd edition. (1st edition 1979). Princeton: Princeton University Press.
Martin, M. (1971). "Referential Variance and Scientific Objectivity." *British Journal for the Philosophy of Science* **22**: 17–26.
Millikan, R. (1984). *Language, Thought and Other Biological Categories: New Foundations for Realism*. Cambridge, Mass.: MIT Press.
Putnam, H. (1975). *Mind, Language and Reality. Philosophical Papers, Volume 2*. Cambridge University Press.
Putnam, H. (1981). *Reason, Truth and History*. Cambridge: Cambridge University Press.
Oberheim, E. and P. Hoyningen-Huene (1997). "Incommensurability, Realism and Meta-Incommensurability." *Theoria* **12**: 447–465.
Quine, W. (1952). *Methods of Logic*. London: Routledge & Kegan Paul.
Quine, W. (1961). *From a Logical Point of View*. 2nd edition. (1st edition 1953). Cambridge, Mass.: Harvard University Press.
Rey, G. (1998). "A Naturalistic *A Priori*." *Philosophical Studies* **92**: 25–43.
Rorty, R. (1979). *Philosophy and the Mirror of Nature*. Princeton: Princeton University Press.
Sankey, H. (1994). *The Incommensurability Thesis*. Aldershot: Avebury.
Sankey, H. (1997). "Incommensurability: The Current State of Play." *Theoria* **12**: 425–445.
Sankey, H. (1998). "Taxonomic Incommensurabilty." *International Studies in the Philosophy of Science* **12**: 7–16.
Stove, D. (1991). *The Plato Cult and Other Philosophical Follies*. Oxford: Basil Blackwell.
Whorf, B. (1956). *Language, Thought and Reality*, ed. and intro. J. Carroll, Cambridge, Mass.: MIT Press.
Wolterstorff, N. (1987). "Are Concept-Users World-Makers?" In J. Tomberlin, ed., *Philosophical Perspectives, 1: Metaphysics, 1987*, pp. 233–267, Atascadero: Ridgeview Publishing Company.

GERALD DOPPELT

INCOMMENSURABILITY AND THE NORMATIVE FOUNDATIONS OF SCIENTIFIC KNOWLEDGE

Abstract. In this paper, I defend the relativity of scientific reasoning and knowledge to scientists' diverse normative allegiances to historically variable standards of theory assessment. My aim is to justify a position of moderate relativism. This position seeks to accommodate the essential role of evidence, reasons, and reasoning in scientific choice/belief; while also giving sociological and historical factors an essential role in explaining and justifying which reasons and choices prove compelling to some scientists but not others who are equally rational and scientific. My strategy of argument is to take up the main criticisms of this sort of position in the work of post-Kuhnian writers such as Laudan, Lakatos, Siegel, Shapere, Scheffler, Sankey, and others. In particular, I examine the following influential rebuttals of relativism: (1) It is self-refuting; (2) It is refuted by the existence of rational debate in science; (3) It is refuted by the existence of neutral, external, universal standards of theory-assessment in science; (4) It is refuted by the fact that there are typically good scientific reasons for preferring some standards to others, which relativists cannot see because they embrace a misguided holistic paradigm of scientific change; (5) It is refuted by normative naturalism which provides an objective, ahistorical, and empirical method for evaluating the effectiveness of standards of scientific knowledge in attaining the aim or aims of scientific inquiry. I acknowledge the insights of these positions. But, I argue, they fail to rebut moderate relativism, and in particular, the relativity of reason and justification to diverse, historically shaped normative commitments to rival conceptions of scientific knowledge. The enduring philosophical contribution of Kuhn lies not in radical relativism, extreme incommensurability, or virulent anti-realism. Rather, it consists in the promise of a more interdisciplinary model of scientific knowledge and the logic of its growth.

Three decades after the ground-breaking work of Thomas Kuhn, there is little agreement among philosophers of science concerning the precise import and significance of his work for the theory of scientific knowledge. In the wake of intense post-Kuhnian debates over incommensurability, relativism, rationality, objectivity, and realism, what sort of conception of scientific knowledge is now most plausible? How should we reformulate the aims and standards of epistemology to accommodate any durable insights that may have emerged from the post-Kuhnian debates? These are the questions I hope to clarify in this paper.

I will argue that the most durable and important insight can be expressed as a thesis which asserts the context-bound normative relativity of scientific knowledge. This thesis holds that the justification of scientific beliefs or theories, thus scientific knowledge itself, always presupposes the normative commitment of some community or group of scientific practitioners to historically particular standards of theory-evaluation. Scientific knowledge is thus relative to the variable normative commitments and changing standards of real scientific groups.

P. Hoyningen-Huene and H. Sankey (eds.). Incommensurability and Related Matters, 159–179.

Secondly, I will argue that this thesis, does *not* justify an *extreme* or *radical* incommensurability, relativism, and irrationalism in epistemology. On the other hand, it *does* justify what I characterize as a moderate relativism concerning theory-choice, justification, and knowledge. Moderate relativism recognizes that scientific theory-choice and the establishment of knowledge is a reason-governed, rational process. But it claims that *which* reasons and *whose* arguments prove to be compelling for scientific practitioners requires both a justification and explanation in historical and sociological terms, not purely philosophical ones. Moderate relativism thus argues that exclusively epistemic or reason-based conceptions of justification cannot, all by themselves, either explain or justify what specific scientific practitioners reasonably take to be justified and known. While I thus firmly reject the extreme relativism/incommensurability criticized by philosophers of science in the post-Kuhnian debate, the moderate relativism I defend still challenges the dominant framework of contemporary philosophy of science. For, if I am right, we should embrace a more *interdisciplinary* model of epistemology and the justification of scientific knowledge

My strategy of argument in this paper is as follows. I develop the thesis of the relativity of scientific knowledge and my argument for moderate relativism. I then critically evaluate the most promising lines of rebuttal of my argument in the work of post-Kuhnian philosophers. Examining various moves and counter-moves, I argue that moderate relativism is the inescapable outcome of the post-Kuhnian dialectic of argument.

1. MODERATE RELATIVISM

Putting the merit of Kuhn's philosophical claims to one side, it is still undeniable that his work has reshaped the terms of debate, and much research, in philosophy of science. In short, his work has given a new centrality and relevance to the history of science, and the examination of specific scientific practices, for philosophers. At one level, the historicity of science is a commonplace and unremarkable feature. The pre-Kuhnian tradition of logical empiricism itself stresses the enormous conceptual innovation and theoretical change at the heart of scientific development. Yet for this tradition, historical change in the content of science is entirely compatible with a timeless, unchanging, unified scientific method that enables philosophers of science to capture "the" logic of scientific testing, confirmation, prediction, explanation, meaning, progress, etc.

Against this background, Kuhn's most important claim is that in the course of its development, science changes not just its conception of nature, but more fundamentally, its conception of its own nature – its conception of the very kinds of evidence, empirical data, theories, explanation, reasoning, and standards which are, and are not, *genuinely* scientific and conducive to *genuine* scientific knowledge. Scientific knowledge involves not just straightforward changes in theoretical and empirical belief, but normative transformations in the very problems, aims, values, and standards taken by scientific practitioners to be essential to scientific knowledge. This Kuhnian claim thus asserts normative variability in the very standards of knowledge to which scientific practitioners are actually committed. From this perspective, scientific controversies may involve disagreement over the proper standards of theory assessment. The resolution of such controversies and the triumph of a new theoretical framework may involve the transfer of allegiance among scientific practitioners to a new set of standards, and a new basis for evaluating claims to scientific knowledge in the relevant discipline or domain of inquiry.

On traditional epistemological models such as logical empiricism, the fact of such normative variability is either denied or held to be irrelevant to theory-assessment. On such models, in typical cases of a rational displacement of theory A by theory B, scientific practitioners eventually abandon A and accept B because B either is, or promises to be, better justified than A, on the basis of universal standards of evidence, explanatory scope, predictive accuracy, simplicity, etc. For moderate relativism, such models are overturned by the thesis of normative variability, taken together with a thesis that asserts the context-bound normative relativity of scientific knowledge to prevailing standards, those to which groups of scientific practitioners are actually committed. Two sets of standards are incompatible if and only if they respectively justify contradictory judgments concerning which among rival theories is better justified.

On the resulting relativist model, one scientific theory B may displace a rival theory A, even though in the context where they compete, B is no better than A. The context is in fact characterized by scientific practitioners' commitments to incompatible standards of theory-assessment. What the proponents of A or B count as good evidence, an important empirical problem, a valid inference, a legitimate type of explanation, or a suitably unifying theory may not count as such for proponents of the rival theory, and vice versa. Furthermore, the relativist argument requires contexts in which each theory satisfies its own set of standards to a greater extent than the rival theory satisfies it. Even assuming the variability of standards, and the relativity of justification to accepted standards, relativism is avoided if all shifts in standards and theories are "cumulative"; cumulative shifts are ones in which the triumphant theory satisfies its own standards, and the standards of its predecessors, to a greater extent than the predecessor theory(s) satisfied either its own set of standards or those of the successor. In this non-relativist scenario, despite the difference in standards, theory B is unequivocally better justified than A because it beats A at its own game, as well as beating A by the standards of B's game.

But in the relativist scenario, many historical contexts are such that each theory is better justified than its rival, on its own standards of theory-assessment. When this situation obtains, it raises the question of how we are to account for the triumph of theory B over A when, *ex hypothesi*, B is not better justified than A. On the moderate relativist model, the answer is that B comes to be better justified than A *only if and when* scientific practitioners transfer their allegiance to a new set of standards, and a new conception of the aims and methods of science. Clearly, once a new set of standards is established, scientific practitioners have a fully rational basis in which to re-evaluate all preceding theories and scientific controversies, up to and including the theory(s) they have overturned or revised, in favor of the current theory. Given the commitment to the new standards, and a record of success in realizing them, such commitments enable us to characterize current scientific theories, or some of their ancestors, as better justified than their predecessors.

In turn, this picture raises the question of how we are to account for and evaluate changes in the standards to which scientific groups are committed. As several post-

Kuhnian critics have argued, relativism is avoided if scientists' embrace of new standards is itself a rational process in which better-justified aims and standards replace discredited or otherwise flawed ones. Or perhaps, there are viable external meta-standards for evaluating the variable standards internal to the practices of past and present scientific groups. The bulk of this paper will argue that such philosophical moves do *not* succeed in undermining moderate relativism. The moderate relativist need not deny that there are reasons underlying the transfer of allegiance to, or the embrace of, new standards of scientific knowledge. Indeed, my moderate relativist can recognize that *some* of these reasons are epistemic, that is reasons for believing that this standard or that is or is not a mark of scientific knowledge. As I argue below, such reasons by themselves rarely establish that one set of standards is or is not better justified, a more reliable guide to genuine scientific knowledge, than a rival set. Rather, the reasons which drive commitments to standards are practical and interest-laden. They require an historical and sociological account of how and why individual scientific practitioners became interested in and identified with a particular practice of inquiry and a particular community of inquirers bound together by relations of trust and credibility denied to outsiders. If moderate relativism is defensible, such practical reasons, interests, and investments explain which epistemic reasons particular scientific practitioners do and do not accept, both for the evaluation of standards, and theories. The resulting picture of science is intended to steer a middle path in the debates between sociologists of scientific knowledge who stress the social, and cultural motives of scientific beliefs or practices, and philosophers of science who stress the reason-driven logic of scientific inquiry and knowledge.

2. THE REBUTTALS OF RELATIVISM

Since I first elaborated and defended a moderate relativism based on a shifts-in-standards argument, various philosophers of science have sought to rebut this position. The most important objections are as follows:

1. Moderate relativism, like relativism generally, is self-refuting, because its defense assumes the existence of the very sorts of standards of justification whose possibility it denies.

2. Moderate relativism admits the existence of rational debate between the advocates of rival theoretical frameworks in science. Rational debate presupposes the existence of neutral, *external* standards of reasoning and theory-assessment. These standards allow us to circumvent any relativism which confines theory-assessment to the historical standards *internal* to particular scientific practices and groups.

3. Quite apart from the above argument concerning rational debate, there are neutral, external, and fairly universal standards of theory-assessment in science: standards of empirical success, predictive accuracy, explanatory scope, simplicity, and problem-solving effectiveness. These standards provide an

unequivocal basis for determining which scientific theories are best justified, in just the way that is denied by the moderate relativist.

4. Moderate relativism ignores the possibility that there are good reasons for preferring some standards to others, that we can learn which standards are most conducive to genuine scientific knowledge. The relativist misses the possibility of rational changes in standards because s/he holds to a mistaken holistic theory of scientific change in which all of science – its theories, aims, concepts, methods, and standards – changes all together and at once, as in Kuhn's doctrine of scientific revolution as global paradigm-shifts. A more accurate gradualist picture of scientific change allows us to see how elements of continuity provide justifications for changes in standards.

5. Moderate relativism can be overcome with the resources of a normative naturalism that objectively evaluates all methods and standards in science as more or less effective means to its aims; based on empirical evidence exhibiting the causal efficacy or inefficacy of this or that method in bringing about the realization of the aims of science, such as the attainment of truth about nature. Such a naturalized epistemology can reveal which standards of theory-assessment are most reliable and most conducive to scientific knowledge. It can thus circumvent a relativism based on scientists' actual methodological commitments and internal standards.

In what follows, I will critically evaluate each of these rebuttals of moderate relativism.

3. THE SELF-REFUTATION REBUTTAL[1]

The moderate relativist seeks to justify, and provide reasons for accepting relativism, to those not already committed to this philosophy, including anti-relativists. This project assumes the existence of general or external standards of justification. But the relativist denies *a priori* the possibility of such general standards. As a result, either the relativist cannot justify this position, or it is self-refuting.

Reply: Moderate relativism, as I construe it, does not express an *a priori* thesis concerning the in principle impossibility of shared, general, or external standards of justification. Part of what makes it a "moderate" relativism is that its thesis concerning the variability of standards is an empirical and historical claim not intended to rule out the possibility and indeed the fact of some shared standards in some contexts of scientific controversy and theory-change. The plausibility of the claim resides in the fact that disagreement over, and variation in, the standards of scientific knowledge is well-documented in the case-studies and research of many historians of science. Similarly, moderate relativism does not assert the *a priori* impossibility of universal, external standards of theory-assessment. Rather it argues that the familiar proposals fail to do the work required of them in the case of scientific evaluation (see below).

As such, the moderate relativist's claim does not apply across the board to all domains of inquiry, including philosophy; rather it needs to be defended case by case.

One would have to independently argue that such variability of standards also applies to the philosophical controversy over relativism concerning science. Thus moderate relativism, as I construe it, is not vulnerable to a gambit of quick, decisive self-refutation.

4. THE RATIONAL DEBATE REBUTTAL[2]

Echoing other critics, Harvey Siegel argues that the existence of rational debate between the exponents of rival theories implies the existence of neutral, external standards – standards shared by the rivals. Moderate relativism, as I construe it, acknowledges the existence of rational debate. If this in turn implies external standards, then these standards furnish a rational basis for non-relativist theory-assessment, and moderate relativism is circumvented.

The idea here is that rational debate requires that the parties to the debate commonly recognize at least some of the considerations offered by each side to be reasons. In turn this requires the existence of some shared, theory-neutral, external standards that enable both sides to recognize certain sorts of considerations as reasons. This defeats moderate relativism.

I have two replies to this line of argument. First of all, there is good reason to reject the view that rational debate implies the acceptance of common standards of justification. In rational debate, each side may recognize that the other side is presenting reasons, which count as good reasons for that side, but quite poor or weak reasons from its own point of view. To identify a consideration as a reason, one only needs to recognize that there is some intelligible and defensible standard and judgment which motivates it. One need not accept the standard as well. Consider a rational debate over whether there is sufficient evidence to find the accused guilty in a criminal trial. One side urges that an array of circumstantial evidence is sufficient proof of guilt beyond a reasonable doubt. The other side sees the relevance of circumstantial evidence, but holds to a higher or different standard of proof requiring either confessions, eyewitness testimony, and/or DNA fingerprinting. Each side can provide reasons for the verdict it favors, even though they subscribe to different standards of proof.

This opens the way to my second and more decisive reply to the rational debate rebuttal of moderate relativism. I have characterized the above situation as one in which the parties to a rational debate subscribe to different standards but can nonetheless provide reasons to one another for the verdict each favors. But, I might better characterize the situation as one in which the parties subscribe to different, but *overlapping*, standards. For though they disagree over the standard of evidence which suffices for a guilty verdict in cases of this sort, they do agree on what is and is not relevant to the matter of guilt or innocence. We may thus presume that they mutually accept some standard of relevant evidence. This overlap in standards can be invoked to account for the possibility of rational debate between groups who otherwise are committed to different standards of proof. Their acceptance of these neutral or external standards of relevant evidence does not suffice to establish which of their incompatible verdicts – guilty beyond a reasonable doubt, or not so – is the better justified, given their commitments to different standards of proof.

In my work and that of other philosophers of science, there are many examples of scientific inquiry, controversy, and change which fit this scenario. In such cases, proponents of rival theoretical frameworks in science are committed to different sets of empirical problems, explanatory concepts, theoretical aims, and standards of theory-assessment as the epistemic center of the discipline. How then can we account for rational debate and the fact that their theories are rivals which make incompatible claims concerning nature? The answer of the moderate relativist consists in recognizing the existence of overlap in problems, language, aims, and standards. As I read Kuhn's key notion of anomaly, it illustrates such overlap – an empirical situation commonly recognized as an unsolved problem for one theoretical tradition which is resolved by its rival. Nonetheless, rival groups may acknowledge that the ability of a theoretical framework to solve a given empirical problem is one relevant measure of its success, and yet systematically disagree concerning its epistemic importance and evidentiary force relative to other problems and aims on which they do not overlap.[3]

In sum, the moderate relativist can account for the existence of rational debate without abandoning his or her claims concerning the incommensurabilty of standards and justification.

5. THE EXTERNAL STANDARDS REBUTTAL[4]

The discussion brings us to the third rebuttal of relativism, which rests on the claim that there are after all, fairly neutral, external, and universal standards of theory-assessment implicit in the aims and conduct of all scientific inquiry: standards of confirmation, empirical success, predictive accuracy, breadth of explanatory scope or unification, simplicity, and/or problem-solving effectiveness. To be sure, different philosophers of science favor different standards depending on how each interprets the aims and attainments of scientific inquiry. I will sidestep these differences here, but return to them later. In any case, on this rebuttal of relativism, such external standards provide an unequivocal basis for determining which scientific theories are best justified, in just the way denied by moderate relativism.

As I construe it, moderate relativism does not deny *a priori* the possibility or existence of such external standards. Indeed, for the sake of argument, it will concede the existence of such external standards and aims, and pose the following dilemma for the externalist. On the one hand, an examination of the history of science reveals that proponents of rival theoretical frameworks often embrace incompatible interpretations of external standards, in effect reducing them to internal standards. If we can establish the relativity of external standards to internal ones, then the existence of external standards poses no threat to moderate relativism. On the other hand, for the sake of argument, suppose that philosophers of science do succeed in characterizing an external standard like empirical success, or problem-solving effectiveness, such that it is genuinely independent of rival theorists' internal standards and yet rigorous enough to secure comparative evaluations of rival theories. If there is no evidence that scientific groups (including contemporary practitioners) were or are actually committed to this standard, what justification is there for taking it to be "the" standard of scientific knowledge? I will elaborate these points.

Undoubtedly, the external standard with the best claim to universality within scientific inquiry is that of empirical success, or the relative ability of theories to "save the phenomena". But "which" phenomena are genuine and most important to save? What sort of theory, and what sort of reasoning or proof is required to genuinely "save" these phenomena? The history of science embodies fundamental disagreements between rival scientific groups concerning how this question should be formulated and answered. Examples of such disagreements are as follows:

1. First, concerning how the domain of phenomena proper to a given science is defined and bounded; for example, many of the observational phenomena thought essential for a chemical theory to explain on the standards of the premodern chemistry of neo-Aristotelians and alchemists are excluded from the domain and replaced in epistemic importance by other sorts of phenomena, on the standards of modern chemistry (Doppelt, 1978, pp. 43–45; Shapere, 1984, pp. 325–336).

2. Secondly, concerning the relative epistemic weight, evidential or probative power, or explanatory importance of one, as against another kind of phenomena, evidence, or empirical achievement; for example, phenomena deducible from a theory that are surprising, previously unknown, or different in kind from those that the theory is designed to explain have special or even unique evidential force in the standards of Herschel and Whewell, which they completely lack on the standards of Mill and others (Laudan, 1981, pp. 127–136).

3. Thirdly, concerning the form(s) of inference that must exist between observational evidence and hypothesis if the latter is to gain rational credibility from the former; for example, for the Newtonians a hypothesis is only knowable on the basis of evidence if it is a strict inductive generalization from that evidence, while on the method of hypothesis endorsed by the ether theorists of the day hypotheses may also be indirectly known on the basis of evidence that it implies or explains but does not inductively generalize (Laudan, 1981, pp. 111–127).

4. Fourthly, concerning what sorts of hypotheses and entities may gain scientific credibility from the evidence or observational phenomena they imply; for example, Newtonian as against the seventeenth- and eighteenth- century ether theorists concerning the legitimacy of unobservable entities (Laudan, 1981, pp. 325–340).

Such examples are characterized as cases of conflict or shift in basic standards of theory evaluation in Doppelt (1978, pp. 33–60), Laudan (1981, pp. 111–141, 163–181; 1984, pp. 55–62), and Shapere (1984, pp. 325–340). In these cases, scientific practitioners may be commonly committed to some "external" standard of maximal empirical success and yet employ different criteria concerning what counts as success, and which successes carry the most evidential weight in establishing claims to scientific knowledge. Consider Laudan's bold attempt to establish an external standard of problem-solving effectiveness and a theory-neutral calculus for identifying, counting, and weighing the various empirical and conceptual problems solved and unsolved in rival or successive scientific research traditions. An impressive and ingenious achievement, his argument nonetheless flounders on a debilitating internal contradiction. On the one hand, his account stresses that rival research traditions embody commitments to different problems, different ways of individuating, counting, and weighing important kinds or types of problems. These features of his view provide powerful support for the relativist thesis that rival scientific traditions embody commitments to different standards of problem-solving effectiveness.

On the other hand, while stressing the historical relativity of what counts as a problem and a solution, Laudan seeks to construct an ahistorical metric of problem-solving effectiveness which enables it to function as a neutral external standard of theory-evaluation. Two problems confound this part of his project. First, it is implausible to hold that there is any ahistorical, absolute way to individuate, count, and weigh more and less important problems, and stronger and weaker problem-solutions. For example, what may count as two types or natural kinds of empirical phenomena/problems in one theoretical tradition may count as one and the same type in a rival tradition, or as a pseudo-problem, a non-phenomenon, in it. To take another example, Laudan's ahistorical metric holds that the epistemic weight of a problem increases if rival theories tackle and solve it. But there are many counter-examples, as in the case of the new chemistry of Lavoisier and Dalton whose most important explanations addressed problems that were unrecognized, for the most part, in the alchemic tradition.

Secondly, for the sake of argument, suppose that we could provide a neutral metric and standard of problem-solving effectiveness external to particular scientific traditions, and capable of generating comparative assessments of their rival theories. Is there any good reason to regard this, or any such, external standard(s) as the standard(s) of scientific knowledge? What justifies a philosophically articulated standard of empirical success as the standard by which all scientific theories and claims to knowledge should be evaluated?

To the moderate relativist, the most convincing answer is that appearances to the contrary, all scientific practitioners are already committed to, or implicitly accept, the external standard. We may assume that there are at least three sorts of evidence relevant to establishing which standards scientific groups are actually committed to: (1) their choices of which theories to accept or reject; (2) their grounds and reasons for these choices; and (3) the standards to which they appeal or which they explicitly reject, in the course of scientific controversies. As I have already argued, as I read this evidence, it establishes the historical variability of standards of theory-assessment. It fails to provide any convincing evidence that all scientific practitioners have accepted one and the same determinate external standard of the sort Laudan's argument, or a confirmation theorist, requires to circumvent relativism.

This leaves open the possibility that an external standard could be justified on the basis of a philosophical analysis or theory of the very nature of empirical inquiry or scientific reasoning itself. Of course, this is the familiar path of traditional, *a priori* epistemology called into question by the historicist and naturalist turns of the last few decades. It is plausible to characterize empirical inquiry as the use of sense experience and reason to produce general and true laws of nature, causal explanations, accurate predictions, problem solutions, simple and unified accounts of disparate phenomena, etc. Nonetheless, we now recognize that among the wide variety of practices of scientific inquiry, there is significant variation in *which* of these aims are embraced and which rejected; and moreover, in how each of these aims is epistemically constituted by different scientific groups to regulate inquiry itself. It now seems doubtful that any *a priori* conceptual analysis of the nature of scientific truth, knowledge, explanation, reasoning, or proof will really justify any contentful external standards or neutral metric of theory-assessment.

In sum, the bare existence or possibility of external standards such as "accept the most empirically successful, or best confirmed, theory" does not provide a convincing rebuttal of moderate relativism. On the one hand, such external standards do not interpret themselves. What counts as empirical success or confirmation varies historically, effectively reducing external to internal standards. On the other hand, perhaps philosophers of science can articulate determinate external standards which are genuinely independent of internal ones; but it is doubtful that there is any way to justify them as the right and proper basis for all theory-assessment. According to moderate relativism, the foundation of justification is normative commitment, the historical fact that some group of scientists is invested in a particular network of standards, aims, and social relations of mutual credibility and trust.

6. THE 'JUSTIFICATION OF STANDARDS' REBUTTAL[5]

This brings us to the most current and promising strategy for the rebuttal of moderate relativism. As I have characterized it up to this point, the moderate relativist grounds the existence of scientific knowledge in the sociological and historical fact that a scientific group is normatively committed to, and invested in, specific standards of knowledge. But suppose that some standards are demonstrably more reasonable to accept than others, and either are embraced or could be for this reason. Against moderate relativism, we could then fully justify and explain changes in standards by appeal to reason(ing) alone. Against the relativization of knowledge to accepted standards, scientific knowledge could be characterized on the basis of the best justified standards. I will evaluate two types of current accounts of the justification of standards which aim to rebut relativism: gradualist, historicist models, and normative naturalist models. I begin with historicist models (Laudan, 1984; Shapere, 1984).

These models begin with a critique of Kuhn's holistic historiography of normal and revolutionary science. Normal paradigm-governed science is supposedly ordered by a wholesale commitment to a specific framework of concepts, problems, standards, methods, aims, etc., all of which inextricably depend on one another. Scientific revolutions are supposed to involve the simultaneous and wholesale abandonment of all the elements of one framework, and the embrace of an entirely different one. For critics, this Kuhnian model of total epistemological breaks sets the stage for incommensurability, irrationalism, and relativism.

They counter this model with a gradualist picture of scientific development which stresses the continuities in scientific language and belief that characterize even large-scale scientific change. Such elements of continuity provide scientific groups of the day with all the resources they require to justify change, including the embrace of new standards. The gradualist historical model of scientific change is thus taken by critics to spell the antidote to relativism.

The moderate relativist, who also stresses overlap, continuity, and rational debate between rival theoretical frameworks, will agree with the gradualist model, up to a point. But for the relativist, the key question is whether or not the elements of continuity and agreement central to the gradualist model provide the basis for establishing that certain standards of scientific knowledge are better justified than rival standards. Put more concretely, the relativist will want to determine whether or not an allegiance to each of the rival standards implies certain epistemic losses and gains different from those implied by the rival, but equally rooted in the elements of continuity, the shared corpus of scientific belief and aim in the context. For, if rival standards carry different epistemic losses and gains for the science of the day, there may be equally good reasons for adhering to an established standard as there are for embracing a new one.

In order to resolve these questions, we need to examine specific accounts by gradualists of the justification of standards. In my work, I have critically examined the accounts provided by Larry Laudan and Dudley Shapere (Doppelt, 1986; 1988). In this paper, I will briefly examine certain problems with Laudan's account, and argue that they reflect general weaknesses with the historicist or gradualist strategy for overcoming moderate relativism.

On Laudan's model in *Science and Values* (1984), scientific inquiry can be characterized as a structure of theoretical and empirical beliefs, cognitive aims, and methodological standards. Changes on any one of these three levels can be justified by elements of continuity and agreement at the other levels, even if we accept the Kuhnian view that there are no sacrosanct and permanent aims or standards in science with which to anchor justification.

In particular, scientific practitioners will have good reason to abandon one set of standards A and embrace another, B, if B proves to be more consistent with the most successful theories of the day than A, and enables scientists to realize cognitive aims that proponents of A have failed to realize. The change in standards is thus justified, first, because it enhances the internal consistency of theory, aim, and standard; and second, because it enables scientific practitioners to make progress in realizing aims and successful theories blocked by adherence to the old standards.

One of Laudan's central historical examples is a fairly well-known case – the decision of scientific practitioners in the 19th century to embrace the method of hypothesis (or, what we call hypothetico-deductive reasoning); and abandon an exclusive commitment to the Newtonian empiricist standard, which excluded the possibility of using observation to establish scientific knowledge of unobservables (Laudan, 1984, pp. 56–59; 1981, pp. 111–141). The empiricist standard was widely thought to be responsible for the great achievements of Newtonianism. But the only successful theories of well-known electrical, chemical, gravitational, and other sorts of observed phenomena in the 18th and 19th centuries posited the existence of an unobservable aether and thus violated the Newtonian inductive-empiricist standard. This situation provoked explicit scientific controversy over the legitimacy of hypothetico-deductive inference as a route to genuine knowledge, and thus over the epistemic status of the aether theories of chemistry, electricity, gravitation, etc., themselves.

On the gradualist model, there are elements of continuity and agreement in the historical context. These provide the basis for a rational change in standards to the method of hypothesis. Both the inductivists and the advocates of hypothetico-deductive reasoning could agree on the desirability and aim of gaining a knowledge of well-known electrical, chemical, gravitational effects. Both groups could see that the then current theories accomplishing this aim violated the inductivist standard and required the method of hypothesis. Moreover, many accepted these theories. For Laudan, these considerations provided good reasons for embracing the method of hypothesis and abandoning inductivism as the sole standard of genuine natural knowledge. As he puts it, "what forced the change was a growing recognition that the explicit axiology of empiricism was fundamentally at odds with the axiology implicit in scientists' theory preferences" (1984, p. 60). Thus, the achievement of consistency, as well as added theoretical success, provided good reasons to accept the method of hypothesis. By this route, the gradualist models of Laudan and others succeed in establishing the existence of reasons for changes in standards. On the other hand, as I will now argue, it does not follow that the gradualist model succeeds in a rebuttal of moderate relativism. Why not?

A closer examination of the above historical example, and other such examples, reveals a more complex epistemological situation and disagreement over standards. Scottish natural philosophers such as Thomas Reid steadfastly continued to defend Newtonian empiricism as the sole standard of genuine knowledge. Were they less consistent, rational, or scientific than those who embraced the method of hypothesis, such as Le Sage, Hartley, and Boscovitch? This is doubtful. First of all, the Scottish natural philosophers did not accept the various aether theories of electricity, chemistry, magnetism, etc., as genuine knowledge and thus preserved consistency between their theory preferences and their commitment to Newtonian inductivism, as against Laudan's account. Secondly, they could also embrace the aim of genuine knowledge of electrical, chemical, and magnetic effects, without sacrificing their consistency. Gaining such knowledge within the constraints of Newtonian inductivism remained a possibility. Indeed, for them it seemed the only possibility for genuine knowledge of these phenomena. Thus, consistency did not provide a compelling or good reason for all the scientific practitioners involved in this controversy to embrace the method of hypothesis.

Thirdly, the historical context provided the Scottish natural philosophers with a perfectly good reason for rejecting the method of hypothesis and recalcitrantly defending the inductivist standard. The method of hypothesis not only lent credibility to the new theories of electrical, magnetic, and chemical phenomena. It also lent legitimacy to Cartesian physics – which by the mid-18th century was widely perceived by scientific practitioners as a system of empty pseudo-knowledge. The method of hypothesis could not distinguish between what was regarded as the very paradigm of scientific knowledge – Newton's laws – and the very paradigm of non-science – the unsavory, *ad hoc*, empirically vacuous hypotheses of Cartesian physics, such as the vortex theories, which could always be arbitrarily manipulated at will to save the phenomena. For the Scottish philosophers, the inductive standard of Newtonian empiricism could demarcate Newtonian from Cartesian physics. Given that this aim mattered in the historical context, it provided a good reason to accept inductivism, and reject the method of hypothesis.

In sum, a closer look at the historical context reveals that there were good reasons on both sides of the methodological controversy. Put differently, given the historical context of continuity and agreement that the gradualist model emphasizes, commitments to old standards or new ones each entails epistemic losses and gains for the science of the day. There were typically several aims of scientific inquiry; for example, gaining knowledge of chemical, gravitational, and electrical phenomena while also maintaining a sharp epistemological demarcation between Newtonian and Cartesian physics, as in the historical context described above. A standard may serve one aim (the epistemic gain) and simultaneously frustrate another (the epistemic loss). Controversies and shifts between standards occur in an ocean of reasons. For the moderate relativist, this raises the question of how to explain *which* reasons, aims, and standards define the allegiance of one scientific group, and why other such reasons, aims, and standards prove more compelling to a rival group. We need to understand how such groups are constituted and reproduced, and the ways their social relations and practices embed an allegiance to certain standards, a certain conception of inquiry, within their identities as knowers of nature. This is the province of an historical sociology, though not one hostile to the role of reasoning and argument in the development of science.

Thus if my argument is correct, gradualist models like Laudan's may succeed in establishing that there are reasons for accepting or rejecting standards of theory-assessment, *without* thereby undermining moderate relativism's contention that such reasons alone do not fully explain or justify commitments to standards, and thus to the theories which embody them. But this is not an *a priori* claim, and requires the kind of argument about historical cases illustrated by my dispute with Laudan concerning the above case. On the other hand, we certainly know that historically, commitments to new conceptions of chemistry, earth science, and electricity entailed epistemic losses, empirical phenomena or problems that could not be explained on the new conception (Laudan, 1977, pp. 147–150).

This brings us to a consideration of normative naturalism, perhaps the most influential current strategy for rebutting relativism. Normative naturalism proposes that we evaluate methods and standards of scientific inquiry in terms of their demonstrable empirical efficiency in realizing the aims of scientific inquiry. The proposal is normative because it provides a way of evaluating standards of knowledge, reasons for accepting, rejecting, or revising historical standards of inquiry. It is "naturalistic" because it bases such evaluations on empirical knowledge of the consequences of employing certain standards, their effects, and thus their effectiveness in bringing about the avowed aims of scientific inquiry. If the naturalist can thus show that certain standards have replaced others because scientific practitioners have learned that there are more effective means to the aims of science, then moderate relativism can be circumvented. Such growing knowledge concerning the best means to advance the ends of science could both fully explain and justify historical shifts in standards, and indeed the current standards of science.

7. NORMATIVE NATURALISM: THE DISUNITY VERSION[6]

There are at least two versions of normative naturalism which I will discuss here. While both positions evaluate standards in terms of their effectiveness in realizing the aims of scientific inquiry, on the first, the aims of science are taken to change and exhibit disunity. On the second version, the fundamental aim(s) of scientific inquiry is taken to remain constant, and to exhibit the unity of scientific inquiry; for example, a view which takes the discovery of truths about lawlike patterns in nature to be the aim of all scientific inquiry. I intend my criticisms of normative naturalism to apply to both versions. But I begin by briefly comparing the strengths and weaknesses of the two positions, which we may characterize as the disunity view and the unity view.

The disunity view has the virtue that it seems to accord better with the history of science, and evident shifts in the avowed aims of scientific practitioners within certain areas of inquiry. For example, the pre-modern alchemists aimed to explain and control the sensible qualities of substances such as ore and gold and to bring imperfect Earth to more qualitative states. Modern chemists from Lavoisier to Dalton redefined the aims of chemistry to stress the explanation and control of changing weight relations and breakdown products of processes such as combustion, and the mixing of substances. While we could say that both of these traditions of chemistry aimed to unlock the secrets or truths of nature, this characterization would mask the fact that they put different questions to nature, and took different aspects of nature, different natural phenomena, as central to the aims of their science. The disunity view captures these dynamics.

On the other hand, the assumption of disunity generates a serious difficulty for normative naturalism. If the aims of scientific inquiry vary, then aim-satisfaction provides a shaky criterion for the naturalists' attempt to evaluate standards. A standard which is effective in realizing one aim may be much less effective, or ineffective in realizing another. The method of hypothesis seemed effective enough in guiding the construction of theories aimed at accounting for chemical, electrical, and magnetic effects. But it seemed less effective than the method of induction in satisfying the aim of demarcating the great Newtonian achievement from the bogus *ad hoc* pseudo-science of Cartesian vortices. Exponents of the disunity version of normative naturalism will need yet another presumably naturalist account of how to evaluate aims themselves. Of course, it is open to the naturalist to argue that scientific practitioners can gain empirical evidence and thus learn that certain aims of inquiry are unrealizable. But in the very nature of scientific inquiry, scientific practitioners typically enjoy some mixture of success and failure in realizing their aims, which is precisely what drives research programs. The judgment that any given scientific aim or research program is unrealizable and bankrupt requires an appeal to standards of epistemic evaluation, which lands us back where the naturalist started, evaluating standards by reference to aims!

8. NORMATIVE NATURALISM: THE UNITY VERSION[7]

The normative naturalist who embraces the unity-of-aim position avoids these difficulties. If there are certain fixed, universal aims of scientific inquiry, then aim-satisfaction can, at least in principle, provide a naturalist criterion for evaluating the relative efficacy of various standards and methods. As I have already suggested, the main difficulty of the unity view is accounting for the apparent variability of the aims evident in the history of science. Suppose, for the sake of argument, we concede that all scientific inquiry aims at discovering the truth about nature. This is a bit like characterizing the goal of all politics as peace and order, without indicating what sort of peace and order, on what or whose terms, at whose expense, etc. Are the truths at which scientific inquiry aims confined to the description of observable phenomena, as scientists in the instrumentalist tradition such as Ernst Mach hold? Or, do they emphasize truths concerning theoretical entities and underlying causal processes, as scientists in the realist tradition maintain? Are the truths of science necessarily explanatory, or predictive, or simple and unifying, as some but not all scientific practitioners have assumed? Is it sufficient for a true scientific theory that it account for and predict well-known phenomena, or must it predict and account for surprising, previously unknown phenomena different in kind from the phenomena that the theory is designed to explain? Should we characterize these differences as differences in aims or rather as different conceptions of one and the same aim – truth, or knowledge about nature?

Both alternatives create problems for the unity version of normative naturalism. Suppose the unity naturalist takes the position that scientists who seek explanatory truths seek two aims not one – truth, and explanation. Then the unity view reduces to the disunity view, with its attendant problems. Standards which are effective means to truth may not be effective means to explanation, and vice versa.

By this route, the unity version is lead to the alternative that the aim of truth (or knowledge) is the one common denomination of all scientific inquiry, irrespective of what scientific groups want to do with the truths they discover, the various uses of truth to explain, predict, or manipulate nature. The best justified standards are simply the ones that have proven to be the most effective means to scientific truth. The ancillary, variable aims or uses of scientific theories (to explain, predict, unify, etc.) simply do not count in the evaluation of standards because the attainment of true theories is the underlying universal aim of scientific inquiry. Following John Rawls, we can characterize truth as the first virtue of empirical claims, as justice is the first virtue of social institutions. On this unity-of-aim version of normative naturalism, standards can be evaluated and justified as more or less empirically reliable means to the attainment of true scientific theories.

The cogency of normative naturalism depends on reading standards in science as hypothetical imperatives. On this reading, standards are maxims or rules which prescribe a certain course of conduct for scientific inquiry, on the (hypothetical) condition that a certain cognitive aim – e.g., true theories – is desired. If standards do indeed function primarily as hypothetical imperatives in scientific inquiry, then it is reasonable to evaluate them on the basis of empirical knowledge which establishes the relevant lawlike connection between following the standard and attaining the end.

My most fundamental criticism of normative naturalism is that it misses an essential role and function of methodological standards in scientific inquiry. If I am right, these standards provide the criteria by which scientific groups identify and determine what counts as scientific knowledge and a true scientific theory, not simply, more or less effective strategies, or means to their attainment (Doppelt, 1990). Basic standards of theory-assessment bear a criteriological and not a merely contingent empirical relationship to the aim(s) of scientific inquiry. Thus within specific practices of scientific inquiry, standards function much more like categorical imperatives than hypothetical imperatives. For scientific practitioners committed to a standard, it unconditionally dictates what sort of relation to evidence, or what sorts of evidence, any genuinely true scientific theory and knowledge must possess. Truth and knowledge presuppose human beings' commitments to some such standards because we have no way of identifying truth or knowledge that is not dependent on evidence, and thus on criteria of evidence, and justification.

If I am right, and standards function for scientific practitioners as unconditional imperatives dictating what counts as genuine evidence for scientific truth, this undermines the normative naturalists' strategy of appealing to empirical evidence to evaluate standards of empirical evidence. Consider the controversy between scientific practitioners in the 19th century between those who accepted the method of hypothesis and those who held to the rule of predesignation (Laudan, 1981, pp. 121–136). For the method of hypothesis, any phenomenon implied or explained by a scientific theory counts as evidence of truth. On the rule of predesignation, only new and surprising phenomena implied or predicted by a theory count as evidence of its truth.

Can the normative naturalist find evidence to determine which of these standards is a more reliable and effective means to produce true theories? To be sure, we can determine which standard leads to the production of more theories which satisfy it, and thus *count* as true *for* scientific practitioners committed to that standard. But this judgment will not determine which standard produces more scientific truth, if I am right and standards function as categorical imperatives which determine what sort of evidence truth requires, and thus what counts as scientific truth. For those committed to the rule of predesignation, the success of the method of hypothesis in generating theories that satisfy it must nonetheless fail to be evidence of the truth of its theories, or its effectiveness as a producer of truth. In a nutshell, empirical evidence cannot adjudicate the effectiveness of different standards in attaining truth, when the standards themselves determine what counts as evidence and truth in science. For this very reason, Newtonian inductive empiricists such as Thomas Reid could recognize the success of the method of hypothesis in generating aether theories of magnetic, electrical, and chemical effects while firmly holding that scientific truth and knowledge could simply not be produced in this way.

It is instructive for my argument to acknowledge the existence of some scientific standards which do function as the normative naturalist requires. Consider current standards which prescribe the double-blind testing of drugs for clinical trials, or controlled experiments, or randomized sampling techniques. The standard of double-blind experiments is rooted in the empirical discovery that such a procedure generates a more reliable knowledge of the therapeutic effects of a drug than obtains in the absence of the procedure. That is, if we want the most reliable knowledge of the effects of a drug in general usage among the population as a whole (call this goal G), then we should follow the method M of double-blind experiments and utilize the evidence from such experiments in making inferences to the effects of the drug.

Note that in such cases the standard or method M provides an empirically established means to the knowledge-goal G and *not* the *sole* evidence or criterion for G. Quite apart from following M, there is an independent empirical check on the attainment of knowledge G, namely evidence of the effects of the drug in general usage, after the clinical trials. For this reason, in such cases the naturalist can gain empirical evidence concerning the effectiveness of M in bringing about G. But with standards of theory-assessment such as the ones central to my argument, there is no way of empirically checking to determine whether or not the goal – scientific knowledge or truth – has been attained, without employing the very standard of evidence itself. Without such an independent check on the attainment of the goal, there is no neutral empirical evidence of the sort the normative naturalist requires in order to determine which among rival standards is, as a matter of fact, most effective in attaining the goal of scientific knowledge or truth.

Indeed this explains the manifest lack of such evidence ironically noted by Laudan, himself a leading defender of normative naturalism, in the following remarks:

> It is generally difficult, *and in certain cases patently impossible*, to exhibit that a particular set of rules is the best possible way for realizing a certain set of values... And if we cannot demonstrate the latter, then we cannot argue for the blanket superiority of those rules over all their conceivable rivals for realizing the values in question ... with respect to such familiar cognitive goals as truth, coherence, simplicity, and predictive fertility, scholars have not managed to show that there is any set of rules of empirical investigation which uniquely conduce to their realization...
>
> For, if there were only one set of rules for realizing any specific set of cognitive aims, we should have to conclude that it was irrational for scientists sharing the same ends or values ever to disagree about the appropriate rules for implementing those values. Yet such disputes are chronic in the history of science and philosophy. Consider, for instance, the 150-year-long (and still ongoing) controversy about the so-called rule of predesignation... All parties to the controversy would, I believe, subscribe to substantially the same cognitive aims. They seek theories that are true, general, simple, and explanatory. Yet no one has been able to show whether the rule of predesignation is the best, *or even appropriate, means for reaching those ends. That failure is entirely typical...* So far as we know, there may be equally viable methods for achieving all the cognitive goals usually associated with science (Laudan, 1984, pp. 35–36, my italics).

To my mind, this failure Laudan describes is best explained by my contention that standards of theory-assessment such as the rule of predesignation are not hypothetical imperatives which can be assessed empirically. Otherwise, why is there no empirical evidence linking the rule to the attainment of scientific truth? In my view, it is because this rule functions as a criterion of scientific truth, on which the capacity of a theory to yield new surprising predictions or discoveries is the only genuine mark of its truth. Given that the method of hypothesis defines different criteria of scientific truth, it is hardly a surprise that the historical record of scientific achievement yields no empirical evidence showing which of these two standards is a more effective means to the aim of scientific truth or knowledge. Once the aim of scientific truth or knowledge has been abstracted from any determinate epistemological criterion or context, there is simply no way to know if and when it has been realized by these standards or any others.

9. INDIRECT REALIST NATURALISM

In a recent paper, Howard Sankey (2000) defends a realist version of normative naturalism which cuts against some of the criticisms I have provided of the naturalists' treatment of standards as empirically demonstrable hypothetical imperatives. He concedes, as I have argued "that there may be no direct empirical evidence that use of a method leads to theoretical truth" (Sankey, 2000, p. 223). On the other hand, this is not a fatal objection for him because there may be "indirect evidence" that the use of certain standards or methods is conducive to scientific truth. In particular, suppose that the use of certain standards is empirically successful in realizing the subsidiary aims of scientific practitioners (e.g., explanation, prediction, accuracy, etc.) and generating theories which meet these standards and aims. In that case, so Sankey argues, the *best explanation* of this empirical success is the truth or approximate truth of the theories, and thus the efficacy of the standards for producing theoretical truth (Sankey, 2000, pp. 224–225). Sankey's argument rests on abduction, what I have called the method of hypothetico-deductive reasoning, or 'inference to the best explanation'. He thus concludes that empirical success (in developing theories, which satisfy standards of theory-assessment) provides 'indirect evidence' that theoretical truth has been attained, and thus that the standards are in fact truth-conducive.

My first problem with this argument is a familiar one, noted by Sankey, of whether or not truth does indeed provide the best explanation of empirical success. For Sankey, it does because the only alternative explanation is sheer luck, which seems implausible (Sankey, 2000, p. 225). But there are surely other more plausible alternatives. I would suggest that empirical success can be explained as follows: through hard, ingenious, socially coordinated, relentlessly disciplined scientific labor, groups of practitioners are able to bend certain parts or aspects of nature to their concepts, aims, standards, etc., and ignore, exclude, or marginalize other parts or aspects of nature which otherwise would, will, or did provide the basis for their downfall, or 'refutation'. This seems to provide a very plausible explanation of the enormous empirical success of theoretical traditions such as 19th-century aetherial theories which we now regard as false. Given that no scientific theories known to us explain everything, or resolve all of their empirical problems, why shouldn't the same explanation of scientific success always be at least as plausible as the assumption of truth? Indeed the social explanation I have provided seems more plausible than Sankey's realist hypothesis, because it explains much more: namely, the empirical failures of scientific theories, as well as their empirical success!

My second, more basic problem with Sankey's strategy derives from my own argument in this paper. His argument seems to ignore the whole problem of rival methodological standards and the disagreements they imply over what constitutes the sort of empirical success required by scientific truth. To make his realist case, Sankey imagines a scientific theory which successfully realizes a wide variety of methodological standards and subsidiary aims (Sankey, 2000, pp. 224–225). I wonder if he is imagining one that succeeds by all known scientific standards and aims, though he does not say this. In any case, recall the original motivation of normative naturalism: to provide *comparative* evaluations of rival methodological standards, in order to determine *which* are most, and *which* least effective, in realizing the aim(s) of scientific inquiry. This motivation reflects the real world of science, including contemporary science, in which different scientific groups are committed to rival standards and subsidiary aims; and rival theories which are successful in satisfying *some* of these at the price of *violating others*, or satisfying them *to a much lesser extent than their rivals*. The naturalist is supposed to have a method for adjudicating between standards in this scenario. How does Sankey's idealized case provide a basis for adjudicating between conflicting standards, subsidiary aims, or theories, to determine their relative truth-conduciveness? Is his view silent on this key issue?

Perhaps his position is that in evaluating rival standards and subsidiary aims, the *most* empirically successful provide the *strongest* indirect evidence of their truth-conduciveness, and thus are the best justified, on the realist's naturalist criterion. If this is Sankey's position, then it doubly depends on standards of empirical evidence and success which some philosophers and scientific groups reasonably reject. First, consider those who reject abductive (or hypothetico-deductive) reasoning on the ground that it is too weak to generate scientific truth and knowledge. For them, Sankey's realist naturalism cannot work even if truth does provide the best explanation of successful standards, subsidiary aims, and theories. It cannot work simply because inference to the best explanation is too weak to establish the truth of the theory which provides the explanation, or 'saves the phenomena'. We know that false theories have, and explain, true consequences. Thus, Sankey's realist hypothesis would need to meet the other standards to which philosophers and scientists are committed. It is not clear that his realist hypothesis can meet these standards or meet them to a greater extent than rival accounts of empirical success.

Thirdly, Sankey's exclusive reliance on the standard of abduction to establish his realist theory is in tension with his requirement that true scientific theories should satisfy a wide variety of methodological standards/virtues. Why doesn't he apply his rigorous methodological requirement concerning true scientific theories to his own realist hypothesis? The fact that he does not generates the possibility that rival non-realist theories of empirical success will satisfy other standards (e.g. predictive accuracy, or greater explanatory scope embracing empirical failure, as well as success) that his realist theory fails to satisfy. Thus his realist theory falls prey to the very moderate relativism I have defended here.

Fourthly, consider scientific groups divided by normative commitments to rival standards and conflicting subsidiary aims. In my arguments in this paper, such groups are often committed to rival criteria of scientific truth. In this context, the 'empirical success' of one group A in realizing its standard(s) is given a realist explanation by Sankey. But A's success will not reasonably count as *genuine* empirical success, providing the sort of evidence required by theoretical truth, for scientists or philosophers who reject the sufficiency of A's standard(s) for theoretical truth. For them Sankey's whole argument is simply a non-starter! Because, on their standards, there is no *genuine* empirical success for A to explain; and truth could not possibly provide the explanation because for them, theoretical truth in science requires standards more rigorous than A's standard(s).

On the other hand, for scientists and philosophers who share standards, even the abductive standard Sankey appeals to, and obtain empirical success, Sankey's realist theory is unnecessary. For their standards and subsidiary aims *directly* tell them that their successful theories are true, without need of Sankey's meta-abduction to realism.

Thus, given the normative commitments of scientists and philosophers to rival standards of theoretical truth in science, Sankey's defense of a realist normative naturalism is not rationally compelling to some, and unnecessary to others, all of whom may nonetheless be realists and take theoretical truth to be the aim and achievement of science. Of course, as I have observed above, my critique may be at cross-purposes with Sankey's realism, if he wishes to restrict it to a purely ideal scientific theory which satisfies many or all the standards and subsidiary aims to which scientists and philosophers have ever been committed. One problem with this ideal picture is that some of the standards and aims to which scientists have been committed are logically incompatible with others. But the more basic problem, as I argue above, is simply that such an idealized defense of realism fails to fulfill the normative naturalists' central aim of adjudicating between the rival methodological commitments, aims, and theories which are the heart of scientific controversy and change.

10. CONCLUDING REMARKS

Finally, the normative naturalist may find my argument much too dependent on the historical standards internal to past scientific practices. Why can't the naturalist defend some external, independent, empirical perspective from which to evaluate the instrumental efficacy of standards? Can anyone really deny, with the benefit of hindsight and new scientific knowledge, that we have learned, for example, that Newtonian inductive empiricism, or the alchemists' standards of what needed to be explained, are simply much less effective than contemporary standards, or plainly counter-productive, for attaining the aim(s) of science? Don't we know that following certain standards is in fact incompatible with modern scientific knowledge?

This is an almost irresistible argument. But it is irresistible *for us* because we assume the triumphs of modern science and evaluate everything past from the embodied standpoint of current standards, aims, and theoretical attainments. Once scientific practitioners are committed to new standards and aims – including whole new domains of phenomena, evidence, and problems which are taken to both require and receive successful scientific explanation, theoretical representation, and/or technical control, then and only then is there sufficient reason for finding older aims and standards to be ineffective and deficient, like outdated or refuted theories themselves.

The moderate relativist can affirm the reasonableness or rationality of this standpoint, from which scientific knowledge, truth, and access to reality or nature flow, on one condition. The condition is that we recognize that the reasonableness of this standpoint is not grounded in reason or reasoning, but rather in socially and historically contingent investments in and allegiance to certain conceptions and practices of scientific knowledge – its standards and aims. Without these normative commitments, reasoning, justification, and knowledge dissolve – as I have argued throughout this paper. The fact that these normative commitments are rooted in cultural and historical process, not the very nature of reason or science itself, does not make them any the less reasonable or scientific. But it does point the way to a more interdisciplinary model of scientific knowledge in which we do not expect science, reason or epistemology to justify or explain themselves. For that we require distinctively social and historical understandings of the normative commitments that make scientific knowledge possible.

University of California, San Diego

NOTES

[1] For discussions of this position, see Doppelt (1980) and Siegel (1980; 1987, pp. 3–32 and 77–92).
[2] On this position, see Doppelt (1978;1980) and Siegel (1980; 1987, pp. 77–92).
[3] This interpretation of Kuhn is elaborated in Doppelt (1978, pp. 39–45, 49–52).
[4] See Doppelt (1978; 1980), Laudan (1977) and Siegel (1987, pp. 77–92).
[5] See Doppelt (1986; 1988), Laudan (1981; 1984) and Shapere (1981; 1984).
[6] See Doppelt (1990) and Laudan (1987).
[7] See Leplin (1990), Rosenberg (1990) and Sankey (2000).

REFERENCES

Doppelt, G. (1978). "Kuhn's Epistemological Relativism: An Interpretation and Defense." *Inquiry* **21**: 33–86; reprinted in J. Meiland and M. Krausz, eds., (1982), *Relativism: Cognitive and Moral*, pp. 113–146, Notre Dame: University of Notre Dame Press.

Doppelt, G. (1980). "A Reply to Siegel on Kuhnian Relativism." *Inquiry* **23**: 117–123.

Doppelt, G. (1983a). "Laudan's Pragmatic Alternative to Positivism and Historicism." *Inquiry* **24**: 253–271.

Doppelt, G. (1983b). "Relativism and Recent Pragmatic Conceptions of Scientific Rationality." In N. Rescher, ed., *Scientific Explanation and Understanding: Essays on Reasoning and Rationality in Science*, pp. 106–142, London: University of Pittsburgh and University Press of America.

Doppelt, G. (1986). "Relativism and the Reticulational Model of Scientific Rationality." *Synthese* **69**: 225–252.

Doppelt, G. (1988). "The Philosophical Requirements for an Adequate Conception of Scientific Rationality." *Philosophy of Science* **55**: 104–133.

Doppelt, G. (1990). "The Naturalist Conception of Methodological Standards." *Philosophy of Science* **57**: 1–19.

Laudan, L. (1977). *Progress and Its Problems*. Berkeley and Los Angeles: University of California Press.

Laudan, L. (1981). *Science and Hypothesis*. Dordrecht: Reidel.

Laudan, L. (1984). *Science and Values*. Berkeley and Los Angeles: University of California Press.

Laudan, L. (1987). "Progress and Rationality? The Prospects for Normative Naturalism." *American Philosophical Quarterly* **24**: 19–31.

Leplin, J. (1990). "Renormalizing Naturalism." *Philosophy of Science* **57**: 20–33.

Rosenberg, A. (1990). "Normative Naturalism and the Role of Philosophy." *Philosophy of Science* **57**: 34–43.

Sankey, H. (2000). "Methodological Pluralism, Normative Naturalism and the Realist Aim of Science." In R. Nola and H. Sankey, eds., *After Popper, Kuhn and Feyerabend: Recent Issues in Theories of Scientific Method*, pp. 211–229, Dordrecht: Kluwer.

Shapere, D. (1982). "The Concept of Observation in Science and Philosophy." *Philosophy of Science* **49**: 485–525.

Shapere, D. (1984). *Reason and the Search for Knowledge. Boston Studies in the Philosophy of Science, Volume 78*. Dordrecht: Reidel.

Siegel, H. (1980). "Epistemological Relativism In Its Latest Form." *Inquiry* **23**: 107–117.

Siegel, H. (1987). *Relativism Refuted*. Dordrecht: Reidel.

DUDLEY SHAPERE

REASONS, RADICAL CHANGE AND INCOMMENSURABILITY IN SCIENCE

Abstract. A view is presented according to which scientific change, including radical change in the most fundamental scientific conceptions, takes place for reasons. In addition to changes in meanings and substantive claims, such change also often involves alterations in standards, goals, and methods of science. Some of the strengths of this view, as contrasted with some major alternative interpretations of science, are sketched. In particular, the view that some ideas, in some theories or traditions, are "incommensurable" with ideas in at least some other scientific theories or traditions is analyzed critically and reinterpreted.

1. THE PROBLEM OF WHAT COUNTS AS A REASON IN SCIENCE

In this paper, I want to outline a view according to which science proceeds by presenting reasons for the research in which it engages and the conclusions which it draws therefrom, and to ask whether calling science an enterprise based on reasons is justified. As I will try to show, this examination gets little help from the great classical tradition in the philosophy of science from Aristotle through logical empiricism, which did focus on reasoning in science, but did so by looking for rules of procedure which were independent of the science to which those rules were applied. It also gets little help from what I will call the "early postclassical" approach to understanding science, represented by the work of Thomas Kuhn and others.

After presenting some of the central aspects of my view, I will analyze what I take to be some of the most fundamental objections to the classical and early postclassical ways of conceiving the scientific enterprise, and will show how the view presented here can be applied to a resolution of those objections. In doing so, I will give special attention to the nest of problems surrounding the topic of incommensurability, since those problems, and the way I will deal with them, will illuminate both, especially when placed in the larger context of developments discussed in the paper. I will end with some remarks on the significance of the new analysis of what counts as a reason in science, enumerating the chief respects in which my view resembles and differs from both the classical and early postclassical attempts to interpret the scientific enterprise.

What constitutes a reason in science can be seen most clearly and explicitly in sophisticated modern research, where scientists have learned how to lay out the reasoning relevant in their research clearly and explicitly. In a number of ways, the solar neutrino experiment serves as a particularly clear example of this sophistication[1]. There we see a host of separate precise considerations brought together and applied in three distinguishable ways: in the conception of the possibility of the experiment, in its

P. Hoyningen-Huene and H. Sankey (eds.). Incommensurability and Related Matters, 181–206.

execution, and in the interpretation of its results. In this paper I will be able to examine only the first two of these (for the third, see Shapere, 1982).

The solar neutrino experiment was conceived in the context of trying to test a theory of how the sun and similar stars produce energy – that is, how they shine. Previous attempts to understand this process had been utter failures, the postulated sources producing energy in completely inadequate amounts[2]. According to the *theory of the source of stellar energy* developed by Bethe (1939) and von Weizsäcker (1938), nuclear reactions could supply sufficient energy to account for the measured quantity of energy released by the sun; this was *the theory to be tested.* Prior knowledge of the mass and composition of the sun made it possible (on the basis of the *theory of stellar structure*) to calculate the temperature prevailing in its central core; and, according to the recently-developed *theory of nuclear reactions* (today subsumed under the theory of the strong interactions), that temperature is high enough to initiate and sustain the reactions in question. Furthermore, Bethe and von Weizsäcker were able to lay out in detail the specific nuclear reactions that would occur in the solar core, and to show that the energy released by those reactions would be sufficient to approximate closely the easily-measurable amount of energy actually released by the sun.

Was it possible to test the Bethe-von Weizsäcker theory by experiment or observation? The surface of the sun could be observed and inferences made about what goes on in its deep interior; but whatever information was carried by such radiation would be highly degraded in its 100,000-year passage from its production-point in the core to the solar surface where it would finally be released into space. Could a more direct test be performed? In the 1950's, a theoretical possibility of such a test was suggested by the *theory of weak interactions*, which had been developed over the decades since the 1930's. The equations developed by Bethe and von Weizsäcker showed that neutrinos, weakly-interacting particles, would be produced in the nuclear reactions occurring in the solar core according to their theory. Because they interact so extremely weakly with other types of matter, these neutrinos could be expected to pass through the entire body of the sun without undergoing any reactions; information carried by them would thus be unaltered in their passage and in their subsequent trip through largely empty space. Therefore, unlike the electromagnetic information received from the surface of the sun, such neutrinos received at the earth would give a direct way of determining if the postulated nuclear reactions do occur.[3]

All the items I have mentioned so far enter into the conception of the experimental test as possible; but was the experiment practically feasible? Could it actually be carried out, "executed"? Again, a widely diverse background of previous beliefs was brought to bear to show that it could; here I can mention only a few of the most striking such items, to illustrate their diversity. It was proposed that a certain specific reaction actually existed which could make the experiment feasible. Again, the theory of nuclear reactions was the basis of the reasoning, this time supplemented by a host of experimental results having to do with reactions whose relevant characteristics could not be deduced from theory, or at least from theory alone. The reaction in question consisted of the capture of a neutrino by a chlorine atom to produce a radioactive isotope of argon, whose decay would be detectable. Its half-life of about 35 days made things convenient for experimenters, since the argon could be collected every 35 days and monitored for the number of decays, and that number compared with the number predicted on the basis of the Bethe-von Weizsäcker theory. Argon, an inert gas, was also convenient because it would not be trapped in the tank by (for example) combining chemically with other atoms, and a highly efficient method was available for separating the 500 or so argon atoms from the enormous tankful of ordinary cleaning fluid (perchloroethylene, C_2Cl_4) that, if the theory were correct, could be expected to be captured in each counting cycle, and a reliable method of detecting their decay was available. The amount of chlorine needed to produce the required number of captures of the very weakly interacting neutrinos could be calculated, and the equivalent of seven railroad tank cars of perchloroethylene would serve as a suitable target in the light of those calculations. Also, possible interferences – processes potentially imitating the capture-and-decay processes relevant – could be listed and dealt with: the experiment would have to be about a mile underground, to protect it from certain cosmic rays which, through known processes, might mimic the relevant chlorine-capture processes, and safeguards instituted against similar mimicking by radioactive decays in the cave walls (those processes being also understood and the safeguards feasible).

2. THE ROLES OF BACKGROUND INFORMATION IN SCIENCE

Such use of a variety of background beliefs in a variety of ways to provide what I have elsewhere called "background information" is typical of a broad range of cases throughout sophisticated science – not only in the physical sciences, but also in biological, geological, neurophysiological, and other sciences, not only for experimental research, but also for theoretical, and also for technological, including medical, applications. Indeed, it is typical, with minor variations, of *most* (and likely all) research in *most* (or all) areas of science. It is typical also of research in earlier science, though there the pattern illustrated by the solar neutrino experiment is not exhibited so clearly and unambiguously; simply because it is early, it is likely to depend on background beliefs which are neither well-established nor in any clear sense "scientific." In all such cases, in whatever field and at whatever epoch, without a great deal of prior theoretical information, the experiment, or the research in general, would have been, in the most literal sense, inconceivable; without a considerable body of background information (sometimes including portions, like weak interaction theory, which are also used in its conception and interpretation), it would not have been executable; and, without some portions of background information, not necessarily the same ones used in the previously-mentioned contexts, it and its results would be uninterpretable. (In this paper I omit discussion of this latter point.)

Certain features of such cases are especially relevant to the incommensurability problem. Note first that this body of background beliefs, in terms of which the experiment is conceived, executed, and interpreted, does not constitute a unity: there is no single background "theory" which governs the research. The background brought to bear is much more of an assortment in which many of the items used are entirely independent of each other – that is, they are not all deducible from some single theory, viewpoint, or set of exemplary methods of approach. Many of them cannot, in any natural sense, even be called "theories." Nor do the same background ideas play roles in every instance of scientific research: in general, the items that are relevant will vary with the needs of the particular research being conducted. The considerations are brought in, so to speak, "piecemeal," as needed for construction of the research. It is true that within particular disciplines, such as stellar astrophysics, high-energy physics, or molecular biology, there will be high-level theories which shape the directions of specific research; the Bethe-von Weizsäcker theory of stellar energy production, for example, in its extended and improved forms, still guides much research not only on that topic, but also on the evolution of stars from birth to death, the abundances of the chemical elements, and much else; similarly, the gauge-theory model of a theory of elementary particles and forces is a major factor guiding research in its field, despite its having been generalized in various ways; similarly, molecular biology techniques guide much research in evolutionary and developmental biology. I will discuss the roles of such high-level theories toward the end of this paper; for now, note that they are not the only source of background reasons relevant in particular research-situations. And apart from them, much of the reasoning is, as I have remarked, "piecemeal" in character, with independent items of background belief being employed as needed.

Finally, it also is not an arbitrary matter which background beliefs are brought to bear. Though there is often room for controversy about the relevance of certain elements of alleged background, it is by no means the case that, as Feyerabend (1975) insists, "anything goes." On the contrary, in the case of a great deal of the background, there is little if any alternative: reasons have to be given in order to cover the specific problems arising in the specific research (think again of why the neutrino detector has to be located underground). It is one business of the philosopher to explain why the reasons given are the ones that are relevant to the research itself, and why other considerations are extraneous and irrelevant.

A further point to be noted is that, in a clear sense, the solar neutrino experiment – its aim, methods, interpretation, and so on – is describable *completely* in the kinds of terms I have mentioned. It *might* be that the selection of these sorts of considerations depend on higher-level assumptions, for instance on a Kuhnian global paradigm or more specific exemplary approaches, or on metaphysical presuppositions, or on metascientific methodological procedures, or on the prevalent cultural, political, economic, or social milieu. But even if it does, what would need to be accounted for by such a claim would be precisely, and in detail, the considerations I have described. It is in this sense that the kind of description I have given is describable completely in terms of the sorts of background beliefs I discussed: whether or not there are *global* determinants which shape not only a particular piece of research but also a whole "tradition" of research, there is what I will call a *local completeness* to the *scientific* reasoning involved in a particular piece of research like the solar neutrino experiment: the immediate scientific motivations, procedures, and interpretations are specifiable precisely and completely in terms of specific items of background information. And any view which maintains that some more global determinant shapes or determines the way the research is conducted must show precisely how *each* item of background reasoning is shaped by that global determinant. This is a stringent requirement, but we should not be satisfied with anything less. No handwaving is permitted here.

A final lesson learned from the solar neutrino example is that the background beliefs shaping the conception, execution, and interpretation of the solar neutrino experiment play the general role of *reasons* for conceiving the experiment, for doing it the way it is done, and for interpreting its results. Detailed examination of many cases in the way I have here looked at the solar neutrino experiment shows that there is a large pool of such background beliefs, from which some items can be selected to function as needed, as reasons in a particular piece of research. But how did the items of belief come to be admitted to the pool in the first place, while others are excluded? And is there any justification for calling them "reasons" when they are applied in research?

3. REASONS IN SCIENTIFIC CHANGE

I will discuss two of the most important ingredients in the scientific conception of reasons which is illustrated in the solar neutrino experiment. The first is the wide adoption, roughly in the sixteenth through eighteenth centuries, of what I will call a *piecemeal approach to the study of nature*.[4] This approach is typified by the work of Galileo on motion, by Newton's *Opticks*, and by the bulk of experimental research conducted on the *Opticks* model on specific bodies or domains of information on heat, electricity, gases, and chemical composition in the eighteenth century. In this piecemeal approach, such specific bodies of information were studied and explained in isolation from one another. Explanations tended to be based on ideas of intelligibility taken from common sense, philosophy, religion, or some other source which would later be far more critically viewed.

The possibility that such "domains" could be studied and explained independently, in isolation from any influences of other domains was of course an assumption: as holistic philosophies have argued throughout history, the connections between things might be too tight and too significant for anything useful to be learned by considering them independently of all else. But it was an assumption of possibility that, in the case of the domains listed, turned out to be realizable: success in explaining domains independently studied and explained – *Domain Success*, as I call it – was actually achieved in a large number of cases, among them those domains I just mentioned. Though it could have failed, the piecemeal approach proved, as a matter of contingent fact, to be able to do what it promised to do, in a number of areas. This higher-level success, of fulfilling its promise with regard to a number of independently-studied domains, I call the *Goal Success* of the piecemeal approach, based on its achievement of many individual domain successes. This general Goal Success gave the piecemeal approach itself, because of its individual domain successes, a higher justification: success in achieving its goal led to the piecemeal approach being considered as a justified approach to inquiry, and results obtained by that approach being considered as reasons for investigating a new idea, for informing us how we are to go about investigating it, and for determining whether or not the idea is acceptable.

The second ingredient in the scientific conception of what count as reasons in science is as follows. The piecemeal approach was gradually extended in various ways, going well beyond its original form, ultimately transforming it in profound ways. Among these extensions was a further approach which came to have particular importance by the 19th and 20th centuries. In it, beliefs are selected for further employment as background information in the light not only of their *success in explaining a particular subject-matter*, but also of their *coherence with other beliefs* that had already come to be selected. One type of such coherence consists in unification of the explanatory accounts given of separate domains. Well-known examples of such unifications, differing from one another in certain respects, include: terrestrial and celestial physics (Newton); electricity and magnetism (Faraday) and the incorporation of light into the resulting electromagnetic theory of light (Maxwell); the unified treatment of matter (chemistry) and light in quantum mechanics; the (partial) unification of the electromagnetic and weak interactions under the electroweak theory; the idea of a Standard Model of elementary particles and forces based on both the electroweak theory and the theory of the strong force, quantum electrodynamics, being gauge theories; and the unification of Big-Bang cosmology and the particle physics of the Standard Model and beyond. (For details of these, see Shapere (1991).) Again, whether such unification or coherence could be achieved was a contingent matter; but Goal Success was once again achieved, and such syntheses provided a set of *reasons* for elevating coherence to the status of an aim of science, and for counting the coherence, or lack thereof, as a reason for accepting, or rejecting, a new idea. Whatever psychological motivations individual scientists may have had to seek unity, there was now *reason* to consider unification to be a goal of science: namely, it had proved possible to *get* unifications – *to fulfill what the attempt promised to do*. The results of unification could therefore be added to the attempts to explain isolated domains and incorporated into the pool of background information guiding further research, at least until specific doubt should arise leading to question their qualification for playing this guiding role.

The principle of Goal Success says that *any approach to inquiry must get its credentials as worthy of further adoption by succeeding in doing what it promises to do as its goal*. It can thus be said to be a "criterion" for selecting among competing approaches to inquiry. But it is not a "criterion" in the sense that judgments made in its terms are made "from a point of view," that is, in terms of assertive presuppositions which the approach being judged might deny. In other words, the principle of Goal Success makes no assertive content, no substantive claims. It relies for its force *only* on the claims of a specific approach to which it is applied, and what that approach promises to do, and on how well it fulfills that promise. The criterion of Goal Success does not, therefore, reject an approach or its results by the question-begging procedure of assuming what that approach denies.

Also, no vicious infinite regress, and no vicious circle, is involved in asserting that what is counted as a *reason* in science is *justified as reasonable*: the process of deciding what to count as a reason is not made by an *a priori* decision as to what is to count so – a decision that might be arbitrary or else require a higher-level decision which would be equally arbitrary. Rather, it is a matter of *building up* what to count as a reason: the process is "bottom-up" rather than "top-down," based on critical experience and not on armchair intuition. What this means is the following. In the history of inquiry, it has turned out, as a matter of contingent fact, that our best hope for gaining knowledge and understanding of the world we experience is by interacting with the specific areas with which inquiry is concerned. This hope is justified by the numerous successes achieved through the piecemeal approach and its extension to the program of unification; that extended piecemeal approach is itself justified because it has been able to do what it promised to do[5]. Such Goal Success has not been achieved by the traditional rival theories of how to gain knowledge, such as rationalism, holism, mysticism, and astrology.

There are other ingredients besides the two I have discussed which serve as "criteria" of what is to count as a good reason, and whose Goal Success entitles them to serve as background specifying aims which science ought (normatively) to try to achieve. Among the more important of these are the following: that the growth of the principle that science must deal with specific details, in a complete and precise way; that the way to achieve aims of precision in description of domains and in explanations of domains is, in many cases at least, through the use of mathematics; and the usability and usefulness of idealizations and similar "conceptual devices," subject in their use to certain constraints.

4. COMPARISON OF THIS VIEW WITH OTHERS

4.1. Classical Approaches to the Interpretation of the Knowledge-Seeking Enterprise

The meaning and significance of the view I have sketched can be seen clearly when we compare it with other views of the scientific enterprise. Two major alternative types of accounts have been given about the nature of the scientific process and its results and applications[6]. The oldest, which I call the *classical view of the knowledge-seeking enterprise,* consisted of two sub-versions. The first, the *traditional version of the classical view* began with the awareness, with Robert Grosseteste and other writers in the fourteenth and fifteenth centuries (Crombie, 1962; Randall, 1961), that Aristotle had failed to specify how the rational soul manages to extract essences from sense-perception without error. This led these writers to seek explicit rules by which essences could be abstracted, and this in turn became the search for explicit rules of scientific method, in particular, ones which would permit valid induction to generalizations, causes, hypotheses, theories, predictions, and other scientific conclusions. This traditional version is generally taken to be exemplified by such writers as Bacon (1939) and J.S. Mill (1874). The most widely-advertized difficulties of this approach, beginning as it does with the "given" data of sense-perception and proceeding to the desired conclusions, are common knowledge, and we only need to recall two of them here. The "given" in experience, which was alleged to provide the first step in the process of scientific reasoning, was never specified; and the rules (usually but not always conceived of as "inductive") were never formulated in ways which could do justice to the reasoning that takes place in actual science. The second – and more modern – major type of classical approach to interpreting the scientific enterprise consisted of the so-called "hypothetico-deductive" versions, in which a hypothesis was supposed to be advanced as a first step in the logic of inquiry, followed by deduction of observable consequences therefrom, and then testing to see whether those deductive predictions were borne out by actual observation. Again, numerous objections have been raised against such views. Here I will focus on one common objection, that, in general, such views alleged that no reasoning whatever is involved in the formulation and presentation of a hypothesis, even though it is not difficult to find massive numbers of cases in historical and contemporary science in which such reasons can be laid out. Indeed, the solar neutrino experiment as I presented it earlier is one such case.

But though such objections are widely accepted as fatal to the two classical approaches, they show only that the programs of those approaches were wrong or misguided. They do not demonstrate that the classical approaches could not be carried out with more careful effort. They do not, in other words, serve as reasons to suppose that the classical approaches were wrong in some fundamental way, *in principle* rather than simply in not following out their own doctrines with sufficient care and thoroughness. There are deeper objections that do in fact show that the program, as conceived, *cannot* be brought to fruition, because it is wrong at a very fundamental level. These have to do with the ubiquitous employment of background information in every area of science – in research, interpretation, and application. Here I will illustrate this kind of objection in detail through one example (Popper), instructive for present purposes because the failure is so blatant and extreme that it is impossible to miss the positive direction in which we must move in our attempt to gain a better understanding of science. After discussing this example, I will make some remarks about the hypothetico-deductive approach more generally, before going on to discuss critiques of the entire classical approach to the attempt to understand science, particularly Kuhn's.

Popper (1959) advanced a version of hypothetico-deductivism in which a hypothesis is tested by seeing whether its predictions are falsified. A hypothesis H is to be tested in terms of observational consequences O which it implies: "If H then O." Then, if O is *not* borne out by observation – if "not O" is the case – hypothesis H is falsified, refuted. The reasoning involved is logically valid: by the rule of *modus tollens* in deductive logic, denial of the consequent of an if-then proposition – e.g., a hypothesis – logically implies denial of the antecedent. *But this negative outcome of the test, the falsification of H, provides the only valid type of reasoning available to science.* For in the opposite case, where O *is* borne out, Popper declares that no valid conclusion about H can be drawn. On the one hand, we cannot say that O *proves* H, because that would be to commit the "fallacy of affirming the consequent", "if H then O, but O, therefore H," a logically invalid argument. Nor, according to further arguments which Popper offers, can one make the weaker claim that O at least provides some degree of *positive evidence* (though not proof) in favor of H. For, according to him, no "logic," no set of formal rules of reasoning, exists or is even possible which could validate claims that a given body of observation "confirms" a hypothesis. Thus there can be, according to him, no such thing as evidence supporting a hypothesis, only evidence refuting it. Falsification exhausts the reasoning available to science.

However, this view faced a potentially fatal objection. As Duhem (1954) had pointed out, hypotheses cannot be tested in isolation from a large body of "auxiliary hypotheses": the observational predictions follow only from the conjunction of the totality of hypotheses, not the one singled out by Popper for test. And when this is recognized, the falsification of the single Popperian hypothesis does not follow logically from the failure of the observational prediction; only the conjunction of all the involved hypotheses is falsified, and this is equivalent to saying that, as far as logic is concerned, any one or more of the conjoined hypotheses can be legitimately rejected.

Popper's response was to introduce the notion of *background knowledge* (Popper, 1962; 1972). To my knowledge, Popper does not offer a systematic and thorough discussion of background knowledge and its various functions, but only gives a few specific examples of such functions, one at a time. The most important one, for our purposes, is his view that background knowledge can be brought to bear in any testing situation, and would (or presumably at least could) on some occasions make it possible to decide which of the propositions in the conjunction should be rejected if the predicted outcome was not observed. Thus stated, his claim is intuitively appealing, and seems to be an anticipation of the notion of background information put forth in the present paper. For his readers naturally suppose that (as in the case of my background information) the background knowledge to be brought to bear is already well-grounded, and they also think that it may be sufficient, at least in some problem-situations, to pick one hypothesis from among all the conjuncts as being the one to reject in the light of a negative observational outcome.[7]

Unfortunately, this intuitive interpretation cannot be what Popper has in mind. For despite his admission that "... we constantly add to our background knowledge" (1962, p. 239), he repeatedly insists that we accept any piece of background knowledge "only temporarily," "for the time being, and for the discussion of this particular problem" (1962, p. 238). Indeed, given his view that we can never, logically, accept any proposition in science by virtue of its confirmation and acceptance on the basis of positive evidence in the past, he could hardly say otherwise. Hence

> ... the falsificationist or fallibilist ... does not *accept* this background knowledge, neither as established nor as fairly certain, nor yet as probable. He knows that even its tentative acceptance is risky, and stresses that every bit of it is open to criticism, even though only in a piecemeal way. (Popper, 1962, p. 238).

Background knowledge for Popper is thus always suspect, not accepted, used only temporarily and tentatively, without even the possibility of any positive support, in the context of a particular discussion, presumably for the sake of argument. Indeed, he also remarks that "admittedly the 'background' of the problem will contain theories and myths" (Popper, 1972, p. 181), and he also equates background "knowledge" with "the common-sense background which must be criticized." (Popper, 1972, p. 33) Such remarks lead to the suspicion that *any* piece of background belief can be used as background "knowledge"; no belief used as background knowledge is any better than any other belief whatever. Indeed, this is what Popper *must* say, given that no background belief can, on his view, have any better credentials than any other.[8]

It thus becomes impossible to see why anyone (except perhaps for an extreme Feyerabendian, a parallel which Popper would not have appreciated) would use such "background knowledge," seriously rely on it, or even call it "knowledge." Further, it becomes impossible to see why the *same* background knowledge is used over and over again in different problem-situations, why it is accepted in a more lasting and secure sense than the immediately temporary and tentative. Clearly, in Popper's most consistent eyes, to do so should be a foolhardy thing to do. Indeed, he *cannot* admit this aspect of background knowledge, since that would contradict his central thesis that there is no such thing as positive support. He is thus deprived of any way of accounting for the patent fact that the background is accepted in a more lasting sense than the immediately temporary. He cannot have any conception of science as building on beliefs that we already have strong reason to accept – that background knowledge is used in the construction of further knowledge, as, in short, a rationale, or at least a guide, of discovery. These features of science are illustrated in the solar neutrino example. There we saw a negative background consisting of the failures of earlier theories to account for the energy produced by the sun (theories which themselves were constructed on the basis of the newly-established thermodynamics and other items, *e.g.*, regarding gravitation, meteors, chemistry, and radioactivity). And we saw positive guides to the conception, execution, and interpretation of research: for example, the theory of nuclear reactions, which itself solved pre-existing problems, and that of radioactive beta decay (subsequently understood as a type of weak interaction), which then themselves served as background for the conception of a theory of the source of stellar energy, and for an experiment which would directly test that theory. All the items of background information utilized also appear in other problem-situations to which they are relevant. Not just any belief is qualified to serve as background information; we have learned to use those which have proved successful in describing and explaining domains (Domain Success) and in achieving coherence between domains and their explanations (Coherence). And those *kinds* of success, as opposed to the individual instances of them, in turn gain their epistemic credentials, as background *information*, from the fact that both piecemeal study of domains and coherent accounts of domains and their explanations have proved capable of doing what they promised to do (Goal Success), despite the fact that their success is only contingent, not guaranteed in advance.

In his futile struggle to avoid the fundamental inconsistency of his attempt to couple falsificationism and the idea of background information, Popper has thus presented us with a hopelessly flawed account of the knowledge-seeking enterprise. Yet other versions of classical empiricist philosophy of science have done no better. In their well-known critique of Mill's methods of experimental inquiry, Cohen and Nagel (1934) apply the hypothetico-deductive approach to show that those methods are unworkable without antecedent hypotheses; but they offer no account of the bases of the needed hypotheses. Though they sometimes refer to the background required for application of the methods as "facts," these are very simple pieces of common sense, beliefs attained from simple sense-perception; they are not the kinds of sophisticated background applied in the solar neutrino experiment. Furthermore – just like Popper – they sometimes refer to them as "assumptions" rather than as "facts." This is far from being a mere figure of speech with no great significance. For a central doctrine of the hypothetico-deductive approach is that "There is no logic of discovery," that is, there are no reasons or reasoning involved in the discovery and presentation of a hypothesis; "logic," reasoning, has a place in science only in the testing of hypotheses, whose source may be psychological, cultural, historical, or any sorts of factors other than reasoning. Yet this viewpoint not only fails to do justice to what happens when, for example, we conceive and construct, execute, and interpret a sophisticated scientific experiment like the solar neutrino one described above; it also contributes (along with other misinterpretations of the knowledge-seeking enterprise) to the antiscientism embodied in movements which consider its ideas as simply "myths," with no stronger basis than any other belief. As traditional classical empiricism failed to recognize any role for background information in science, the modern hypothetico-deductive version of empiricism, which did recognize that hypotheses are necessary as antecedent background for scientific investigation, botched the job miserably. In the first place, it provided meager understanding of a few simple roles played by antecedent hypotheses, focussing primarily on the antecedent hypothesis' specification of what is relevant to the problem at hand. Further, and worse, it eschewed, as a matter of basic and explicit principle (that there is no logic of discovery), any view that those hypotheses are developed rationally, and therefore gave no analysis of the *process* of science which showed how that process is a rational one.

This objection is closely related to another one. As we saw, the classical approach to understanding the scientific enterprise came about as a search for explicit rules of scientific method, stemming from Aristotle's claim that we can just intuitively grasp, through sense-perception, memory, and comparison, the essences of things. The search for explicit methodological rules continued to be central to the entire classical approach, inductivist or hypothetico-deductivist. And that approach came to embody the doctrine that the explicit rules of scientific method (discovered, at least according to earlier classical approaches, by someone like Galileo) are to be applied in science, but are never subject to change in the light of the science which applies them. They were, that is, to be "metascientific" rules, rather than scientific; absolute and unchanging, rather than contingent. By the twentieth century, the metascientific rules were required to be formulated largely in terms of formal logic (deductive at least, inductive also if such a logic could be constructed): metascientific concepts were to be defined, wherever possible, in logical terms.[9] Logic was conceived as dealing with "all possibilities"; and therefore to understand science – what any possible science would be, what the "meaning" of the term 'science' was – we must provide our analysis in logical terms. Again, there was a deep mistake. I will return later to the errors concerned with the concept of "meaning"; for the present, note that, if logic deals with "all possibilities," then it does too much to understand the scientific enterprise, its methods and its substantive concepts and their implications. Science concerns *this* world, *this* universe, not all possible ones: to illustrate the point by what is a pure metaphor, it is concerned with *this* particular line of the universal truth table; and so, a true understanding of science must focus centrally on what science does in trying to understand this world, and to what extent it is able to do so, and why.[10]

In the flush of enthusiasm among the scientific revolutionaries of the seventeenth century, there were, initially, a number of rival interpretations of what science was to be. Descartes (1983) believed that the basic laws of nature were deducible by pure reason alone, from the nature of God. Newton seems to have held that, through application of "rules of reasoning in philosophy" (Newton, 1964, p. 324) laws of nature could be deduced from experiment and experience. But to those who witnessed the great successes (or apparent successes) of the piecemeal approach in such domains as motion, gases, specific types of matter, heat, electricity, and others, the interpretation of science in terms of the central doctrine of traditional classical empiricism, inductivism, seemed an increasingly obvious view to take of science. Galileo's (1914) abstraction and idealization of the phenomena of motion in order to study those phenomena in isolation, and Newton's *Opticks* (1952) – despite its deductivist tone – served as models of this interpretation, strongly suggesting that scientific conclusions are to be drawn from sense-experience by a process of "inductive reasoning."[11] But as science developed further, especially in the nineteenth century, this view became more and more difficult to sustain. Increasingly, the frequency and depth of the grand unifications achieved, appealing more and more to explanatory concepts which, *prima facie* at least, seemed to appeal to the unobservable, and the increasingly evident way in which hypotheses seemed necessary, led to the replacement of the traditional inductivist interpretation of science to the hypothetico-deductive one. These tendencies were exacerbated by the twentieth-century discoveries of relativity and quantum mechanics and their offshoots.

But the modern classical philosophy of science was itself doomed. In the first place, the attempt to formulate a "metascience," in the form of definitions and rules of method which were presupposed by scientific investigation, and which were not subject to revision in the light of scientific results, had been unsuccessful. The realization of this failure was connected with a second development: deeper studies of the history of science, a subject professionalized after World War II, suggested that a concentration on "the logic of science," with no attention to scientific change, was profoundly inadequate as a means of understanding science. Third, the failure to specify what is "purely given" in sense-perception made a mockery of the classical empiricist doctrine that all our knowledge rests on sense-perception. Again, this point was associated with a fourth one, namely, that the patent appeal of the new sciences to evidence and processes beyond sense-perception made it clear that, if empiricism were to remain credible at all, it could do so only if the notion of "empirical evidence" was reinterpreted in such a way that such evidence went beyond mere sense-perception, by broadening what counts as observation (or observational evidence). And finally, as a consequence of all four of the preceding points, was a fifth: the complexity and sophistication of scientific reasoning made it increasingly clear that prior knowledge was used in vastly more complex and multifarious ways than had been conceived by the hypothetico-deductive tradition: more attention needed to be paid to actual science and its reasoning, and less to logical formalization. Primarily due to the grasping of these five points, some thinkers by about 1960 had come to see that a radically new approach was required if we are to understand the nature of the scientific enterprise and its results. That understanding, growing out of the five points mentioned, was put into the form of what I call "early postclassical" approaches to the interpretation of science, and I will illustrate it through the work of Thomas Kuhn.

4.2. Early Postclassical Approaches to the Interpretation of Science: The Kuhnian Version

Many aspects of these five points were recognized and incorporated in Kuhn's important 1962 book, *The Structure of Scientific Revolutions* (Kuhn, 1970b), which thereby became the most influential work of the "early postclassical" approach to understanding science[12]. As a physicist himself, he saw clearly the need for closer attention to actual science, and his experience in teaching under Conant at Harvard convinced him that actual science must be understood through its historical development, not through formal logic or a set of explicit rules which are independent of what is accepted in science. Kuhn saw, too, that the classical empiricist notion of a "given" in experience was a sham, and that what is counted, within a tradition or community, as "observation" depended on some prior belief. And finally and most centrally, he recognized that thought and activity in science requires prior beliefs, and that that requirement went far deeper than the modern classical empiricists realized, far deeper, that is, than the almost exclusive emphasis that most of those writers placed on the doctrine that hypotheses determine what is relevant in a scientific investigation.

Despite these insights, the position advocated by Kuhn was ambiguous and unclear, and under any interpretation subject to serious objections. His original presentation of his views, in *The Structure of Scientific Revolutions* (Kuhn, 1970b), contained a great many statements that seemed to imply that "paradigms" are monolithic global determinants (or at least profound shapers) of everything in a scientific tradition – not merely what is relevant to a particular inquiry, but also, far more broadly, its methods, standards, goals, the meanings of its terms, what it took to be facts, reasons, or evidence, *etc.* Indeed, this paradigm determination was so pervasive that different paradigm traditions or communities would, in general, not be comparable with one another in any way. Without paradigm-transcendent standards in terms of which two paradigms could be judged as to adequacy, without paradigm-transcendent reasons in terms of which one paradigm could be rationally selected over another or a new one introduced, it was difficult to make sense of the notion of "progress" in science, despite the unconvincing attempts in that direction made by Kuhn in the final chapter of the original *Structure* to argue that "a sort of progress" would exist if that "progress" were an advance over past error (1970b, p.169). For if there were no paradigm-transcendent reasons for believing that one paradigm was better or more true than another, how could one paradigm be judged to have achieved an advance over the errors of past ones? Such a view was strongly criticized for, among other things, leading to an extreme relativism. (*E.g.*, Shapere, 1964; Scheffler, 1967; Masterman, 1970).

In Kuhn (1970a) and in the *Postscript* to (1970b, esp. pp. 182–186), he attempted to avoid at least some of these objections by retreating from these implications through an analysis of a close-knit "disciplinary matrix" as consisting of four distinguishable types of components: "symbolic generalizations," "metaphysical paradigms," "values," and "exemplars," the latter being "concrete puzzle-solutions" which served as models for scientific research. Thus, whereas the earlier work had emphasized the unity of a paradigm tradition's or community's perspective, the revised version called attention to the diversity of the factors governing the tradition or community. In the eyes of many of his followers (and of Kuhn himself), this move was on the right track to answer the criticisms, and was even said (again, even by Kuhn himself) to have been his real view all along, even in *Structure*.

But despite its efforts to come to grips with the possibility of rational comparison and judgment of theories or paradigms, and with the sense in which progress is possible in science, the 1970 move toward emphasis on diversity was not really a successful answer to the criticisms[13]. In the first place, the diversity interpretation of (1970a and 1970b) was quite distinct from that of the original *Structure*. It is certainly true that, scattered through that book, Kuhn mentions particular governing ideas which can, all too often only by rather Procrustean means, be classified as "symbolic generalizations," "metaphysical paradigms," "values," and "exemplars." But insofar as they are discussed in that book, these various components are seen as all *exhibiting* some common theme, as *illustrating* or *manifesting* some underlying unity, the paradigm. According to the original Kuhn, *it is this unifying theme, common to all the various fundamental models, and not the models themselves, that provides the coherence of the traditions and communities that are taken to be the basic entities in terms of which science is to be interpreted.* Close reading of the texts reveals that the majority of statements concerning paradigms fit this reading.

In the concentration on evading the criticisms of *Structure*'s unified paradigm concept, the 1970 papers emphasize the diversity of the disciplinary matrix, and the further diversity of its components. Yet in those papers and in further work, Kuhn and his followers continued to maintain that, when we analyze the kinds of research conducted in science, the problems raised, the standards adopted, the goals sought, and so forth, it is the communities and traditions which we are to view as the basic units to be examined. Yet what, under the diversity interpretation, guarantees the coherence of those communities and traditions? What now produces the unity underlying the diversity of disciplinary matrices and their constituent parts – the enormous diversity of values, for example, which different members of the same tradition may be in disagreement about, and even where there is agreement, say about "simplicity" or "beauty," there might be widely divergent interpretations? In these 1970 and later papers, and in many of Kuhn's followers, the unity is portrayed not as a monolithic internal or intellectual unity, but rather as a sociological one, due not to any substantive unity of beliefs, but rather of a community or tradition whose members presumably might disagree on a number of very basic views of what are appropriate "symbolic generalizations," "metaphysical paradigms," "values," and "exemplars," or interpretations thereof.

The truth seems to be (if I can borrow one of Kuhn's own expressions) that there is a tension, even an essential one, at least among his adherents, between the need to recognize both unity and diversity in his interpretation of science. On the one hand, there is the insistence of much of the Kuhnian school on the existence and analytic fundamentality of unified traditions and communities in science: they are the basic objects to be examined in the attempt to understand science. This strain is explicit in *Structure*, and is most often exhibited in the idea that there is a common intellectual thread running through or presupposed by all the diversity. In the1970 and later papers, and in many of Kuhn's followers, the unity is portrayed not as a monolithic intellectual (pre-intellectual?) unity, but rather as a sociological one, due not to any unity of beliefs but of a community or tradition of more or less overlapping interests which are described as the (far from unitary) "disciplinary matrix."

On the other hand, there is the second strain, the emphasis on diversity. It is important to note that this strain builds on one of the five major roots of early postclassical philosophy noted above: the need to make room for the diversity involved in the multifarious types of background beliefs employed in science, and in turn the variety of functions performed in different problem-situations by each of those types. The Kuhn of 1970 was indeed on the right track in emphasizing diversity. Only he took that path for the wrong reasons, to attempt to dispel the objections that had been hurled at *Structure*. He supposed that a classification of types of diversity within the total disciplinary matrix would be sufficient to remove those objections. They were not, as I have argued in previous work. However, two further objections are relevant in the context of the present paper. The first has to do with the fact that *neither in Structure nor in the 1970 papers was there any need whatever for a unity underlying the diversity.* In fact, instead of searching for a new kind of unity, most often conceived as sociological rather than an internal unity of the disciplinary matrix, Kuhn would have done better to have abandoned any claim of unity whatsoever. After all, the real source of the relativism of which Kuhn has been accused, and of his alleged failure to account for scientific progress, lay precisely in the apparent internal unity of the paradigm or disciplinary matrix, in its function of shaping or determining what counted as a reason (and everything else) within the paradigm. Whether or not this function could be immunized from such objections by substituting a sociological for an internal unity is a moot question; but even if it could, such a unity is needed *only if* "communities" or "traditions" are taken as the basic items for the analysis of science. Nor, if we reject the conception of the unity as an internal unity of the diverse elements of the disciplinary matrix, is it necessary to substitute a sociological conception. For such a unity is as unnecessary as an internal one. As the analysis of the solar neutrino experiment shows, the scientific aspects of the research are wholly and adequately describable in terms of the patently scientific background relevant in its conception, execution, and interpretation. This is not to say that sociological aspects are not relevant in certain ways: they certainly are in some aspects – for example, in matters of funding, or of personalities, or of local interest in some problems rather than others. But such matters are *about* the science: they do not constitute it.[14]

The categorization of diverse components of a disciplinary matrix laid out by Kuhn in the *Postscript* to (1970b) fails to provide a clear and unambiguous picture of the variety of components, and the variety of each type of component's functions, that we find in actual scientific cases. Fitting the many components of the solar neutrino experiment into Kuhn's categories of "symbolic generalizations," "metaphysical paradigms," "values," and "exemplars"is a frustrating and fruitless occupation, a Procrustean bed into which scientific processes can be fitted only by stretching or narrowing them to make them fit. As just one example, where in these pigeonholes do we put prior knowledge of how to clean the tank? The guarding of the experiment from contamination by cosmic muon decays by burying the experiment a certain distance underground? As to metaphysical paradigms, as far as I can understand what Kuhn had in mind, these are to be found in clear incarnations in earlier, less sophisticated science[15]; but in modern sophisticated science, one has to get away from such primitive ideas as "matter" and "causality" as prescientifically given truths, and find out what "metaphysics" to propose and perhaps to accept. Background information is too diverse, applied in too many ways, for too many purposes, for a simplistic classification into four fundamental types to be truly illuminating.

So diverse are the components of background information and their functions that we may begin, perhaps prematurely, to suspect that there is no unity, either internal or sociological, underlying particular scientific traditions or communities. But what we have found is this: that there *are* very broad "guiding theories" in much of science; the mechanistic philosophy of nature was an early example of an "exemplary" viewpoint which provided a model of what solutions to problems over a broad range of domains ought to be like. A more modern example is that of gauge theories, which, once their power and applicability beyond the strong and electromagnetic interactions became understood, provided a contingent (but not categorical) imperative for the way to try to unify the fundamental forces.[16] But it is a mistake to think that only such broad theories serve as background for research. In all cases in modern sophisticated science (and indeed perhaps constitutive of what it is for research to be "sophisticated") a large number of different theories, some broad, some narrow, play roles, often more than one role; and the background consists not only of what are naturally conceived of as "theories," but also includes very specific facts (how deep a hole is needed for the experiment) and practical knowledge (how to clean the tank). And as to "exemplars," there is not one small set of models on which science proceeds; there are many, and the choice depends on the problem-situation at hand. Science builds by relying on a combination of diverse background, each item performing its own functions, and sometimes performing different functions in different circumstances. A classification of types of background information, and of the types of functions performed by the various types, and even an underlying unity of some sort, *might* well emerge from such investigations as that of the solar neutrino experiment. Nevertheless, the search for such classifications (whether put forth in advance of such investigations or as a result thereof), or for such unity, is not the centrally important task for the philosophy of science. It is understanding how this process works, rather than classification, that must be our primary aim.

This point brings up one further danger resulting from the search for a unifying factor. Not only is such a factor unneeded; not only does it not exist; the search for it is actually a red herring, diverting attention from the real problems of understanding what science is – from the understanding of how we build on what we know (or take to know), conceiving new problems, new lines of research, new methods, new concepts, new theories, new standards, and new goals.

This discussion of the need for concentrating on the diversity in scientific reasoning, and not on some unifying factor behind the diversity, fits in perfectly with that in the discussion of the failures of classical empiricism in its various guises (Subsection 4.1). The essential point underlying the other major failures of classical empiricism was its inability to come fully to grips with the pervasiveness and variety of the roles of background information, and the way it shapes what is counted as observational and as reasons for or against belief. Traditional classical empiricists refused (in word if not in deed) to admit the use of background beliefs in inquiry. Modern classical empiricists admitted the need for background "hypotheses" or "conjectures," but either (as with Popper) as a means of defending a deeper dogma, falsificationism, with which the idea of background knowledge was ultimately inconsistent, or else (as with the more standard hypothetico-deductivists) holding that reasoning existed only in the process of justification, but not in its discovery or presentation. In all these versions of classical empiricism, then, the processes of science were poorly served. The present subsection (4.2) shows that Kuhn, for all his insights into the failures of classical empiricism, did no better. He too failed to grasp the significance of the use of background beliefs. On one interpretation of his earlier work, he supposed there to be one monolithic background paradigm determining everything about a particular tradition in science. This view led inevitably, despite his struggles, to an extreme relativism which failed to account for scientific progress. In his later work he began to emphasize the diversity of science while suggesting a sociological unity underlying that diversity. In doing so, he gave a categorization of types of components of a disciplinary matrix which was so simplistic as to be almost completely unhelpful in understanding what happens in science, and which did not in any case provide a reason why *these* types of components were *justifiably* usable as background. (Why these values rather than any others? Why these metaphysical paradigms in particular? Why these specific exemplars and these symbolic generalizations?)

5. APPLICATION TO THE PROBLEM OF INCOMMENSURABILITY

The account of scientific reasoning that I have proposed in this paper has immediate application to problems associated with the possibility of comparing radically different theories or general approaches. Since they were brought to wide attention by Kuhn (1970b) and Feyerabend (1975), such problems have been grouped together under the general "problem of incommensurability" (Sankey, 1994; Hoyningen-Huene, 1993; Shapere, 1998). However, their formulation and treatment have undergone considerable evolution since then. Kuhn, as we have seen, gradually moved away from initial statements which seemed in many cases to imply that "paradigms" or "high-level background theories" are global determinants of everything in a scientific tradition, so that the meanings of terms in different traditions, could not be compared in any way, and he moved in the direction of considering the determinants of a tradition in terms of a close-knit combination of a number of distinguishable factors (Kuhn, 1970a). Later still, he and his sympathizers came to see the problem of incommensurability as more "local," confined to only some of the terms in a radically new theory rather than to all,[17] and though that problem is still often approached through doctrines deriving from linguistic philosophy, the problems are no longer seen only in terms of the comparison of "meanings," but are sometimes taken as arising from incommensurability of reference.[18] Again, incommensurability is occasionally proposed to be something less than a denial of the possibility of strictly accurate translation, with the (somewhat unspecified) suggestion that more complex description of resemblances between two "incommensurable" uses may be possible.

But in all this evolution, there has been surprisingly little attention to a possibility that seems quite appealing: namely, that two uses of the same term in two distinct theories (for example, in an earlier and a later theory) may be quite different, and yet be related by a chain of reasoning which explains why the term's meaning (and/or reference) had to be changed.[19] The source of this neglect is perhaps evident in the views from which the problem of comparison stems: for if what is counted as a reason in a particular scientific community or tradition is dependent on a "paradigm," or on more specific and distinct exemplary models of problem-solving, the concept of a reason is derivative, and perhaps even varies (perhaps in a radically incomparable way) from one community or tradition to another. I think this has been a mistake: the important task for philosophy of science is not to make it a branch of linguistic philosophy, but to examine what is certainly the most important claim made by science, that its methods and conclusions are based on reasons.

If linguistic analysis is thus committed to the flames as the fundamental way of understanding what happens in inquiry, and if scientists come to their conclusions in cases of major change of theory and/or theoretical perspective by employing relevant prior knowledge as reasons in particular inquiries, then the problems of incommensurability, like so many others, disappear. There will still remain differences, often radical, between two theories and their uses of similar or identical expressions. But is that not what we expect of science, which continually revises not only its meanings and references, but also its conception of what it studies (its domains), the kinds of explanatory accounts to give of what it studies, its methods, standards, goals, and so much else? The view I have outlined in this paper recognizes the depth of such change, and attempts to explain how it takes place in terms of reason and evidence, while allowing both of those also to be subject to change in the light of what we learn. Viewed from this perspective, we can construct a model of scientific change in terms of which some versions of the problems of incommensurability can be solved, and other aspects of the problems be cast in a new way. I will present this model in such a way that it is specifically relevant to the problem of the comparability of terms used in successive theories, but in radically different ways. It is easy to think of this as a form of *conceptual change*, where, for the sake of example, the conceptual change is of the usage of a single term (it is easily extended to cases involving a number of terms or to changes of more complex forms, such as changes of theories.) Though for the sake of simplicity (and for the sake of formulating the issue in the way it would be by Kuhnians) I speak of "theoretical contexts" (paradigms, disciplinary matrices); nevertheless, from the perspective of the view developed in this paper, the context determining the usage at each successive stage must be understood as a diverse body of background information. With this qualification, the model of change of usage of single terms is the following.

Usage	*Theoretical Context*	*Properties (none essential)*
U_1	T_1	ABCD
↓	↓	↓
usage change	theory change	reasons for abandoning D
↓	↓	↓
U_2	T_2	ABC
↓	↓	↓
usage change	theory change	reasons for adding F
↓	↓	↓
U_3	T_3	ABCF
↓	↓	↓
usage change	theory change	reasons for replacing C with G
↓	↓	↓
U_4	T_4	ABGF
↓	↓	↓
...	...	...
U_k	T_k	HKGF

Table 1: Model of change of usage of single terms

The notion of reasons responsible for the successive substitutions of features of usage here is precisely what it has been the purpose of the view outlined in this paper to clarify.[20]

The model can be summarized as follows: If, for two usages U_1 and U_k respectively determined by "theoretical contexts" T_1 and T_k, there is a chain of reasoning in terms of which we can understand why certain properties ascribed in usage U_1 and its successors up to and including U_k were abandoned, altered, or replaced, then that chain-of-reasoning connection explains the possibility of comparing the two usages and their theoretical contexts despite the fact that the final stage U_k involves few or even, in principle, no features in common with the initial usage U_1.

This single-term model can be illustrated by the transition in the concept of mass from Newton to Einstein, used as an example of incommensurability in (Kuhn, 1970b, esp. p. 102). Einstein found that there were reasons, in the sense of 'reason' developed in this paper, why mass had to vary (despite its invariance in Newtonian physics) with choice of reference-frame in the case of high velocity with respect to that frame. In other words, he gave the chain-of-reasoning connection (in the sense of 'reason' portrayed here) between the usages. These reasons included larger-scale ones than those having to do with the single term itself (which is why it is no great matter to extend the model to more profound cases of radical scientific change). Among these reasons were the then-prevailing contradiction in the way the world was treated in classical mechanics and electromagnetic theory, and on Einstein's analysis of how we measure spatial and temporal quantities.[21]

The incommensurability problem in most of its variations has been all too closely identified with a problem concerning two objects (whether "meanings," "referents," or whatnot) which are to be juxtaposed and compared, in isolation from all other considerations. And naturally, when the issue is approached in this way, and two different usages U_1 and U_k of, say, a term like 'mass' are considered, the two usages look so different as to raise a problem of how they could possibly have the same meaning or reference. It makes it appear that they may be "incommensurable" in some extreme sense that somehow has to be given an explanation, usually hoped for in terms of linguistic features of their usage, or, as a second line of defense, in terms of the features of some common containing language which makes feasible a rougher, less precise, comparison. Such a view of the matter ignores what happens in science: the real issue there is not one of side-by-side comparison, but of tracing the reasons for adoption of successive alterations which result in the appearance of radical, "incommensurable" change. It is a matter of why a transition takes place, particularly in cases of radical change.

6. CONCLUSIONS AND IMPLICATIONS

From this perspective, the problems of incommensurability are simply one bundle of instances of the fundamental issue in the attempt to understand science: to make sense of the notion of reasoned change. This is what the view developed in this paper has tried to do, in terms of the notion of background information. In the view I have sketched, linguistic categories are derivative, reasons fundamental – reasons in terms of which we decide how we should go about inquiry, the ways we think and talk about what we are studying, and even the goals we seek in inquiry.

In this paper, I have argued that, in the process of inquiry, we have, as a matter of contingent fact, been able to build a pool of beliefs which qualify for use as background in the knowledge-seeking enterprise. The beliefs which so qualify include not only factual and theoretical beliefs about the universe, but also vocabulary by which we describe and delineate the subject-matters of our inquiry, the methods we employ in the search for knowledge, the normative standards by which we conduct our inquiry and set its goals. In no case does any *one* idea ("theory," "paradigm") determine all aspects of research; there is always a diversity of background information, and different items of background information function in different problem-situations and in different ways. Thus any distinction between "normal science" and that conducted under the aegis of a unified paradigm proves unrealistic, and leads to the problems discussed earlier.

Thus, as the great achievements of science in the last four centuries have shown, we have to *learn how to learn* about the world of experience or nature, about what it is we study and the problems we must deal with; and in doing so, we *learn how to think and talk* about that experience or nature, how to refer to what we have learned we have to refer to, and how to judge the results of our efforts.

The demarcation between science and non-science is not given, or discoverable by armchair logic, but develops with the growth of a distinction between beliefs which are qualified to serve as background information, and those that are not. This distinction is a contingent one: it is a matter of fact that this is the way we have learned, by building on what we have learned already, or at least on what we have the best reason to believe we have learned.

In this sense, the view presented in this paper is a rational descendant of empiricism. But whereas classical empiricism maintained that all our knowledge comes from (or is tested by) sense-perception, the view offered here says that all our knowledge comes from (and is tested by) observation, in the sense in which we have learned to use that term, to include also instruments extending our range of interaction with nature beyond what is accessible to sense-perception. We learn about nature by interacting with it. In that sense, if the present view requires a name, it might be called "interactional empiricism," where what counts as an interaction (observation) itself evolves as a product of inquiry.

The quest began with a piecemeal approach, which was gradually extended to include an approach in which separate domains and domain explanations were found to cohere. These approaches gradually led to the formation of a body of background information – beliefs which, among other things, had proved successful in describing and explaining domains, often in a coherent and unified way. The possibility of achieving success in taking such approaches was a contingent one, and required justification. But as a matter of contingent fact, Domain Success and Coherence were achieved in many cases – they were successful in achieving, in many cases, what they set out to do. That is, they received a higher justification as approaches to inquiry, through their achievement of Goal Success, and thus became elevated to the (contingent) status of normative principles guiding scientific research. This "criterion" of Goal Success is ultimate, and itself needs no further justification. It makes no substantive claims, but relies solely on the internal characteristics (the promise) of the approaches it measures. Any approach to knowledge and its seeking must be judged by its Goal Success. Because, as a matter of contingent fact, the extended piecemeal approach we call "science" did achieve what it promised to do (its Goal Success), the scientific approach won out over rivals which also claimed to give access to knowledge, among them rationalism, holism, mysticism, and others.[22]

The present view is perfectly compatible with the possibility that the background information available at a given stage of development of an area of science is not complete enough, by itself, to solve the problem at hand; with the possibility that we may, at given stages, be mistaken in our belief that an item employed as background information will turn out not to be so qualified; and with the possibility that this *may in principle* happen to *any* particular item accepted as background information in the light of prior investigation. Though all the reasons available up to a certain point may have been favorable to the item, and none unfavorable, it is always possible, *in principle*, that specific reasons for doubt may arise which will force us to reject it, to stop using the item as background information in our search for further knowledge. But this in-principle possibility that doubt may always arise with regard to any of our beliefs – a possibility of doubt that applies *equally* to *every* belief, including its negation – is not itself a reason for doubt.[23]

The present view is also compatible with the possibility of learning truth about the way things are in nature. The terms "success" and "coherence" which have figured so prominently in my account have not been chosen without full realization of the fact that they have also been used to refer to two traditional types of theories of truth. Only, a chief objection against such theories has always been that they give *reasons* for believing, whereas something can be true even if the reasons for believing are false. Success and coherence have been connected here, as they should be, with reasons. But they are also essential ingredients in assigning a firm basis for the possibility of knowledge which "corresponds to reality." (This topic is dealt with in detail in Shapere, forthcoming b.)

The present approach avoids both the absolutism of classical philosophy of science and the relativism of extreme versions of postclassical philosophy of science. That it is able to do so should occasion no surprise, for what I have argued is very simple, even obvious. In all problems, even the most everyday ones, prior knowledge has to be applied: the more we know about the situation we confront, the better able we are to make a decision, whether about how to act or what to believe. Even our Pleistocene ancestors had to know what stones made the best tools, and where to find them, and what plants were edible and when and where they could be found. The piecemeal approach is simply the attention to details, nothing more. Goal Success has always been central in inquiry and action; how else could we proceed? By ignoring details? Surely that is not a way to go. By dismissing coherence of our ideas? Anyone who does so would be in danger of not lasting very long. We learned, long before the modern scientific piecemeal approach gained its credentials, that we must learn by attending to details of what we are trying to learn about, and we learn about those details by applying the best of what we have learned already. That simple lesson is all there is to the piecemeal approach, even in its most sophisticated applications in science.

Wake Forest University

NOTES

[1] See Shapere (1982) for a detailed discussion of the solar neutrino experiment from the point of view discussed in the present paper, but with emphasis on the concept of observation rather than of reasons.

[2] With the development of thermodynamics, three main alternative types of theories were developed in the nineteenth century: (1) the sun and stars produce energy by chemical burning, (2) the energy is produced by gravitational collapse, and (3) the sun is constantly impacted by meteors. All of these theories were quickly shown to produce energy in woefully inadequate amounts. The discovery of radioactivity in the early twentieth century led to a theory in which the energy was produced by the radioactive decay of primordial heavy elements; however, no combination of elements decaying radioactively seemed capable of matching the measured release of solar energy. Though Eddington (1926) suggested that the energy must be produced by "subatomic" processes, it was not until the discovery of the neutron in 1932, and the consequent development of nuclear physics, that it became possible for Bethe and von Weizsäcker to conceive and present their theories.

[3] This "direct way of determining" amounts to an extension of the term 'observation': *information emitted from the source (the core of the sun) is transmitted without interference or alteration from the source to a receptor*. The receptor is an instrument, in particular the neutrino detector to be described shortly, rather than a human sense organ. The development of new receptors led, by rational extension, to the entire electromagnetic spectrum being considered "observable" by means of receptors designed to capture a relevant range of wavelengths in that spectrum; the human eye becomes one sort of such receptor, designed by natural selection rather than by technological innovation. A further extension goes beyond the electromagnetic interactions, to the weak interactions: the neutrino, a weakly-interacting particle, travels unhindered from its source in the energy-producing solar core to the detector on earth. The relevant concept of "information," too, is specified here by currently well-founded physics, which asserts the existence of four fundamental types of interaction, of which the electromagnetic and weak are two. (A further extension is in the offing if it proves possible to "observe," in the extended sense, gravitational waves. Success in that effort would involve observation by means of a third fundamental force, the gravitational, and would amount to observation of a fundamental facet of spacetime.) This provides an example of a change of "meaning" which can be accounted for as being based on reasons; see Shapere (1982) for details.

[4] The use of the term "piecemeal" in this connection is related to the use of the same term in connection with application of background information discussed in the previous section. Both typify the concern with details which characterizes so much of sophisticated (and also much unsophisticated) reasoning.

[5] In this sense, therefore, empiricism, in the form of the extended piecemeal approach, is not simply one approach among others, such as rationalism (cf. Hoyningen-Huene, 1993, pp. 249–252). It gains legitimacy by its goal success, as opposed to rationalism, which has not achieved such success.

[6] The views I discuss here are only versions of empiricism, and should really be referred to as two versions of *classical empiricism* (traditional and modern), rather than as two versions of the philosophy of science (traditional and modern). However, the chief rivals of empiricist approaches – rationalism, for example – were, until the 1960's, no longer influential after the nineteenth century. I will discuss the reasons for their abandonment shortly.

[7] Whether the background knowledge brought to bear can always – if ever – be *sufficient* to narrow the choice to one and only one of the conjuncts is a further question, which is dealt with in forthcoming work (Shapere, forthcoming b).

[8] Popper (1959) has written that beliefs can be *corroborated* even though they cannot be *confirmed*. Whether or not this distinction can be made sense of, Popper's closely-related theory of verisimilitude does seem to have been decisively criticized by David Miller (1974); for useful further discussion, see O'Hear (1982, pp. 49–56, 106).

[9] Logic had to be gone beyond in certain specific respects, however: the purely formal analyses of concepts such as 'explanation,' 'theory,' *etc.*, had to be connected to experience by means of correspondence rules, which themselves might or might not be formulable to at least some extent in logical terms.

[10] A detailed analysis of the roots and ultimate bankruptcy of the central role accorded to logic by twentieth-century logical empiricism, together with a parallel examination of what I have here called the "early postclassical" trend (Kuhn, Feyerabend, and others) in later thought about science in that century, can be found in Shapere (forthcoming a).

[11] For further discussion of Newton's deductivism, however, see Worrall (2000) and references therein.

[12] Some of Kuhn's views were anticipated, to varying extents, in Toulmin (1953 and 1961) and Hanson (1958).

[13] For detailed criticisms of (1970a) and (1970b), see Shapere (1971). I still stand by the objections raised in that review, as I do also for those raised in (1964). However, the objections raised in the present paper differ fundamentally from those found in (1971).

[14] If sociological factors are to be shown to be genuinely inseparable from the science and determinative of it, that claim must be shown by careful and strict deduction of the scientific factors from the alleged sociological causes. Many attempts have been made to do so; the work of Pickering (1984) on the history of quark theory, and Pinch (1980; 1981; 1986) on the solar neutrino experiment, are typical. Neither of these writers is successful in showing that the scientific aspects of the cases they consider are "social constructions." The arguments demonstrating their failure will be found in (Shapere, forthcoming b).

[15] For a discussion of the roots of metaphysics and its influence on early science, see Shapere (1990; 1996; and forthcoming c).

[16] As always, there were reasons why the gauge-theoretical approach became a guiding one. Its original use, in Yang and Mills (1954), was confined to strong interactions, where it seemed to be interesting but useless because it assigned zero masses to the particles (bosons) which carried the force. However, Higgs (1964) proposed a mechanism by which such particles could gain mass. Combining the Yang-Mills gauge theory with the Higgs mechanism, and building on already-known indications of deep relations between the electromagnetic and weak forces, Weinberg (1967) and Salam (1968) brought forward an "electroweak theory" in which, at high enough energies, the two forces were identical. (The unification achieved was not, however, complete, since two separate groups were required rather than a single one.) Even so, the theory was not known to be exempt from infinities ("renormalizable") which would make useful calculations impossible. With 't Hooft's (1971) compelling arguments that gauge theories were renormalizable – that the infinities could be made to disappear – the gauge-theoretical approach became seen by many as "the way to go," and very shortly a gauge theory of the strong interactions was constructed. On the basis of their mathematical analogy, the electroweak theory and the theory of the strong interaction, quantum chromodynamics, were juxtaposed in what has come to be known as the "Standard Model." This, together with hypothetical extensions which would truly fuse the electromagnetic, weak, and strong forces by interpreting them in terms of a single group rather than as a juxtaposition of three separate but mathematically similar ones, led to the possibility of making calculations regarding what happened during the first fractions of a second after the Big Bang, so that particle physics (quantum field theory) and cosmology (general relativity) were also brought together to at least an extent.

However, in these as in other cases where a certain approach becomes a "guiding" one in an area of science, there are also rival approaches, dealing in general with the same subject-matter (domain). In the case of gauge theories, for example, the gauge approach, seeking a single group which would unify all four forces, is not the only contender for a fundamental theory. Many rivals, each "guiding" much research, exist: for example, superstring theory and a large variety of "quantum gravity" approaches. Too, background ideas of lesser scope also combine in various ways with such high-level approaches in various ways to guide specific research; this point is illustrated in the survey given of the solar neutrino experiment.

[17] The problem of comparing two radically different theories is thus allegedly solved by confining the incommensurable parts of the theory to a small "localized" set of concepts which are embedded in a broader language common to both of the theories to be compared. This solution has received considerable favorable attention in the literature and at the conference where this paper was presented. However, it assumes that the local concepts in the two theories, which are not commensurable with each other, are commensurable with the containing language. This assumption must be shown to be justified.

[18] See Sankey (1994) for discussion of the referential approach. For an argument that despite the multiple-facet view of paradigms, they must still be understood as a "holistic" unity, see Hoyningen-Huene (1993), p. 221.

[19] Hoyningen-Huene (1993, chapter 7) gives an interesting discussion of the role of reasons in choosing between a radically new and an older paradigm (but only after the new paradigm has been introduced in the face of anomalies which led to the rejection of the older one). However, despite the value of the discussion,

it remains unclear how the author's designation of these considerations as "reasons" can be made sense of as such within the context of a Kuhnian view as he interprets it.

[20] For the similarities to Wittgenstein's notion of "family resemblances," see Shapere (1989), which also criticizes Wittgenstein's use of this concept in that work.

[21] The contradiction was there for all to see, but was understood in its implications only by Einstein and, perhaps to a lesser degree, by a few others like Poincaré. This point is important: the body of scientific background information exists independently of whether anyone notices all the features of it – its contradictions and implications, for instance. This is the important point behind Popper's (1972) view of a "Third World" of "objective knowledge," Ong's (1982) delineation of the stage of "literacy," and its differences from the preceding "orality," a transition made possible by the invention of alphabetic writing, and Donald's (1991) idea of an "information-storage" stage in the development of thought which replaced the "mythical" stage. This objectivity of knowledge explains why I speak of "science" rather than of "scientists" as the object of the investigation made in this paper. It also suggests the direction in which the view presented in this paper can be subsumed under a far broader picture of the career of the human species and its ancestors.

[22] But though all knowledge is contingent, even this proposition itself may turn out to be false, when and if we fully understand the universe in which we live (or, alternatively, when and if we discover a room in which it turns out that everything we think of turns out to be true).

[23] The kind of doubts we have learned to respect in the knowledge-seeking enterprise are *specific doubts*: ones which are directed specifically at a specific proposition but not at its negation. We have learned this (or should by now have learned it) through, among other things, the failure of Cartesian rationalism, which demanded that even *universal* doubts, such as "A demon might be deceiving me," or "I may be dreaming," be considered as legitimate (philosophical) reasons for doubt. Such universal or philosophical doubts apply equally to every proposition, including its negation, and thus do not, or more precisely cannot, discriminate between beliefs. But the mere possibility of doubt is not a reason for doubt; it amounts only (but importantly) to a reminder that all our knowledge is contingent, at the very least in the sense that there is always the possibility that specific reasons for doubt may arise. The contingency of knowledge does not preclude our legitimately calling a belief "knowledge," if it meets certain conditions (among them, that all available evidence or reasons favor it, and there exist no specific reasons for doubting it). If specific compelling doubt does arise, we simply cease calling it "knowledge" or "information," and cease using it as such.

Thus the failure of the solar neutrino experiment to capture as many neutrinos from the sun as calculations predicted led to a more careful study of all the background information employed in the experiment. As of this writing, the culprit is strongly suspected to be weak interaction theory, which had to be revised accordingly (Bahcall, Davis, Parker, Smirnov, and Ulrich, 1994, pp. 294–337).

REFERE NCES

Aristotle. (1991). *Posterior Analytics*. In J. Barnes, ed., *The Complete Works of Aristotle, Volume I*, pp. 114–166, Princeton: Princeton University Press.

Bacon, F. (1939). *Novum Organum*. In E. Burtt, ed., *The English Philosophers from Bacon to Mill*, pp. 24–123, New York: Modern Library.

Bahcall, J., R. Davis, Jr., P. Parker, A. Smirnov, and R. Ulrich, eds. (1995). *Solar Neutrinos: The First Thirty Years*. Reading: Addison-Wesley.

Bethe, H. (1939). "Energy Production in Stars." *Physical Review* **55**: 434–456.

Cohen, M. and E. Nagel. (1934). *An Introduction to Logic and Scientific Method*. New York: Harcourt Brace and Company.

Crombie, A. (1962). *Robert Grosseteste and the Origins of Experimental Science*. Oxford: Oxford University Press.

Descartes, R. (1983). *Principles of Philosophy*. Dordrecht: Reidel.

Donald, M. (1991). *Origins of the Modern Mind*. Cambridge: Harvard University Press.

Duhem, P. (1954). *The Aim and Structure of Physical Theory*. Princeton: Princeton University Press.

Eddington, A. (1926). *The Internal Constitution of the Stars*. Cambridge: Cambridge University Press.

Feyerabend, P. (1975). *Against Method*. London: New Left Books.

Galilei, G. (1914). *Dialogues Concerning Two New Sciences*. New York: MacMillan.

Gross, D. and F. Wilczek. (1973). "Ultraviolet Behavior of Non-Abelian Gauge Theories." *Physical Review Letters* **30**: 1343–1346.

Hanson, N. (1958). *Patterns of Discovery*. Cambridge: Cambridge University Press.
Higgs, P. (1964). "Broken Symmetries, Massless Particles and Gauge Fields ." *Physical Review Letters* **12**: 132–133.
Hoyningen-Huene, P. (1993). *Reconstructing Scientific Revolutions: Thomas S. Kuhn's Philosophy of Science*, trans. A. Levine. Chicago: University of Chicago Press.
Kuhn, T. (1970a). "Reflections on My Critics." In I. Lakatos and A. Musgrave, eds., *Criticism and the Growth of Knowledge,* pp. 231–278, Cambridge: Cambridge University Press.
Kuhn, T. (1970b). *The Structure of Scientific Revolutions*. 2nd edition. Chicago: University of Chicago Press.
Masterman, M. (1970). "The Nature of a Paradigm." In I. Lakatos and A. Musgrave, eds., *Criticism and the Growth of Knowledge*, pp. 59–89, Cambridge: Cambridge University Press.
Mill, J. (1874). *A System of Logic*. New York: Harper & Brothers.
Miller, D. (1974). "Popper's Qualitative Theory of Verisimilitude." *British Journal for the Philosophy of Science* **25:** 166–177.
Newton, I. (1952). *Opticks*. New York: Dover.
Newton, I. (1964). *The Mathematical Principles of Natural Philosophy*. New York: Citadel Press.
O'Hear, A. (1982). *Karl Popper*. London: Routledge & Kegan Paul.
Ong, W. (1982). *Orality and Literacy*. New York: Methuen.
Pickering, A. (1984). *Constructing Quarks: A Sociological History of Particle Physics*. Chicago: University of Chicago Press.
Pinch, T. (1980). "Theoreticians and the Production of Experimental Anomaly: The Case of Solar Neutrinos." In K. Knorr, R. Krohn, and R. Whitley, eds., *The Social Process of Scientific Investigation, Sociology of the Sciences, Volume 4*, pp. 77–106, Dordrecht: Reidel.
Pinch, T. (1981). "The Sun Set: The Presentation of Certainty in Scientific Life." *Social Studies of Science* **11**: 131–158.
Pinch, T. (1986). *Confronting Nature*. Dordrecht: Reidel.
Politzer, H. (1973). "Reliable Perturbative Results for Strong Interactions." *Physical Review Letters* **30**: 1346–1349.
Popper, K. (1959). *The Logic of Scientific Discovery*. New York: Basic Books.
Popper, K. (1962). *Conjectures and Refutations*. New York: Basic Books.
Popper, K. (1972). *Objective Knowledge*. Oxford: Clarendon.
Randall, J. (1961). *The School of Padua and the Emergence of Modern Science*. Padova: Editrice Antenore.
Salam, A. (1968). "Weak and Electromagnetic Interactions." In N. Svartholm, ed., *Proceedings of the Eighth Nobel Symposium on Elementary Particle Theory*, pp. 367–377, New York: Interscience.
Sankey, H. (1994). *The Incommensurability Thesis*. Aldershot: Avebury.
Scheffler, I. (1967). *Science and Subjectivity*. New York: Bobbs-Merrill.
Shapere, D. (1964). "*The Structure of Scientific Revolutions*." *Philosophical Review*, **49**, 383–394.
Shapere, D. (1971). "The Paradigm Concept." *Science* **172**: 706–709.
Shapere, D. (1982). "The Concept of Observation in Science and Philosophy." *Philosophy of Science* **49**: 485–525.
Shapere, D. (1989). "Evolution and Continuity in Scientific Change." *Philosophy of Science* **56**: 419–437.
Shapere, D. (1990). "The Origin and Nature of Metaphysics." *Philosophical Topics* **18**: 163–174.
Shapere, D. (1991). "The Universe of Modern Science and Its Philosophical Exploration." In E. Agazzi and A. Cordero, eds., *Philosophy and the Origin and Evolution of the Universe*, pp. 87–202, Dordrecht: Kluwer.
Shapere, D. (1995). "On the Introduction of New Ideas in Science." In J. Leplin, ed., *The Creation of Ideas in Physics*, pp. 189–222, Dordrecht: Kluwer.
Shapere, D. (1996). "The Origin and Nature of Time." *Philosophia Scientiae, Entretiens de la session 1994 de l'Académie Internationale de Philosophie des Sciences* **1**: 197–220.
Shapere, D. (1998). "Incommensurability." In E. Craig, ed., *Routledge Encyclopedia of Philosophy, Volume 4*, pp. 732–736, New York: Routledge.
Shapere, D. (forthcoming a). "Logic and the Analysis of Science." To appear in a book, edited by E. Agazzi and P. Weingartner, containing the proceedings of a conference held at Salzburg, Austria, in May, 1999, on "Logic and the Future of Science."
Shapere, D. (forthcoming b). *The Rational Dynamics of Science*. Oxford: Oxford University Press.

Shapere, D. (forthcoming c). "The Ultimate Furniture of the Universe." Forthcoming in Proceedings of a conference on "Concepts of Nature in Ancient Greece and Modern Science," San Sebastian, Spain, October 1998.

't Hooft, G. (1971). "Renormalizable Lagrangians for Massive Yong-Mills Fields." *Nuclear Physics* **B35**: 167–188.

Toulmin, S. (1953). *The Philosophy of Science*. London: Hutchinson.

Toulmin, S. (1961). *Foresight and Understanding*. Bloomington: Indiana University Press.

Weinberg, S. (1967). "A Model of Leptons." *Physical Review Letters* **19**: 1264–1266.

Weinberg, S. (1972). *Gravitation and Cosmology: Principles and Applications of the General Theory of Relativity*. New York: Wiley.

von Weizsäcker, C. (1938). "Element Transformation Inside Stars." *Physikalische Zeitschrift* **38**: I, 633; II: 633–646.

Wittgenstein, L. (1952). *Philosophical Investigations*. Oxford: Blackwell.

Worrall, J. (2000). "The Scope, Limits, and Distinctiveness of the Method of 'Deduction from the Phenomena': Some Lessons from Newton's 'Demonstration' in Optics." *British Journal for the Philosophy of Science* **51**: 45–80.

Yang, C. and R. Mills. (1954). "Conservation of Isotopic Spin and Isotopic Gauge Invariance." *Physical Review* **96**: 191–195.

HARVEY SIEGEL

INCOMMENSURABILITY, RATIONALITY AND RELATIVISM: IN SCIENCE, CULTURE AND SCIENCE EDUCATION

Abstract. In this paper I revisit some of the old debates concerning incommensurability, rationality and relativism, and argue that no relativistic or irrationalistic conclusions can be legitimately drawn from the *ur*-arguments concerning incommensurability. I then consider incommensurability understood more broadly than it is usually understood in philosophy of science, as involving not two or more incomparable scientific theories, but rather fundamentally divergent cultures or world views. I relate this broad understanding of incommensurability to contemporary interest in *multiculturalism*, and consider its ramifications for a subject that philosophers of science unfortunately tend to ignore: that of *science education*. Attending to science education, I urge, allows us to see from a new angle what is at stake in our philosophical musings concerning incommensurability.

INTRODUCTION

The idea that scientific theories, and perhaps other cultural artifacts, are *incommensurable* – i.e., (as a first approximation) that they share no common scale of measurement, and can be evaluated only from within particular paradigms, conceptual schemes, ways of life, world views, historical traditions, or cultures – was at or near the center of attention in the philosophy of science in the years following the early work of Thomas S. Kuhn and Paul Feyerabend.[1] Work on incommensurability dealt, in particular, with questions concerning its genuineness – were theories really incommensurable? – and with related epistemological questions concerning the rationality of theory choice/preference and the relativity of scientific knowledge. These latter questions arose because incommensurability seemed to many to entail that theories admit of evaluation only relative to one or another paradigm or scheme (hence *relativism*); and, since theories (if incommensurable) could not be fairly evaluated across such overarching conceptual frameworks, that the rational evaluation of the respective merits of such alternative frameworks – and the theories judged from within their inevitably limited perspectives – must be impossible in principle (hence *irrationalism*). These issues dominated philosophy of science for a couple of decades, after which they were largely relegated to the back burner and replaced by the realism/antirealism controversy and other issues, especially those arising in the special sciences. Despite their relative neglect during the last two decades, the basic questions concerning incommensurability and their ramifications for the epistemology of science have never been far below the surface; as the papers collected in this volume make clear, the issues remain both fundamental and (to a significant extent) unresolved.

P. Hoyningen-Huene and H. Sankey (eds.). Incommensurability and Related Matters, 207–224.

The philosophical situation in which we find ourselves today is rather different from that which faced philosophers of science in the years after Kuhn and Feyerabend put incommensurability on the philosophical agenda. A dominant intellectual force on the present scene, mainly absent in the heyday of the study of incommensurability, is that of *multiculturalism*. Multiculturalism is readily understood as relevant not only to the cluster of issues concerning incommensurability in the philosophy of science, but also to work at one of the most important places at which science impacts culture: that of *science education*. For, as science educators who travel under the banner of multiculturalism stress, different cultures often have alternative, incompatible conceptions of science, nature, and appropriate modes of investigation. If those conceptions can be rightly said to be incommensurable, the ramifications for science education –concerning, e.g., what it is legitimate to include in science curricula, what can legitimately be said about 'the nature of science,' what can legitimately be presumed concerning (the nature of) nature itself, how scientific method is to be understood and presented to students, etc. – appear to be significant indeed.

In what follows my main aim is to address the impact of incommensurability on recent controversy concerning what might be called the 'multiculturalist movement' in science education. In order to set the stage for that discussion, however, I first review some of the basic philosophical questions concerning incommensurability which have occupied philosophers of science. I begin at the beginning, by trying to get clear on the nature of incommensurability.[2]

1. WHAT IS 'INCOMMENSURABILITY'?

The central idea of incommensurability, as both the etymology of the word and its use in mathematics suggest, is that incommensurables share no common standard of measure. But how this idea is applicable to scientific theories, and what the philosophical ramifications of its applicability might be, are less than clear. Two questions are fundamental: a) Why think that scientific theories are incommensurable?, and b) Does incommensurability preclude comparison and thus rational evaluation? I take these in turn.

1.1 Why Think that Scientific Theories Are Incommensurable?

Kuhn famously argued for the incommensurability of paradigms, from which the incommensurability of theories seems straightforwardly to follow.[3] In his enormously influential book, Kuhn (1970) provided several reasons for regarding rival paradigms as incommensurable: they differ in the meanings of key theoretical terms/concepts; they differ in methods; and they differ in standards, so that the constitution of both legitimate problems and of adequate solutions to them depend upon the paradigm from which such problems and proposed solutions are judged.[4] Kuhn's work developed these reasons with subtlety and not a small amount of controversy; the proper interpretation of the Kuhnian corpus remains the subject of energetic scholarly commentary and debate. Additional reasons offered in support of the 'incommensurability thesis'[5] can be found in Kuhn's writings and in the work of many others, including claims concerning the impossibility of translation and/or reference, and the impossibility of reconciling alternative standards of explanatory adequacy (and attendant 'conceptions of science'), across rival paradigms.[6] It remains controversial whether Kuhn's arguments against the existence of common standards in accordance with which rival scientific theories can be 'measured' – either in fact, in actual historical episodes of theory change,[7] or in principle, in virtue of the features of paradigms just mentioned – succeed. Nevertheless, insofar as 'incommensurability' refers to a lack of a common scale of measurement, so that to claim that theories are incommensurable is to claim that there is no "single scale of units of value" according to which rival theories "can be precisely measured" (Chang, 1997b, p. 2), incommensurability seems an innocuous enough claim. In particular, understanding the term in this way has no untoward consequences concerning the comparison and rational evaluation of incommensurables.

But some – including Kuhn, at least in some passages – argue from incommensurability to incomparability, and claim that because rival theories are incommensurable they cannot be rationally compared and evaluated. This claim is, I believe, untenable.[8]

1.2 Does Incommensurability Preclude Comparison and Thus Rational Evaluation?

Does 'incommensurability' entail 'incomparability'? It is plausible enough to suppose that, if rival theories were to fail to admit of evaluation in accordance with evaluative standards fairly applicable to both, there would be no possibility of fair, neutral, non-question-begging comparative evaluation. However, that rival theories do so fail is i) not the sense of 'incommensurability' just discussed and at issue here; and ii) clearly false.

i) I just granted the plausibility of incommensurability, understood as the thesis that there does not exist a single scale of units of value in accordance with which incommensurable theories can be precisely measured. However, while rival theories might well be incommensurable in this sense, they might nevertheless *not* be incommensurable in the sense that either there are not in fact, or could not be in principle, evaluative standards applicable to both, in terms of which their respective merits can be fairly and appropriately assessed. This second sense of 'incommensurability,' according to which to claim that rival theories are incommensurable is to claim that their respective merits cannot be neutrally, fairly, or non-question-beggingly assessed – that is, that they are *incomparable* and so do not admit of rational evaluation – is far stronger, and more contentious, than the first. Such a potent philosophical claim (i.e. incomparability) cannot be established by defining 'incommensurability' in terms of incomparability, so that the issue is settled by definitional fiat. We need rather to consider whether incommensurability, in its first sense, somehow entails or provides good reasons for embracing the contentious claim concerning the impossibility of comparing, and rationally evaluating, incommensurable theories.

ii) It is perhaps worth pointing out that Kuhn himself denied that incommensurability either was synonymous with or entailed incomparability:

> Most readers of my text have supposed that when I spoke of theories as incommensurable, I meant that they could not be compared. But 'incommensurability' is a term borrowed from mathematics, and there it has no such implication. The hypotenuse of an isosceles right triangle is incommensurable with its side, but the two can be compared to any

> required degree of precision. What is lacking is not comparability but a unit of length in terms of which both can be measured directly and exactly. (Kuhn, 1976, pp. 190–191)

Kuhn did, however, reintroduce incomparability – at least, 'point-by-point' incomparability – in his famous following sentence:

> In applying the term 'incommensurability' to theories, I had intended only to insist that there was no common language within which both could be fully expressed and which could therefore be used in a point-by-point comparison between them. (*ibid.*, p. 191)

Whether this Kuhnian maneuver renders Kuhn's overall position concerning incommensurability and comparability coherent I leave the reader to judge.[9] Rather than engage in hermeneutical interpretation of the Kuhnian corpus, I will instead take the cited passage at face value and, following Kuhn, reject the claim that incommensurability entails incomparability. I argue next that incommensurability does not provide good reasons for thinking that rival theories are incomparable, and moreover that the incomparability claim is false.

Several of Kuhn's early critics, e.g. Shapere (1964), Scheffler (1967), and Kordig (1971), challenged what they took to be Kuhn's case for incomparability. Rightly or wrongly, they took Kuhn to be defending incomparability on the basis of his claim that paradigms include as proper parts of themselves all relevant criteria in accordance with which all relevant paradigms – for any given paradigm, both itself and its rivals – could be properly evaluated, so that evaluation would inevitably question-beggingly favor the paradigm from which the evaluation was made. That is, they took him to be claiming that paradigms bring with them their own criteria of paradigm evaluation – criteria which are internal to paradigms, but to which appeal must nevertheless be made in evaluating rival paradigms during episodes of revolutionary science. They seized on this point in arguing against incomparability (and what they took to be Kuhn's attendant relativism[10]). They argued, among other things, that paradigm-neutral criteria of theory evaluation were available – even if they were not such as to permit 'point-by-point' comparison in terms of precise units of value – and that rival theories, even if properly understood as parts of rival paradigms, could be compared and evaluated in terms of second-order criteria, i.e., criteria not question-beggingly internal to the paradigms involved, such as predictive power, explanatory power, problem-solving ability, testability, precision, scope, fruitfulness, simplicity, and depth.[11] Incomparability, they further claimed, was simply false: rival theories could in fact be fairly and objectively (albeit fallibly, of course) compared and evaluated.

Both these critical points are, I believe, correct. First, even if it is true that a particular pair of rival theories are incommensurable in that there does not exist a single scale of units of value in accordance with which they can be precisely measured (and so compared 'point-by-point'), that fact provides little reason for thinking that they cannot be fairly compared and evaluated. We cannot, for example, precisely measure the units of evidential support or explanatory power enjoyed by rival theories – say, Copernican vs. Ptolemaic astronomy, or contemporary neo-Darwinian biology vs. vitalism, or contemporary quantum chemistry vs. either Daltonian or pre-Daltonian chemistry – but these rival theories can nevertheless be straightforwardly assessed in terms of their evidential support and explanatory power. 'A single scale of units of value in accordance with which they can be precisely measured' is simply unnecessary for the fair, non-question-begging evaluation of rival, incommensurable theories. Consequently, that rival theories are incommensurable, even if true, provides little or no reason to think that they are incomparable.[12]

Second, and perhaps more importantly, it is simply false that incommensurable theories are incomparable. To say that they are is to say that there is no respect in which one can be evaluated as better than, worse than, equally as good as, or (perhaps) roughly as good as, the other.[13] But, for any pair of incommensurable theories, there are many respects in which they can be compared: Ptolemy's theory is *older than* Copernicus'; Newton's theory is *simpler than* Ptolemy's; Einstein's relativity theories *better explain the totality of the relevant evidence* than Newton's; Dalton's theory *solves problems more fundamental to chemistry* than pre-Daltonian theory;[14] Darwin's theory is *easier to grasp without a detailed knowledge of mathematics and statistics* than either contemporary population genetics or quantum mechanics; General Relativity (GR) theory *has made more surprising but confirmed predictions than* Copernicus' theory; Ptolemy's theory is *easier to reconcile with a 'realist' metaphysics than* quantum mechanics; etc. While some of these respects may not involve values philosophers of science have traditionally taken to be important for theory evaluation (e.g. age, ease of understanding, ease of reconciliation with some particular metaphysics, etc.), some involve exactly the sorts of values philosophers have emphasized (e.g. simplicity, explanatory power, problem-solving ability, predictive power, etc). Moreover, they are not made from the perspective of these theories' associated paradigms; one needn't be committed to the Ptolemaic paradigm in order to judge its (greater) compatibility with a realist metaphysics (than quantum mechanics), or to the paradigm associated with Einstein's theories in order to judge that GR has made more surprising but confirmed predictions than Copernicus' theory. Of course one could deny that these are in fact examples of incommensurable theories, but then it would be unclear what pair of theories would count as incommensurable. In short: there is ample reason to think, not only that incommensurability provides no good reason for embracing incomparability, but that the incomparability thesis is false: rival theories, even incommensurable ones, *can* be compared.

Moreover, if they can be compared, incommensurable scientific theories can also be rationally evaluated: the choice of one theory over its rivals can be straightforwardly justified in terms of relevant comparative evaluation.[15] Of course I do not claim that such choice is always or often easy. Sometimes it is unclear whether the reasons favoring one theory are sufficiently strong that they outweigh competing reasons that favor its rival. It must be admitted, further, that it might in some cases also be unclear how to weigh the various respects in which theories can be compared, when a theory fares better than its rival in one respect but worse in another. Still further, there are many other problems not yet mentioned, concerning underdetermination, theory-ladenness, and the like, which a full treatment of incommensurability must address.[16] Still, comparability in terms of criteria such as those listed above secures the possibility of rational evaluation of, and choice between, incommensurable theories. Not only does incommensurability not entail the impossibility of rational evaluation of rival theories and so the irrationality of any choice between or among them; comparability secures the in principle possibility, and the routine actuality, of such rational choice.

To sum up the discussion thus far: First, we must distinguish between two different incommensurability theses, which utilize alternative senses of 'incommensurability,' which have been too often conflated in earlier discussions (including my own). The first, according to which incommensurable theories do not share a single scale of units of value in accordance with which they can be precisely measured, is plausible, but has no tendency to establish that such theories cannot be compared, and, consequently, no tendency to establish either that theory evaluation must be relative or that theory choice must be irrational. The second, according to which incommensurable theories are indeed incomparable, does have these relativistic and irrationalistic consequences, but is clearly false in that rival theories, incommensurable or not, are clearly comparable. Of course this is not to say either that theory choice and evaluation is always easy, or that there is always at hand evidence sufficient to resolve such choice or guide such evaluation. Still, neither sense of 'incommensurability' provides us with any reason to embrace the untoward epistemological consequences from which Kuhn's critics have recoiled.

One might think, however, that this discussion is too limited, in that its target class of incommensurables – scientific theories – is too narrow to be of general philosophical interest. Perhaps scientific theories which are parts of incompatible paradigms can be compared and rationally, non-relativistically evaluated, but what about more challengingly different sorts of things, like chalk and cheese, poetry and pushpin, or the destruction of the world and the scratching of my finger? This question pushes us far beyond the bounds of philosophy of science, but if we are to come to terms with incommensurability, we must endeavor to come to terms with it. I cannot attempt such a large philosophical project here.[17] I try next to take a small step in this direction, by considering the case of putatively incommensurable *cultures*. In considering *multiculturalism* and putatively incommensurable aspects of culture, I hope to broaden the perspective from which the phenomenon of incommensurability might usefully be viewed. I concentrate on those aspects of culture involving science and, in particular, *science education*.

2. INCOMMENSURABILITY BEYOND SCIENCE: ARE DIFFERENT CULTURES (OR FUNDAMENTAL CULTURAL VALUES) INCOMMENSURABLE? THE DEMANDS OF MULTICULTURALISM

It is a truism that cultures differ fundamentally with respect to basic beliefs and values. Some value conformity and obedience to authority; others creativity and independence. Some embrace a fierce individualism; others communalism. Some value wealth and technological advance more highly than spiritual development; others the reverse. Some exploit nature; others revere it. Some approve of, and (its members) regularly engage in, behaviors that others reject as racist, sexist, or worse. Cultural differences are of course as broad as they are deep; it would be folly to try systematically to categorize such differences here. But one sort of cultural difference is of particular interest in the present context: namely, cultural differences concerning the nature of science and of the natural world itself. It is this sort of difference on which I want to focus in what follows. In particular, I want to consider the ways in which science educators ought to conceive and deal with cultural differences concerning science and nature. I consider these matters in the context of contemporary interest, among science educators and science education scholars, in *multiculturalism*.

Science education scholars, like those in many other branches of the academy, often embrace multiculturalism. How this difficult notion is to be understood is far from clear, but it is I believe most plausibly interpreted in *moral* terms, as calling for the just, respectful treatment of all cultures and their members, in particular those cultures which have historically been the victims of hegemonic domination and oppression. (Siegel, 1999b; 1997a, chapters 9–12) Supposing that multiculturalism, so understood, is accepted, what might its implications for science education be? Advocates of multicultural science education (e.g., Hodson (1993), Stanley and Brickhouse (1994; 2000), and Snively and Corsiglia (2000)) – henceforth *MSE* – sometimes suggest serious revisions of the theory and conventional practice of science education on the basis of a commitment to multiculturalism. Their claims can be instructively viewed through the lens of incommensurability.

The most fundamental claim made in this context is that embracing the dictates of multiculturalism, and treating science students with respect and dignity in the science classroom, requires teaching the 'local,' 'ethnic' or 'indigenous' sciences of various cultures – especially those of the students in the class – as accounts of the natural world, and of proper modes of investigating and understanding it, which are as legitimate, and as entitled to a place in the science curriculum, as the more standard science curricula commonly taught in schools in the industrialized West. It likewise requires the abandonment of any pretense to a 'universal' science, i.e. an account of the natural world that is indifferent to, and pretends to be independent of, the vagaries of local, culturally-bound accounts of that world.

Those who, in the name of multiculturalism, reject any such universalist conception of science/nature in science education, typically contrast it with 'local,' 'ethnic' or 'indigenous' science: understandings of the natural world which reside only within particular cultural contexts. For example, Hodson (1993) contrasts the allegedly universalistic science characteristic of 'Western' cultures with African sciences, which are characterized by the "use of mytho-magical explanations, alternative conceptions of time, and the view that nature is benevolent" (p. 686) and Maori science, which emphasizes "[a]ppropriate respect and recognition of spirituality (of forests, mountains, and the sea...)" (p. 703); he details the alternative views of rationality and objectivity embraced by Western, Islamic, Maori, Alaskan native, and many African cultures embodied in their local sciences.[18] Snively and Corsiglia view "every culture as having its own science" (2000, ms. p. 2), and discuss several examples of 'indigenous science' (e.g., of the Nisga'a people of northwest Canada, the Yupiaq of southwest Alaska, the Aborigines of northern Australia, and others), and how these contrast with Western science. They emphasize their view that, as Stanley and Brickhouse put it, "the universalist position is not universal at all but only one perspective among others" (1994, p. 395; see also Harding, 1998). Of particular note is Snively and Corsiglia's discussion of the spiritual, magical or animistic character of many of these local sciences, and the disparity between that character and the character of the dominant Western alternative.[19]

These advocates of MSE often suggest that cultural differences of the sort just mentioned preclude the rational comparison and evaluation of alternative, culturally-specific views of science/nature: different cultures have different views of the constitution of nature and/or good science, and the imposition of any putatively 'universal' view of these matters on other cultures constitutes an objectionable cultural hegemony which multiculturalists should reject. The very idea of a universalistic conception of science should be abandoned, it is claimed, as incompatible with the reality of the culturally diverse views of science/nature which multicultural approaches to science education emphasize, and with the respectful treatment of students from such cultures as well. As Kawagley, Norris-Tull, and Norris-Tull (1998, p. 134) put it, a universalistic conception of science "not only diminishes the legitimacy of knowledge derived through generations of naturalistic observation and insight, it simultaneously devalues those cultures which traditionally rely heavily on naturalistic observation and insight." This line of argument, and the common thread in the multiculturalist science education literature that 'universalistic' science is only one among a number of culturally-specific sciences, raises the following issue: Is there any way in which these alternative conceptions of nature/science can be fairly compared and evaluated, with respect to either their scientific merits or their appropriateness for the science curriculum? For example, can any good reason be given for teaching modern biology, chemistry or physics in the multicultural high school science classroom, rather than one or more of the many extant 'local' alternatives, which attribute to the natural world magical, animistic, or spiritual qualities which their modern, Western counterpart eschews, and which eschew experimental techniques which that counterpart embraces as central to its endeavor?

When articulated in these terms, the issue can be seen as a special case of putative incommensurability. Are these alternative, culturally-specific conceptions of science/nature rightly regarded as incommensurable (or incomparable)?

There are two reasons suggested in the literature for thinking so. First, it might be thought that incommensurability follows from the local, non-universalistic character of all conceptions of science and nature. Second, it might be thought that incommensurability follows from the facts that alternative culturally-local views of science and nature incorporate their own local criteria concerning what counts as good science (i.e., a good account of the natural world), and that these criteria cannot themselves be comparatively evaluated in any fair or neutral way. The key premises of the two basic arguments for the incommensurability of 'local,' culturally-specific accounts of science and/or nature, then, are that (a) all such accounts are 'local' and not universal, and only the latter sort of account (which we cannot have) could preclude incommensurability; and (b) the criteria in accordance with which such accounts might be evaluated are themselves incommensurable in that they do not admit of non-question-begging comparative evaluation. I consider these in turn.[20]

2.1 Are All Accounts of Science/Nature Inevitably 'Local'? Are Universalistic Accounts Impossible, In Principle?

Is 'universalistic,' Western science – i.e. the biology, chemistry, physics, etc. taught in high school and university science departments in the industrialized West – just another local science? Are all views of the natural world, and of appropriate ways of investigating and understanding it, inevitably 'local,' i.e. culturally specific?

We must, I believe, answer these questions in the affirmative: all understandings of nature are undeniably local in that they represent the views of scientists who live and work in particular cultural (and historical, etc.) locations. No scientific theory is issued from some heavenly office, laboratory or research team; none represents a God's-eye view of the natural world. There simply is no perspective (available to the likes of us, at any rate) from which such a view can be had. For humans, at least, all science is local.

However, from the inevitably local character of our scientific musings, the impossibility of an appropriately 'universal' science does not follow. The argument under consideration, which moves from the premise that all views of nature (and of science) are local to the conclusion that a universal view of the natural world is impossible, is a *non-sequitur*. The conclusion follows from the premise only on the assumption of an additional premise, to the effect that 'local' and 'universal' are *contraries* (or *contradictories*). If all views are indeed local, and further, if that which is local cannot also be universal, then the conclusion follows. But the additional premise is false: the local can also be, and often is, universal.[21]

I have argued for this claim elsewhere;[22] here I will content myself with offering what I hope you will agree are compelling examples of scientific claims which are both local and universal. Such examples are readily available. Consider claims made in science classes concerning the speed of light or sound, the process of cell division, the convertibility of mass and energy, the mechanisms guiding the migratory habits of birds, the characteristics of sea-floor spreading and of continental drift, the increased risk of contracting sexually transmitted diseases through unprotected sex, or virtually anything else that is routinely taught in 'Western' science courses. All such claims are local in the sense that they were first conceived by particular scientists who lived and worked in particular historical/cultural contexts, and they are recognized, embraced, and taught in equally particular contexts. Not all children learn about these matters in (the cultural equivalent of) their science classes. (The African and Maori students discussed by Hodson, and the Nisga'a and Yupiaq students discussed by Snively and Corsiglia, presumably do not, except when they leave their own cultural contexts and venture out to participate in dominant-culture-sponsored schools.) But the claims are as true of the light, cells, birds, continents, and diseases found in these non-Western cultural contexts as they are of these things existing in Western cultural contexts; they are as applicable to the members of these 'minority' cultures, and their local environments, as they are to those of 'majority' cultures.

This point can be understood in terms of what it is to make a claim, scientific or otherwise. It can also be understood as a point about the 'absoluteness' of truth, or about the 'universal' character of natural facts, or the uniformity of nature. Whatever philosophical tack one takes here, the key is that light, cells, diseases, continents, birds and most other objects of scientific investigation have the properties they do independently of the cultural contexts in which people study them. Those cultural contexts do not determine the character of the properties of the objects and mechanisms which scientists study (although the former clearly influence the scientific accounts we construct concerning the latter). If the aim of science is to shed light on those properties, the cultural locatedness of scientists must be understood to be completely compatible with the universality of their claims. If this is right, the fact that Western science is just one local science among others – that, as Stanley and Brickhouse say, "the universalist position is not universal at all but only one perspective among others" – has no tendency to undermine the 'universal' character of science.

My suggestion is that the 'local/universal' dichotomy, which the argument on the table presupposes, is defective. In particular, at least many of the claims of modern 'Western' science are both. If so, advocates of MSE cannot get from the true premise that all science is local to the conclusions either that a universal science is impossible, or that local sciences are incommensurable in the sense that they cannot be fairly comparatively evaluated with respect to either their scientific worth or their appropriateness for the science curriculum.[23]

However, my argument makes some important assumptions. First, it assumes that the aim of science is to shed light on the actual properties of the objects and phenomena of the natural world. It also assumes that such properties are best studied and understood in particular ways, i.e. the ways recommended and exemplified by 'Western' methodological principles and practices. It assumes, finally, that the criteria relevant to the evaluation of alternative accounts of nature are uncontroversial. All of these assumptions are challengeable, and have indeed been challenged by advocates of MSE. The deepest of these challenges is best put, I believe, in terms of *criteria*: viz., that the criteria in terms of which alternative accounts of nature are evaluated (both for their scientific adequacy and for their appropriateness for the science curriculum) are themselves 'local'; that, being such, they cannot themselves be fairly evaluated in terms of some overarching meta-criteria – that is, it is the criteria, rather than (or not only) the alternative accounts of nature, that are incommensurable; and therefore that the presumption of 'Western' criteria constitutes not only an unjustifed presumption but an objectionable act of cultural hegemony. I turn to this difficulty next.

2.2 Are 'Local' Criteria of Good Science Incommensurable?

The suggestion that the incommensurability of alternative views of science/nature flows from the bounded, limited-to-home-culture applicability of relevant evaluative criteria will no doubt remind readers of the philosophy of science debates concerning incommensurability and incomparability reviewed above. There I suggested that it is a mistake to regard evaluative criteria as solely internal to paradigms, and so inadequate to evaluate alternative paradigms or their associated theories. Rather, the availability of criteria external to paradigms establishes the possibility of non-question-begging comparative evaluation of the scientific merits – i.e., the comparability – of rival paradigms and theories. Can the parallel argument be made here? That is, are there evaluative criteria, not solely internal to local, culturally-specific views of nature/science but in the relevant sense external to such local views, in accordance with which they can be fairly evaluated, in terms of both their scientific merits and the appropriateness of their place in the science classroom? I believe that this parallel argument *can* be made; I try to make it next.

Examples of criteria external to paradigms offered above included those of predictive power, explanatory power, problem-solving ability, testability, precision, scope, fruitfulness, simplicity, and depth. These criteria are also relevant here, in the following way: being criteria in terms of which scientific merit can be judged, they can also appropriately serve as criteria in accordance with which suitability for inclusion in the science curriculum can be determined. That is, these 'good-making features'[24] of scientific theories are also the features of local accounts of nature in terms of which they earn entry into the curriculum.[25]

Of course there are other factors relevant to such curricular decisions; scientific quality is not the only thing that matters educationally. The age and cognitive development of students, their linguistic capacities, their prior educational experiences, the resources available for their science education, etc., are also relevant. Still, given the constraints imposed by these other considerations, science teachers ought to teach the *best* accounts of the natural world available, as determined by these criteria of scientific quality, to all their students. As we have seen, all accounts are local; nevertheless, some are better than others, and it is the best ones that should be taught. Why? Because teaching students what we take to be the best accounts of the natural world is what is required if we are concerned to treat them as independent centers of consciousness, capable of exercising independent judgment and deserving of dignified and respectful treatment. This is what 'treating students with dignity and respect' requires (Siegel 1988; 1997a). The reason for embracing MSE in the first place – the moral requirement to treat all students, particularly those from minority cultures, with justice, dignity and respect, such respect requiring the respecting of students' beliefs and the fostering of their independent, critical judgment – is equally a reason for thinking that science educators must endeavor to evaluate alternative accounts of aspects of the natural world, and determine curricular content accordingly. Given this moral requirement, and the uncontroversial idea that science concerns the understanding of the natural world, the only defensible option available to science educators is to base the curriculum on what we take to be good science. The standard 'good-making' criteria are the ones we do, and should, utilize in order to evaluate alternative accounts of the natural world and determine which accounts are indeed the best accounts available to us at a given time. These criteria are – in virtue of the moral requirement to treat students with respect, and the fact that so treating them requires teaching them what we regard as worthy of their attention, with full attention to the development of their critical abilities to evaluate that which is taught – the criteria to be embraced by science educators committed to multiculturalism. I take it as uncontroversial that 'Western' accounts of particular aspects of the natural world – e.g., biochemical and epidemiological accounts of aspects of the AIDS epidemic, physical accounts of the panoply of astronomical and technological phenomena, geological accounts of earthquakes and volcanoes, meteorological accounts of hurricanes, etc. – satisfy these criteria to a greater degree than (many, though not necessarily all) non-Western, 'indigenous' accounts which involve animism, magic or spiritualism. To the extent that they do, it is the former, rather than the latter, which should occupy center stage in the science curriculum.

Moreover, a central aim of science education is the development in students of the abilities to grasp and apply these criteria, in order to make their own independent judgments of the value of alternative scientific accounts of the natural world, and, eventually, of alternative criteria of good science themselves. If the multiculturalist imperative in education is rightly understood as requiring the respectful and dignified treatment of students from 'minority' cultures (Siegel, 1997a; 1999b), this requirement is operative in science education as a special case. Treating students with dignity and respect requires striving to impart these abilities to them; if science educators fail to endeavor to so equip them, that failure virtually guarantees students' inability to critically assess the scientific beliefs, and conception of science and criteria of good science, of their own cultures. This surely fails to treat them with respect.[26]

This blunt judgment – that treating students with dignity and respect requires helping them to become able to critically assess the scientific beliefs, and conceptions of science and criteria of good science, of their own cultures – places a high value on independent, critical thought. That value is itself often regarded as 'Western' (and 'male'), and it cannot be denied that some cultures place a lower value on such thought, and might disagree that its development is rightly regarded as a central aim of science education (or education more generally). It seems, then, that the line I have been taking requires treating cultures in ways they might reject, which itself seems to constitute objectionable cultural hegemony of the sort MSE aims to rectify.

This is a deep problem, which deserves a more serious response than I can give it here. However, I believe it can be effectively answered, and elsewhere I have tried to answer it in some detail. First, the justification of MSE, and of multicultural educational initiatives more generally, must be taken seriously. If MSE is not justified in the moral terms offered above, on what basis should it be thought to be justified? Why should (science) educators take the demands of multiculturalism seriously? Here the character of multiculturalism, and the case for it, must be articulated and evaluated. My claim here is that the moral (and political) justification offered above is the only successful justification of multiculturalism available; and, further, that 'imposing' *some* values, including both those of independent, critical thought, and those of multiculturalism itself, does *not* constitute objectionable cultural hegemony but is rather an inevitable and unobjectionable aspect of education (Siegel, 1997a; 1999b).[27]

Second, the constitution of *respectful* treatment must be clarified: why does striving to enable students to engage in critical evaluation count as treating them with respect, while failing to so strive does not, even in the case in which they and their culture do not value such critical evaluation? Here difficult conceptual and moral theoretical questions concerning respect must be addressed, but I believe the case can be made (along roughly Kantian lines) that a key dimension of the respectful treatment of others, and one particularly relevant in educational contexts, involves respecting (and, in education, striving to foster) their independent judgment. Basically, the case is this: to fail to do so amounts, in effect, to a disrespectful form of paternalism: 'We know what you should believe; your own (culturally determined) beliefs are of no importance here.'[28] Surely such a cavalier attitude toward others' culturally informed beliefs is disrespectful. How can it be altered so as to be more respectful? First, it must regard those beliefs as *worth critical scrutiny*: perhaps the 'native,' 'local' beliefs are stronger, epistemically, than the cavalier attitude suggests. But this can only be determined by evaluating them, and establishing their epistemic merits, in terms of relevant criteria. Second, it must regard the 'local' believers as themselves capable, at least in principle, of participating in that evaluation; and if they are students, educators must strive to make them actually so capable (as it must for all students). Otherwise, they are treated as 'permanent children,' cognitively defective, or worse – surely the antithesis of respectful treatment. In this way respectful treatment can be seen to require the fostering of critical judgment (Siegel, 1988; 1997a, in progress). Of course, one might think that this disrespectful attitude is the very one I have been advocating: in effect, 'We know what the criteria of good science are; you must simply accept them.' But the universalistic view of science I am advocating makes no such claim. Rather, its advocates should say to science students: 'We believe that the aforementioned criteria of good science are themselves appropriate criteria, but we hold that belief fallibly. Learn what they are and how to apply them; learn our reasons for thinking them appropriate; learn about the philosophical problems concerning them that philosophers of science routinely address; develop the ability and disposition to engage in critical scrutiny of them and the reasons for them; subject them to your own independent judgment; join in the project of critically evaluating them and improving them.' By endeavoring to enable science students to do all this – in effect, to become scientists and philosophers of science themselves – science educators treat their students with respect.

Of course much more needs to be said here. If the 'local believers' reject the epistemic authority or legitimacy of critical judgment, for example, difficult questions concerning the justification of rationality must be faced (Siegel, 1997a, chapter 5). A given culture might also reject the vision of education (as involving fundamentally the fostering of critical thinking) I am here assuming; in this case, fundamental questions concerning the nature and aims of education must be addressed (Siegel, 1988; 1997a). Other difficult problems also remain. Nevertheless, I believe that these two points taken together effectively blunt the challenge here being considered, by suggesting that multiculturalism can be coherently understood and advocated only in the moral terms advanced here, and that such advocacy requires that educators strive to enhance the critical thinking of their students, even when their cultures do not value such thinking highly, by (among other things) directing their critical attention toward that cultural value itself.

I have been arguing that curricular and other decisions made by science educators should be governed primarily by the standard, 'Western' criteria by reference to which scientific quality is normally judged. But it must immediately be granted that there are other legitimate criteria – e.g., Does this view help or hinder environmental health? Does it foster cultural continuity and cohesiveness? Does it speak to deep human needs concerning spirituality, psychological well-being, or community? – which are vitally important, and in terms of which alternative views of science/nature can be judged. But criteria such as these, important though they are, are not criteria relevant either to the assessment of the specifically *scientific* merits of alternative local accounts of nature, or to the determination of the science curriculum. Science education should concern itself not with the general assessment of the human worth or value of cultures generally, but rather with the task of helping students to develop a decent understanding of the natural world (as determined by the 'standard' criteria mentioned above), one which will enable students to appreciate, and in their turn enhance, that understanding. This is not the only sort of understanding of nature, or of science, that is of value to students. But it is the sort that the *science* educator should strive to impart.

I said above that the scientific accounts of the natural world offered by Western science ought to take center stage in the science curriculum, but I do not mean to suggest that all other approaches to the study of the natural world should be denied entry into it. On the contrary, it is essential that we recognize that other 'local' sciences do have an important place in science education. For one thing, their inclusion in the science curriculum makes available to students alternative perspectives from which to view mainstream theories and methods. That availability can sensitize students to hitherto unsuspected assumptions and biases, thereby enabling a more *critical* understanding of standard, 'Western' curriculum content and methodology. It can, moreover, enable students to critically evaluate the criteria by which scientific worthiness is judged, thus providing the opportunity for a deeper and more critical understanding of the very science being studied. In all these ways that inclusion can serve to bring to the fore important *philosophical* questions concerning science, questions whose explicit consideration is essential to the deeper understanding of science which science education aims to achieve (Siegel, 1988, chapter 6). Last but certainly not least, the inclusion of other local sciences in the science curriculum can bring to light the scientific strengths – as judged by 'Western' criteria – of those non-Western accounts of nature. For example, Snively and Corsiglia (2000) detail an impressive array of local ecological knowledge that would handsomely complement (and perhaps in some instances replace) the content of many existing 'Western' courses. The place of other local sciences in the standard curriculum is (or should be) secure. But that place is compatible with the universalistic view of science defended here, according to which 'universalist' criteria of good science can legitimately be appealed to in the formulation and execution of the science curriculum.

I have endeavored to discuss these matters without attempting to rely upon any particular definition of 'science,' or to demarcate science from non-science. The history of this 'demarcation problem' is well-known to philosophers of science; that history does not give us any reason to expect a clear principle of demarcation. Fortunately, such a principle is not required for the resolution of the issues here before us. The key question confronting science education is not 'what is science?' but rather 'what is *good* science?' If my arguments are successful, the standard criteria invoked above, and the contrast between them and other criteria which, however important, are not relevant to the science curriculum questions we have been concerned with here, allows an answer to the latter question which can adequately and legitimately guide science educators.[29]

Finally, let me connect this discussion of criteria directly to the matter of incommensurability. If the account of criteria offered here is correct, incommensurability – or rather, incomparability – is avoided.[30] So is the relativism associated with it, which, to varying degrees, advocates of MSE embrace (Snively and Corsiglia, 2000; Stanley and Brickhouse, 1994; 2000). Local sciences can be compared, and comparatively evaluated, both with one another and with the 'universal' Western alternative. The criteria which Western science utilizes to pick out good science are indeed local, but their use in evaluating non-Western local alternatives is not problematic. That use is not problematic because (a) science, *qua* science – whether Western or any other – seeks the understanding of the natural world, as manifested in the accounts of that world offered by the world's scientists; (b) these criteria enable us to judge the scientific worthiness of such accounts – in that (for example) an account which explains some phenomenon more adequately is a better account, *ceteris paribus*, than one which does not (and parallel claims can be made for the other good-making features captured by the criteria); and (c) the multiculturalist imperative concerning treating students with dignity and respect requires that students be taught the best science available. The availability of such criteria is itself a crucial dimension of an adequate universal conception of science.[31]

3. CONCLUSION

Rather than reviewing my conclusions concerning incommensurability here, I will close by urging philosophers of science to consider science education in their work. Contemplation of standard philosophical issues concerning science is typically conducted without regard to science education. This is I think regrettable for two reasons. First, science educators and scholars of science education need the help that philosophers of science are especially well placed to provide. Second, fundamental philosophical issues concerning science can be illuminated by contemplating them in the context of science education. I hope to have shown that incommensurability is one such issue.

University of Miami

ACKNOWLEDGEMENTS

This paper was written for and presented at the conference on "Incommensurability (and related matters)" organized by Paul Hoyningen-Huene and Howard Sankey, in Hannover, Germany, June 1999. I want to express my appreciation to Hoyningen-Huene and Sankey for conceiving and executing such a fine conference, for drawing philosophical attention once again to the cluster of issues involving incommensurability, and for their excellent editorial suggestions. I am grateful to my commentator at the Hannover conference, Hugh Lacey, and to Edward Erwin and Bruce B. Suttle, for helpful criticisms of an earlier draft.

NOTES

1 Kuhn's publications are listed in the bibliography of Hoyningen-Huene (1993, pp. 273–278, 302). Several of Feyerabend's important discussions of incommensurability are listed in the same bibliography, p. 284.
2 My discussion is limited to what might be called 'epistemological' or 'methodological' (as opposed to 'semantic') incommensurability. Thanks to Howard Sankey for introducing this very useful distinction in his excellent introduction to this volume.
3 Kuhn regards paradigms as including "theory, methods and standards ... usually in an extricable mixture." (1970, p. 109). In what follows I focus mainly on incommensurable theories, in order to side-step needless complications concerning the difficult notion of paradigm.
4 Kuhn's discussions of incommensurability, as well as those of his defenders and critics, are reviewed in Siegel (1987, chapters 3–5).
5 As we shall see, there is more than one such thesis.
6 For discussion of translation and reference, see Hoyningen-Huene (1993) and Sankey (1997); for discussion of standards of explanatory adequacy and attendant conceptions or "definitions" (Kuhn, 1970, p. 148) of science, see (in addition to Kuhn's writings) especially Gerald Doppelt's several papers on this theme (discussed in Siegel, 1987, chapter 4).
7 As Doppelt interprets Kuhn; for discussion see Siegel (1987, pp. 77–92).
8 The complex interweaving of stronger and weaker readings of incommensurability in Kuhn's writings is discussed in Siegel (1987, chapter 3). I must acknowledge that in that chapter I do not clearly distinguish

between the two senses of 'incommensurability' in play here: the first, which denies that incommensurables enjoy a single scale of units of value in terms of which they can be precisely measured; and the second, labelled here 'incomparability,' according to which incommensurability precludes comparability. My claim in the chapter (pp. 60 ff.) that the former precludes the latter clearly depends upon understanding incommensurability in this second sense. Here, under the influence of Chang (1997b), I am understanding it in the first sense. This I hope explains the apparent discrepancy between the chapter and my claims here. On matters of substance, and in particular on the questions of relativism and rationality, the two efforts are I believe completely consistent. In drawing a sharp distinction between incommensurability and incomparability, and in denying both that the former entails the latter and that incommensurable theories are in fact incomparable and so do not admit of rational evaluation, I am following Chang's insightful discussion.

[9] I argue that it does not in Siegel (1987, pp. 60 ff). But see Hoyningen-Huene (1993) for a more sympathetic reading of Kuhn.

[10] For discussion of Kuhn and relativism, see Siegel (in press).

[11] I do not mean to suggest that this list is exhaustive.

[12] Kuhn can be interpreted in such a way that this point is accepted, but its implications for comparability denied, by noting that such second-order criteria function only as 'values' which systematically vary across paradigms. For discussion see Siegel (1987, pp. 56ff).

[13] Here too I am indebted to Chang's insightful analysis (1997b). What I am here calling 'respects' she calls 'covering values.'

[14] I concede that this example is more problematic than the others; I include it only to goad my friend and disputant Gerry Doppelt.

[15] Note that such a choice can, but needn't, involve the judgment that the chosen theory is true. Comparative evaluation can justify this sort of judgment, but also judgments that the superior theory is probably true, worth further investigation, etc. There are many possible doxastic attitudes that can be taken towards theories; my claims concerning comparison and rational evaluation in the text apply indifferently to all such attitudes.

[16] I have not addressed here the many more specific arguments for incommensurability just mentioned; my concern has been rather on the relationship between incommensurability and comparability. For discussion of some of those other arguments, see Siegel (1987). Thanks here to Ed Erwin.

[17] For a partial snapshot of the current state of play, see the essays in Chang, (1997a).

[18] For discussion of Hodson's position, see Siegel (1997b).

[19] For critical discussion, cf. Matthews (1994, pp. 187 ff).

[20] A third reason is definitional: Many would reject magical, spiritual or animistic accounts of nature as (by definition) unscientific; their MSE opponents would counter that the presumed definition of 'science' is in fact contentious in ways not acknowledged by most contemporary science education (in the industrialized West). While there is I think some insight to be gained from pursuing the controversy in these definitional terms, I cannot do so here except in passing. For further discussion of it see Siegel (in progress).

[21] In saying that local, culturally-specific accounts of the natural world can also be 'universal,' I intend to claim simply that the truth and applicability of such accounts are not inevitably restricted to the particular cultural locations in which they are articulated and embraced. For further clarification of this problematic notion, see Siegel (in progress).

[22] Siegel (1997a, chapters 10–12, esp. pp. 174–178; 1997b; 1999a, pp. 192–194; 1999b).

[23] It is perhaps worthwhile to note a telling resemblance between the local/universal dichotomy and the traditional distinction between the contexts of discovery and justification. According to the latter, how a scientist discovers a thesis, theory or claim is irrelevant to its justificatory status: however irrational a given process of discovery might be, the resulting theory might nevertheless be perfectly justified by relevant evidence, subsequent testing, etc. Similarly, a theory's 'locatedness' is irrelevant to its possible 'universal' status: whatever the historical/cultural location in which it is articulated and embraced, it might nevertheless be universal in the sense that it is true, or applicable, in other locations – and indeed, even in all such locations. Thanks to the editors for calling my attention to this resemblance and inviting this note.

[24] I borrow this felicitous expression from Newton-Smith (1981, chapter 5).

[25] Of course these criteria are not themselves transparent; much ink has been spilled concerning the proper understanding of virtually all of them. But these disputes – e.g., that concerning the character of explanation, or of simplicity – are for the most part only tangentially relevant to the science education issues being considered here. As indicated below, the philosophical issues concerning these criteria have an important place in the science classroom.

[26] It would be worth considering, in this context, the views of science education advocated by the 'philosophers of incommensurability,' Kuhn and Feyerabend. I regret that I cannot do this in detail here. See Siegel (1988, chapter 6) for critical discussion of Kuhn's treatment of science education.

[27] No doubt there is a sense in which *education itself* can be hegemonic; cultures, I happily grant, are free to reject it as incompatible with their basic values. But our topic here is science education; we are contemplating the proper character of science education within cultures that are already engaged in, and value, the science education of their children.

[28] Notice that this formulation covers both the case in which what 'we know' *contradicts* local cultural beliefs, and the case in which the former *consists in* those local beliefs. Both cases, whether pro- or anti-existing local beliefs, fail equally to treat students with respect.

[29] For further discussion of the claim that what is needed for science education is not a criterion of demarcation between science and non-science, but rather criteria in terms of which alternative theories can be evaluated as better or worse, see Kitcher (1984) and the essays by Laudan, Quinn and Ruse in Ruse (1988, Part Four).

[30] I hasten to acknowledge that at least some of the advocates of MSE cited above explicitly disavow "radical" incommensurability in their critique of 'universalist' conceptions of science. (Stanley and Brickhouse, 1994, p. 390) I discuss this paper in Siegel (1997b).

[31] There is a further, political point worth making here: if we are genuinely concerned with the just treatment of oppressed, dominated or marginalized peoples, we need to insure that they are able to draw upon the resources of modern science in their struggles for justice. Here we have a very different sort of reason for utilizing Western criteria of scientific quality in determining the content of the science curriculum. See here the telling remark of Noretta Koertge (1981, p. 354).

REFERENCES

Chang, R., ed. (1997a). *Incommensurability, Incomparability, and Practical Reason*. Cambridge: Harvard University Press.

Chang, R. (1997b). "Introduction." In R. Chang, ed., 1997, pp. 1–34.

Coburn, W. and C. Loving. (2000). "Defining 'Science' in a Multicultural World: Implications for Science Education." *Science Education* (in press).

Harding, S. (1998). *Is Science Multicultural? Postcolonialisms, Feminisms, and Epistemologies*. Bloomington: Indiana University Press.

Hodson, D. (1993). "In Search of a Rationale for Multicultural Science Education." *Science Education* **77**: 685–711.

Hoyningen-Huene, P. (1993). *Reconstructing Scientific Revolutions: Thomas S. Kuhn's Philosophy of Science*, trans. A. Levine. Chicago: University of Chicago Press.

Kawagley, A., D. Norris-Tull and R. Norris-Tull. (1998). "The Indigenous Worldview of Yupiaq Culture: Its Scientific Nature and Relevance to the Practice and Teaching of Science." *Journal of Research in Science Teaching* **35**: 133–144.

Kitcher, P. (1984). "Good Science, Bad Science, Dreadful Science, and Pseudoscience." *Journal of College Science Teaching* **14**: 168–173.

Koertge, N. (1981). "Methodology, Ideology and Feminist Critiques of Science." In P. Asquith and R. Giere, eds., *PSA 1980: Proceedings of the 1980 Biennial Meeting of the Philosophy of Science Association, Volume 2*, pp. 346–359, East Lansing: Philosophy of Science Association.

Kordig, C. (1971). *The Justification of Scientific Change*. Dordrecht: D. Reidel.

Kuhn, T. (1970). *The Structure of Scientific Revolutions*. 2nd edition. Chicago: University of Chicago Press.

Kuhn, T. (1976). "Theory-Change as Structure-Change: Comments on the Sneed Formalism." *Erkenntnis* **10**: 179–199.

Matthews, M. (1994). *Science Teaching: The Role of History and Philosophy of Science*. New York: Routledge.

Newton-Smith, W. (1981). *The Rationality of Science*. London: Routledge & Kegan Paul.

Ruse, M., ed., (1988). *But Is It Science? The Philosophical Question in the Creation/Evolution Controversy*. Buffalo: Prometheus Books.

Sankey, H. (1997). *Rationality, Relativism and Incommensurability*. Aldershot: Ashgate.

Scheffler, I. (1967). *Science and Subjectivity*. Indianapolis: Bobbs-Merrill. Second edition Indianapolis: Hackett, 1982.

Shapere, D. (1964). "The Structure of Scientific Revolutions." *Philosophical Review* **73**: 383–394.
Siegel, H. (1987). *Relativism Refuted: A Critique of Contemporary Epistemological Relativism*. Dordrecht: D. Reidel.
Siegel, H. (1988). *Educating Reason: Rationality, Critical Thinking, and Education*. New York: Routledge.
Siegel, H. (1997a). *Rationality Redeemed?: Further Dialogues on an Educational Ideal*. New York: Routledge.
Siegel, H. (1997b). "Science Education: Multicultural *and* Universal." *Interchange* **28**: 97–108.
Siegel, H. (1999a). "Argument Quality and Cultural Difference." *Argumentation* **13**: 183–201.
Siegel, H. (1999b). "Multiculturalism and the Possibility of Transcultural Educational and Philosophical Ideals." *Philosophy* **74**: 387–409.
Siegel, H. (in press). "Relativism." In I. Niiniluoto, M. Sintonen and J. Wolenski, eds., *Handbook of Epistemology*. Dordrecht: Kluwer.
Siegel, H. (in progress). "Multiculturalism, Universalism, and Science Education: In Search of Common Ground."
Snively, G. and J. Corsiglia. (2000). "Rediscovering Indigenous Science: Implications for Science Education." *Science Education* (in press).
Stanley, W. and N. Brickhouse. (1994). "Multiculturalism, Universalism, and Science Education." *Science Education* **78**: 387–398.
Stanley, W. and N. Brickhouse. (2000). "Teaching Sciences: The Multicultural Question Revisited." *Science Education* (in press).

HUGH LACEY

INCOMMENSURABILITY AND "MULTICULTURAL SCIENCE"

Abstract. Responding to criticisms made by Siegel of Kuhn's views on incommensurability, I argue that "multicultural science" is possible. More exactly, values derived from different cultures underlie the salience of questions (e.g., about seeds) that arise when phenomena are considered explicitly as objects of human experience and social value, and that are open to empirical address but side-lined in mainstream modern scientific inquiry. Attention to such values thus points to the potential importance of identifying alternative approaches to systematic empirical inquiry that may involve interesting developments of approaches deployed in gaining "traditional" knowledge, e.g., in agriculture – as we see when we contrast mainstream approaches to agricultural research (those, e.g., that use biotechnological methods) with practically incompatible competing agroecological approaches.

Proponents of *multicultural science* – that is, of legitimate culture-based variations in approaches to scientific practice – often draw upon Kuhn's views about the incommensurability of theories developed within incompatible paradigms. It is fitting, then, that Siegel (this volume) attempts to dispose of strong claims made in the name of multicultural science using the same argument that he deploys to rebut Kuhn on incommensurability. In contrast, I will move a little beyond Kuhn's own views in order to open the door (at least a little bit) to the genuine possibility of multicultural science.

1. A MODEL FOR THEORY CHOICE

Siegel and Kuhn agree that there are multiple criteria for assessing the cognitive value (rational preferability) of theories: empirical adequacy, explanatory power, etc. Like Kuhn I regard these criteria as *values* (Kuhn, 1970, Postscript; 1977) – *cognitive values* (Lacey, 1999a [*SVF*], chapter 3). Then, although rational choice among competing theories cannot be made in the light of only one cognitive value, it may be made according to the model [MODEL 1]: where T_1 and T_2 conflict, T_1 is rationally preferable to T_2 if and only if it manifests all (most?) of the cognitive values more highly than T_2 does.

Siegel seems to endorse MODEL 1 as the unique model of rational theory choice and to see no obstacle to its applicability in principle to resolve all theoretical conflicts. Kuhn, on the other hand, holds that there are conflicts among theories developed within different paradigms which cannot be resolved by applying MODEL 1. Siegel takes him, thereby, to deny in principle the rational comparability of such theories, even suggesting that Kuhn sometimes made the error of inferring such principled incomparability from the existence of multiple criteria (Siegel, this volume, p. 210). For Siegel, recognizing that there are multiple criteria of rational theory choice, a set of cognitive values whose

P. Hoyningen-Huene and H. Sankey (eds.). Incommensurability and Related Matters, 225–239.

items – taken singly – may point *ceteris paribus* towards different theories, takes care of all that is sound in Kuhn's discussions of "incommensurability." Not so for Kuhn!

2. COMPETING STRATEGIES

The disagreement between Kuhn and Siegel derives from certain features that Kuhn attributes to "paradigms" ("disciplinary matrices"). In order to lay the groundwork for further discussion of theory choice (section 3) and for the possibility of multicultural science (section 5), it will suffice to elaborate only one of these features, *viz.* that a paradigm incorporates a *strategy*. A strategy prescribes *constraints* upon the kind of theory that may be entertained in the research practices conducted within a paradigm, in part by prescribing the use of a particular structured lexicon. Reciprocally, because many of the cognitive values (e.g., empirical adequacy, explanatory power)[1] involve relations between theories and relevant empirical data, it *selects* the kinds of empirical data to be sought (and thus, the phenomena and the relevant aspects of them to be observed and experimented upon) and fitted by an acceptable theory, and the appropriate categories to use in order to report them. Adopting a strategy is inseparable from delimiting the realm of phenomena considered of interest for investigation and the kinds of possibilities desired to be encapsulated; and so the reasons for its adoption may include metaphysical and value commitments.[2]

For example, under *materialist strategies* theories are constrained so that phenomena are represented in terms of being generated from *underlying structure, process, interaction and law*, abstracting from any place they may have in relation to social arrangements, human lives and experience, from any link with value (thus deploying no teleological, intentional or sensory categories) – so as to encapsulate the *material possibilities* of things, those possibilities that can be represented as generable from the underlying order in abstraction from whatever social, human and ecological possibilities may also be open to them. Reciprocally, empirical data are selected such that their descriptive categories contain mostly quantitative terms, applicable by virtue of measurement, instrumental and experimental operations.

Aristotelian strategies, in contrast, involve relating phenomena to their places in the cosmos, and their theoretical categories reflect organic metaphors; while relevant data may deploy (e.g.) common sensory and teleological categories. *Agroecological strategies* (the key to my discussion of incommensurability and multicultural science) are also different. Quoting from a foremost exponent:

> The agroecological approach regards farm systems as the fundamental units of study, and in these systems, mineral cycles, energy transformations, biological processes and socioeconomic relationships are investigated and analyzed as a whole. Thus, agroecological research is concerned not with maximizing production of a particular system, but rather with optimizing the agroecosystem as a whole. This approach shifts the emphasis in agricultural research ... toward complex interactions among and between people, crops, soil and livestock (Altieri, 1987, pp. xiv–xv).

The strategies of agroecology do not abstract from the social, human and ecological dimensions of things (*SVF*, chapter 8). Their focus is upon objects – productive and sustainable agroecosystems, and their constituents (seeds, plants, etc) – whose possibilities cannot be reduced to material possibilities (at least to those that can be encapsulated under currently available theories). At the same time, under agroecological strategies, knowledge gained under materialist strategies is drawn upon, e.g., to enable the identification of constituents of agroecosystems, such as micro-organisms and chemical nutrients, that are not directly observable.

"Materialist strategies" is the term I use to refer to strategies, that admit of varieties not all of which involve physicalist reductionism (*SVF*, pp. 68–69), that will readily be recognized as the ones virtually exclusively adopted in modern science. Their virtually exclusive adoption is closely linked with commitment to materialist metaphysics: that all possibilities are material possibilities (*SVF*, chapter 6), but adopting them to investigate particular classes of phenomena and possibilities does not imply any metaphysical commitment (McMullin, 1999; Lacey, 1999b). Indeed, I doubt that there would be any strategies entertained today, including those implicated in my argument for the possibility of "multicultural science," that would not allow an important, albeit subordinate or circumscribed role within them to some uses of materialist strategies (*SVF*, pp. 240–243). In common idiom "science" tends to be applied to inquiry conducted under materialist strategies. I "stretch" the term to apply to any systematic inquiry in which claims about causation and about what is possible are settled effectively by appeal to empirical data and the cognitive values, *and* that in principle can (where appropriate) draw upon knowledge gained under materialist strategies (*SVF*, p. 241).

The single scientific objective to gain theories that manifest the cognitive values highly – or to gain understanding of the world – provides no clear direction for research (*SVF*, chapters 5, 7) so that without adopting a strategy we cannot address coherently and systematically: what questions to pose, what puzzles to resolve, what classes of possibilities to attempt to identify, what kinds of explanations to explore, what phenomena to observe, measure and experiment upon, what procedures to use? When a strategy is adopted answers to these questions are stipulated, but different strategies stipulate different answers, each of which (if successful in consolidating theories that manifest the cognitive values highly) enables us to identify particular classes of possibilities. Under the strategies of agroecology, e.g., the possibilities of producing crops so that people in the region of the production will gain access to a well balanced diet in a context that enhances local agency and well-being and that sustains the environment; and under those of biotechnology[3] (instances of materialist strategies) the possibilities of maximizing crop production under conditions – use of fertilizers, pest and weed management techniques, water, machinery, strains of seeds, etc – that can be widely replicated. In general the possibilities encapsulated by theories developed under different strategies at most overlap, suggesting that a "complete" understanding of the world cannot be obtained (even in principle) under one kind of strategies.

Different strategies require different "playing fields" – Kuhn calls them 'worlds'.[4] Their prescriptions (concerning constraints on theories and selection of data) are practically incompatible; they structure "incompatible modes of [scientific] community life" (Kuhn, 1970, p. 94). They cannot be followed simultaneously within the same practice: the application of the structured lexicon of one (e.g., one that deploys the categories of quantity and law) may preclude that of another (e.g., one that deploys sensory and teleological categories); or interactions with natural objects needed to obtain prescribed kinds of data for one may interfere with the conditions needed for obtaining the kinds of data prescribed by the other (e.g., the interactions involved in investigating crop production as a function of widely replicable methods interfere with maintaining local ecological stability; and engaging in experiment involves modification of "natural" settings, the observation of which provides prescribed data under some strategies). This *practical incompatibility* of strategies, I suggest, is what Kuhn had in mind when he referred to "the incommensurability of competing paradigms" (Kuhn, 1970, p. 150; cf. Hoyningen-Huene, 1993, pp. 208–212). It is also related to claims about the "incommensurability" of *theories*. Only where strategies compete does the issue of the incommensurability of theories arise; theories such as those of quantum mechanics and evolutionary biology, that clearly are about different things, are not said to be incommensurable (Hoyningen-Huene, 1993, pp. 218–221; Carrier, this volume). Some strategies are simply different; the ones implicated in incommensurability *compete* – their "playing fields" impinge on one another threatening to undermine the "games" being played; their respective scientific 'worlds' clash in the shared social "world" that nourishes them (*SVF*, chapter 7).

There are several (not mutually exclusive) ways in which strategies may compete. (1) *Concerning application* to *phenomena of our shared social world*: Numerous interesting phenomena are described using the lexicons of common idiom which is shared by most people regardless of theoretical orientation. In order to explain them recourse is had to available scientific theories; theories are applied *to* them for explanatory and predictive ends; in doing so, objects described in common idiom are identified with objects articulated theoretically. Contradictions may be generated (in the lexicon of common idiom when it incorporates the theoretical lexicons) from attempts to apply theories developed under different strategies to the same phenomena of our common experience (*SVF*, pp. 161–167; cf. Hoyningen-Huene, 1993, p. 221, on Kuhn's recognition that incommensurable theories may be compared in their "empirical potentials"). (2) *Concerning application* in *practical projects*: Theories produced under different strategies may encapsulate possibilities of the same objects (e.g., seeds) – e.g., those of agroecology and those of biotechnology – that cannot be co-realized (to any significant extent) in our shared social world; *on application* they may inform human practices in fundamentally different ways serving incompatible value-outlooks. Then there is competition about which strategy's knowledge will be applied; and, in turn, since research is dependent upon the availability of relevant material and social conditions, there will also be competition for the resources and conditions needed for research under the strategies – the outcome of which is strongly influenced by the social values and forces with interests in the applications. Social competition almost inevitably leads to certain strategies remaining underdeveloped, or to competitors to predominant strategies not being entertained. (3) *With roots in conflicting worldviews*: Far-reaching assumptions (about the general character and possibilities of things) that are drawn upon to support adoption of strategies, or to legitimate applications of their theoretical products *in* practical projects, may be inconsistent. (a) Conflicting metaphysical assumptions, e.g., were involved in seventeenth century competition between materialist and aristotelian strategies. (b) More immediately, there are assumptions that reflect fundamental value commitments of dominant contemporary social forces and institutions (rather than the outcomes of empirical inquiry), and that have become deeply ingrained in the common consciousness of our times, that serve both to legitimate the expanded role of biotechnology in agriculture and to support adoption of biotechnological rather than agroecological strategies in research – assumptions like "the inevitability of economic globalization" and "feeding the world's rapidly expanding population requires the development and implementation of biotechnology-informed agriculture."[5]

The practical incompatibility of competing strategies includes "semantic incommensurability" of theories as one of its features: Posits in theories developed with the structured lexicons of different strategies, and their negations, may bear no entailment relations with one another, since the lexicons may not be isomorphic and thus relevant items of the lexicons cannot be inter-translated. How widespread the phenomenon of semantic incommensurability is, where key theoretical terms from different lexicons actually have different meanings, is a matter of dispute. Kuhn himself narrowed the scope of his claims considerably in writings after *Structure* (Kuhn, 1970). (See Sankey, 1997 and Hoyningen-Huene, 1993 on the developments in Kuhn's views over the years.) Sankey also argues convincingly that semantically incommensurable theories may be about the same objects – thus showing that Kuhn's view, that theories developed under different strategies are about objects in different 'worlds', does not follow directly from semantic incommensurability.[6]

Moreover, the practical incompatibility of competing strategies is the source of "methodological incommensurability": The degrees of manifestation of the cognitive values in theories developed under competing strategies cannot (generally) be compared, so that MODEL 1 is not (generally) applicable to resolving theoretical conflicts across strategies; and perhaps: Choice among competing strategies is under-determined by considerations that rest upon the cognitive values. I think this explicates what is sound in the notion of "methodological incommensurability". Others propose to explicate it more or less as follows: The *methodological standards* for appraising theories are different within different paradigms – where, under "standards," cognitive values and strategic prescriptions tend to be grouped together indiscriminately. Often this indiscriminate grouping occurs as part of an argument that the adoption of methodological standards, including cognitive values, reflects social and cultural contingencies. This is a mistake. Cognitive values (allowing for some historical development) may and ought to be shared across strategies, so that adopting them is not open to the same kind of social and cultural explanation that can be appropriate for strategic prescriptions. Adopting incompatible strategies is consistent with holding that theories developed under all strategies should be appraised for acceptability (though not necessarily for being provisionally entertained or being considered worthy of further investigation or of application – *SVF*, pp. 13–16) in the light of the same set of cognitive values. It is important to separate the (logical) moments of adoption of strategy where social and cultural values have legitimate play, *and* of theory acceptance which rests only on the data and the cognitive values (*SVF*, chapter 10; Lacey, forthcoming).[7] This permits that theories developed under strategies that compete, especially in the second way, may encapsulate different classes of possibilities of the same things, so that strategic competition need not always produce theoretical competition (section 4).

3. A SECOND MODEL FOR THEORY CHOICE

To get at the limits of applicability of MODEL 1, consider empirical adequacy. It gains a concrete interpretation, that enables the degree of its manifestation in a theory to be "measured," only in the context of a strategy that prescribes what kinds of data should be brought into contact with what kinds of theories. A theory is not simply *empirically adequate* but *empirically adequate with respect to specified kinds of data.* But the specified kinds vary with strategy. Thus, across strategies, we cannot (generally) compare empirical adequacy, for the only *relevant* comparison occurs when the data that one theory (should) fit constitute a subset of those that the other (should) fit.[8] Similar conclusions can be drawn about the other cognitive values that concern relations between theory and data. Thus, MODEL 1 applies (generally) only to the comparison of theories developed under the same strategy.

Another model [MODEL 2] can apply to competing fundamental theories T_1 and T_2, generated respectively under competing strategies S_1 and S_2: T_1 is rationally chosen if and only if T_1 manifests highly the cognitive values interpreted according to the (constraint/selection) prescriptions of S_1, and T_2 manifests (has come to manifest) many of them to a low degree interpreted according to the prescriptions of S_2; and S_2 has ceased to be very *fruitful*, i.e., efforts under S_2 – supported by appropriate material and social conditions – to generate theories that manifest the cognitive values highly are not (have ceased to be) successful (*SVF*, p. 229). Where MODEL 2 applies, accepting T_1 is inseparable from rejecting S_2 as a strategy rationally worthy of adoption (and thus inseparable from ruling out hope of finding a "better" successor to T_2 through further research under S_2).

For Kuhn a theory choice made according to MODEL 2 is rational (cf., the analysis of Kuhn's account of the rationality of the theory choices that end a "revolutionary" period offered in Hoyningen-Huene, 1993, pp. 241–243). There is, however, no assurance that a comparison of theories developed under competing strategies can always be made according to it – if, e.g., both strategies are fruitful to some extent, as Kuhn suggests tends to be the case *during* a "revolutionary" period. (Carrier's article, this volume, clarifies the logic of MODEL 2; for further elaboration see *SVF*, ch 7.)

In the light of both MODELs 1 and 2, accepting a theory is portrayed as choosing it rather than a competitor. In the light of MODEL 2, it is also implicated in rejecting the strategy under which the latter has developed; but (contrary to Kuhn – see next section) not necessarily in adopting (for ongoing research) the strategy under which the former has developed. I said above that adopting a strategy is inseparable from identifying the kinds of possibilities desired to be encapsulated in theories. When a theory is soundly accepted, i.e., when it manifests the cognitive values to a high degree of certain domains of phenomena, according to either model it successfully encapsulates relevant possibilities of these domains. That remains untouched, even if the theory is later displaced by a theory that manifests some of the cognitive values more highly.[9] Sound judgments made according to MODELs 1 and 2 are "universalistic" (in Siegel's sense; Siegel, this volume, pp. 219 ff); their grounds are independent (logically) of cultural, moral and social values.[10] Nevertheless, it does not follow, where MODEL 2 is operative, that the strategy under which the chosen theory has developed should be adopted "universalistically",[11] for there may be yet another strategy (S_3) which remains fruitful (or would if research under it were supported with appropriate material and social conditions) and which leads (or would lead) to understanding deemed more *significant* (of more *social value*): more applicable *to* valued objects and *in* valued social activities (Anderson, 1995; *SVF*, pp. 15–16). Accepting a theory according to either of MODELs 1 and 2 does not imply its significance "universally," regardless of cultural, moral and social values held.

4. A MODEL FOR CHOICE OF STRATEGY

When adopting a strategy it is (rationally) proper to consider both fruitfulness and significance. For Kuhn fruitfulness is sufficient, since he maintains that research (in a given field) normally is conducted under a single strategy adopted in the aftermath of theory choices made according to MODEL 2, accompanying which a non-fruitful strategy is rejected, leaving one fruitful strategy in place to be pursued until its fruitfulness is exhausted (see Note 5). It is indeed true that fruitfulness suffices to rationalize the adoption of materialist *rather than* Aristotelian strategies, but not necessarily rather than *any other* strategies (*SVF*, chapter 7). Perhaps at the time of "the scientific revolution" the pertinence of significance was not apparent (though it seems to have been for Bacon), for then – in the light of the emerging values of modernity – materialist strategies were widely deemed highly significant, as they still are (so that the issue of possible alternative strategies, other than the discarded aristotelian ones, simply did not arise): their theoretical products are applied *in* the projects of leading social forces and apply *to* the technological objects that have come to figure so centrally in our lives. In general, theories developed under materialist strategies are highly significant wherever modern technological and economic values, the values of "modernization",[12] are widely held. I have in mind particularly the distinctive role played by control of natural objects and the way in which it is valued in modernity: its scope, centrality in daily life, relative insubordination to other social values, and the deep sense that control is the characteristic human stance towards natural objects, so that the expansion of technologies into more and more spheres of life and into becoming the means for the solution of more and more problems is highly valued (Lacey 1999c, forthcoming; *SVF*, chapter 6). In modernity, then, both fruitfulness and significance seem to point towards adopting materialist strategies (*SVF*: chapters 6 and 7). But they can be pried apart.

The values of modernization (unless highly qualified) are not held universally – for some, because of the social and ecological devastation that modernization is perceived to have borne in its wake; for others because these values contribute to undermine "traditional" values and important dimensions of human well-being that might provide the basis for alternative forms of "development" (*SVF*, chapter 8; cf. Feyerabend, 1999, Part 2, chapter 9). For those who question modernization, the general significance of understanding gained under materialist strategies is compromised (though the fruitfulness of materialist strategies remains established and alternative strategies may freely draw upon the positive knowledge gained under them). That is a good reason for them to seek out alternative strategies that may be able to generate acceptable theories that are more significant in light of their values, theories that might be applicable *in* their preferred ways of life and *to* important phenomena and possibilities in them.[13] (Their quest, however, is not guaranteed success.) There may be two fruitful strategies that enable the identification of possibilities of the same objects, but different classes of them: (e.g.) their possibilities as relevant to technological control, and those connected with furthering the manifestation of competing social values (to which technological implementations are considered subordinate).[14]

Many objects – including experimental and technological phenomena – whose material possibilities are well grasped under materialist strategies are also social objects, objects of social value. Understanding them *fully* (*SVF*, pp. 96–100) requires reference to the human/social descriptions that can also be given of their boundary and initial conditions and of their effects – thus grasping the possibilities that these objects have in virtue of their relations with human beings, social conditions and (more broadly) systems of things. Certain material possibilities cannot be actualized (in historical context) without also actualizing particular social possibilities (e.g., furthering modernization) and undermining others. To focus on the material and leave aside the social possibilities is to abstract; and to limit the domain of science to the former is just an arbitrary stipulation when systematic empirical inquiry that involves the latter can also be conducted.

For example, the possibilities identified by recent biotechnological research on seeds cannot be actualized (under current conditions) without furthering the social process of turning seeds into commodities[15] – which not only changes the character of farming but also has profound ecological and social implications (Lewontin, 1998). That certain kinds of biotechnological possibilities have been identified makes a "universal" claim; that seeds become turned more completely into commodities *also* does.[16] (Both are soundly accepted claims in the light of available data and the cognitive values – cf. the examples in Feyerabend, 1999, p. 251.) But at least the latter does not represent a "universal" value, and so *ceteris paribus* the research practices that increase our grasp of the possibilities of biotechnology are not universally considered of high social value.

Consider the two kinds of possibilities described in section 1 about crop production, the first about it serving local well-being and sustaining the environment, the second about maximizing it under widely replicable conditions. Those who adopt materialist strategies (including those of biotechnology) exclusively are effectively able to address questions about the second, but those about the first cannot be addressed if one abstracts from the social and ecological contexts. Questions about the first, however, will have greater salience to those whose values are in conflict with "modernization". Thus, since they are open to empirical investigation under agroecological strategies, these strategies will rightly take precedence for them; rationality does not require that they wait until the fruitfulness of the competing kinds of materialist strategies has been exhausted. Competing strategies, if they were to gain the material and social conditions necessary to develop, might all be fruitful; they might encapsulate different classes of possibilities of the same objects (e.g., seeds).

Moving beyond Kuhn, I propose yet another model [MODEL 3]: Only adopt a (potentially) fruitful strategy; but if competing strategies S_1 and S_2 are both (potentially) fruitful, adopt the one that may produce understanding of significance for the cultural, moral and social values one holds. MODEL 3 has clear affinities with several of Feyerabend's themes: the methodological importance of proliferation of theories (Feyerabend, 1981); the multiplicity of scientific traditions many of which involve interaction and continuity with "traditional" forms of knowledge; that such proliferation and multiplicity serve to gain access to possibilities that are important for "humanitarian" and "democratic" reasons, but which are otherwise inaccessible (1993); that scientific theories are appropriately evaluated for their ethical value [significance] or their place in a way of life (1999); and that the dominance of a single science [conducted under materialist strategies] threatens human well being (1993, 1999).[17]

MODEL 3, of course, is not a model for theory choice. Its use does not support judgments that theories that might be developed under a discarded (non-adopted) strategy are false, or that theories that fit the constraints of the favored strategy are acceptable – only MODELs 1 and 2 can properly underlie such judgments. Neither does the use of MODEL 3 lead to the definitive rejection of one of two competing strategies; and investigators, who hold different value outlooks, may use it to support adopting different strategies – then the dispute between them is located at the level of moral, social and cultural values and not in judgments involving cognitive values.

Making use of MODEL 3 confronts serious difficulties in practice. In section 2 I listed three ways in which strategies may compete. When they compete concerning application *in* practical projects, the accompanying competition for resources may leave one of the strategies significantly underdeveloped (of interest only to marginalized groups), so much so that its potential for fruitfulness may not be discernible – and so one strategy may be followed as if it has no legitimate competitors, thus disguising the role that values play in supporting its adoption. This situation is reinforced when the competing strategies also draw upon conflicting views, one set of which consists of presuppositions of the value commitments of dominant social forces and institutions.

The combination of the two ways of competing may render proposals developed under strategies that compete with the dominant strategies virtually unintelligible. Think of those forms of agroecology that maintain explicit continuity with traditional agricultural practices in some Third World countries (Altieri, 1987; 1990; Kloppenburg, 1991; Shiva, 1991; *SVF*, chapter 8, and references there). On the one hand, if agroecological research gained the material and social conditions needed to explore fully and systematically the possibilities of crop production serving local well-being, and it were successful, that would threaten the increasing dominance of agribusiness in agriculture (Lewontin, 1998) and challenge some of the presuppositions of "globalization," all of which are well served by scientific developments (e.g., in biotechnology) under materialist strategies. (There is an increasing amount of evidence that such research might be very fruitful: theoretical – Lewontin & Berlan, 1990; experimental – Tilman, 1998; in agricultural practice – Altieri, 1999; Altieri, Rosset & Nicholls, 1997). On the other hand, the categories deployed with materialist strategies have gained such a grip on contemporary imaginative and conceptual powers that the idea that there might be serious alternatives to research under materialist strategies, under which soundly accepted theories might be consolidated, simply tends not to be considered (*SVF*, 126–130). Then, in effect, the prescriptions of materialist strategies become treated as if they were cognitive values.

The tendency to mistake dominant strategic prescriptions for cognitive values partly accounts for the lack of recognition of the role that MODEL 3 should play. This mistake may be reinforced by the fact that rarely is a strategy adopted as a result of explicit deliberation, but rather in the first instance an investigator generally enters into research activities under established strategies, so that a novice (e.g.) learns how to follow strategic prescriptions and how to estimate the degrees of manifestation of the cognitive values at one and the same time. Thus we should be wary of defenses of dominant strategies of the "It's the only game in town" type. It may be the only game in town because (whether deliberately or without awareness) competitors have been suppressed, not recognized, or denied resources for their development. The reasons that support adopting a strategy generally are articulated *post hoc*; then the role of MODEL 3 may become apparent, for empirical inquiry under any one type of strategies is clearly inadequate to establish that all possibilities are of the kind that can be encapsulated under these strategies.

Grounding the adoption of a strategy with arguments that deploy MODEL 3 explicitly is likely to lead to the recognition that under it various classes of possibilities will be left unexplored, and that assumptions about some of the unexplored possibilities are among those that support both the value of adopting the strategy and the legitimacy of applying its products. For example, innovations with transgenic crops – and indeed adopting the strategies of biotechnology research itself – are often legitimated by appeal to the assumption "It's necessary to feed the world's rapidly growing population" (Nuffield Council, 1999) which in turn assumes that other agricultural approaches, including those of agroecology, cannot produce the required food (Lacey, forthcoming). These assumptions are not the outcome of research under materialist strategies, and cannot be; they could only be established with research that involved the significant development of strategies such as those of agroecology (*SVF*, chapters 8, 10). Note that it is not enough to show that biotechnology provides a means to increase crop production, for that is compatible with agroecology being a comparable or superior competitor (and also with the harmful side-effects of applying biotechnology technologies outweighing their benefits).

Recognizing these things should engender, according to traditional notions of scientific *neutrality* (*SVF*, chapter 10), a tolerance to provide as much social space as possible to explore alternative strategies, and to provide (to the extent possible) conditions for the development of multiple strategies for scientific (systematic empirical) research, (say) for those of agroecological as well as biotechnological strategies. It should also give rise to a community of investigators that endorses the methodological legitimacy and the possible fruitfulness of a multiplicity of strategies, that regards disputes about competing strategies as a normal part of scientific activity, that permits applications to be informed fully by knowledge gained under various strategies, and in which investigators temper commitments they have to the values that are linked to the strategies they adopt with humility and tolerance for other approaches (both scientific and socioeconomic). This is the context for the proper use of MODEL 3. (I do not suggest that this context can be brought about simply by choices and judgments made within the scientific community, without broader changes in socioeconomic ordering.)

No paradoxical kind of relativism is hereby implied. It is just that under the different strategies (in principle) different classes of possibilities are encapsulated that may not be able to be co-actualized in the same places and projects. Furthermore, one may endorse multiple strategies in the scientific community, while choosing to adopt particular ones oneself, and while recognizing that research conducted under materialist strategies is capable of indefinite expansion – in part because it is of the nature of experiment and technology to create phenomena and the spaces in which certain phenomena are encountered, understanding of which is produced under these strategies. But indefinite expansion of material possibilities is compatible with not all possibilities reducing to material ones. I foresee, for example, no limit to the realm of possibilities that might be uncovered by research in biotechnology, but that provides no reason to hold that the possibilities open to things in virtue of their locations in agroecological systems are reducible to the material possibilities of the constituents of these systems.

5. "MULTICULTURAL SCIENCE"

Thus the door is opened to "multicultural science": values derived from different cultures and embodied in radically different forms of life point to the salience of questions, open to empirical address, about material objects (e.g., seeds) not abstracted from their place in human experience and social structures, that are side-lined by the mainstream of modern science, and thus point to the potential value of identifying alternative strategies which may involve rich developments of the strategies deployed in gaining "traditional" knowledge (*SVF*, chapter 8 and references there; Altieri, 1987; 1990; 1999; Altieri, Rosset & Nicholls, 1997; Kloppenburg, 1991; Lewontin & Berlan, 1990). In principle, there is no reason why there should not be strategies, incompatible with *uniquely privileged* materialist strategies and competing with them concerning application in practical projects, that are fruitful. Furthermore, matters of social justice, informed by the values of various cultures, endorse the potential significance of products they might generate; e.g., they might successfully identify novel possibilities about crop production and local well-being.

I do not know how far the claims of multicultural science can be pushed. That needs to be established case by case in the light of careful investigation. *Prima facie* that a culture's cosmovision includes a view like "that nature is benevolent" (Siegel, this volume, p. 213) is not *per se* a reason to dismiss the products of the systematic empirical (scientific) inquiry associated with it – unless the view is shown to be inconsistent with a theory soundly accepted under materialist (or other) strategies. It is not enough that it is inconsistent with materialist metaphysics, a view often (falsely) thought to be a presupposition of modern science. In this article, I just want to get the camel's nose of multicultural science under the tent – or into the classroom. The occupants of the tent have some big sticks and perhaps they will succeed in beating the camel back. But the camel, I think, smells of reason and social justice. It also carries an unambiguous record of fruitfulness: the seeds developed in the course of traditional agroecology have become the *sine qua non* for the development of transgenic seeds (Kloppenburg, 1987).

6. CONCLUDING REMARKS

What lies behind the disagreement between Siegel and me? It is easy enough to pinpoint. Siegel holds the research conducted under different strategies has a common object: "the natural world" (Siegel, this volume, pp, 216, 220) or parts of it, or objects in it "that have the properties they do independently of the cultural contexts in which people study them, [where] such properties are best studied and understood in ... the ways recommended and exemplified by 'Western' methodological principles and practices (p. 216)" [i.e., under materialist strategies]. Kuhn, of course, disagrees with this. For him the objects investigated under different strategies are largely different, occupants of 'different worlds,' which are partly constituted by and in the research activities themselves.

I, not departing very far much from Kuhn, have suggested that the object of inquiry under materialist strategies may be considered to be the material possibilities of things (*SVF*, chapter 6; Lacey, 1999c, forthcoming). In all cultures some of the material possibilities of things are salient (since no possibilities are realized without also realizing certain material possibilities), and that is why I expect overlap between materialist and any other strategies concerning what the object of inquiry is. But materialist possibilities do not exhaust the possibilities of things insofar as they make contact with our lives (though perhaps they do, in principle, of the possibilities of technological control). Furthermore, theories developed under different strategies may compete in their accounts of what possibilities are realizable in the "world" of daily life by and in interaction with the objects encountered in the actual social world.

In multicultural science – more exactly, under a range of strategies reflective of a variety of cultural outlooks – material objects may be investigated for the social (including agroecological) possibilities in which they are implicated. If science is systematic empirical inquiry that enables us to grasp "the world in which we live" and that serves to inform our practical activities, there is a certain urgency to exploring concretely the potential scope of multicultural science, and no reason to accept that a unique strategy will pass the test of fruitfulness. Of course, given that strategies linked with different cultural values will compete with materialist strategies concerning application *in* practical projects, major barriers to such exploration will remain in place.

That there are practically incompatible competing strategies, all of which may be fruitful, is the key to my argument. They are the source of the methodological incommensurability of competing theories developed under different strategies (section 2). But fruitful competing strategies need not generate competing theories; their respective soundly accepted theories may encapsulate well different classes of possibilities of the same phenomena, whose realization in practical applications may serve different cultural, moral and social values. Thence, competing strategies are also the source of the possibility of multicultural science. Reflecting on this common source of methodological incommensurability and the possibility of multicultural science also enables us to discern the impediments to the latter's growth and even to the recognition of its intelligibility.

Swarthmore College

ACKNOWLEDGEMENTS

I wish to thank the National Science Foundation (SES–9905945) for partial support in preparing this chapter, Harvey Siegel for comments, and the editors for very helpful suggestions.

NOTES

[1] Other cognitive values concern relations among theories: inter-theoretic consistency, consilience. Others still concern features of theories themselves: consistency, simplicity (*SVF*, pp. 58–62). Shapere's analysis of reasons for scientific change could be usefully applied to enrich the discussion of cognitive values (Shapere, this volume).

[2] While the term "strategy" is my own (*SVF*, pp. 21–22, chapters 5, 8, 9), the idea derives from Kuhn's view

that research is always conducted in the light of commitments about "legitimate methods, problems and standards of solution" (Kuhn, 1970, p. 48, also pp. 103–104; cf. the "second type of component of the disciplinary matrix" – "Postscript," p. 184; see also Lacey, 1990). Here I emphasize the roles of strategies in constraining theories and selecting data. Further roles involve a variety of heuristic tactics: when and how to deploy *ad hoc* hypotheses, sources of analogies and metaphors, etc.

[3] "Biotechnology" sometimes is used to encompass any systematic harnessing of a biological process in the production of goods. It also, in many recent discussions, has a much narrower usage in which it refers only to applications of advanced molecular and cell biology, and specifically of the genetic modification of organisms (Smith, 1996). In this paper I am employing this narrower usage.

[4] The interpretation of Kuhn's use of the metaphor of "different worlds" has been much disputed. I, more concerned with explicating a Kuhn-inspired view that I regard as sound than with the details of Kuhnian scholarship, emphasize that a scientific 'world' is linked with a "form of life," its required skills (habits, expectations and sense of what is possible), its organizing structures, and its ways of actively engaging in research (cf. Rouse, 1987); and that different 'worlds' are all located in the shared sociohistorical "world," whose phenomena they may enable us to understand and whose practices they may inform (*SVF*, 149–154; Lacey, 1999d). Hoyningen-Huene (1993), on the other hand, offers a neo-Kantian interpretation, identifying scientific 'worlds' with "phenomenal worlds" and "the world" with the epistemically inaccessible "world-in-itself".

[5] Kuhn considers only competition of types (1) and (3a). This reflects his views that competition among strategies occurs (or should occur) only at times of "scientific revolution," and that social factors, including those relating to applied science – though necessary for investigation to proceed and perhaps to motivate the initial exploration of certain strategies – are not among the grounds for adoption of the strategies that guide "normal science" (see section 4).

[6] Note also: First, even where the lexicons of observational reports (produced under different strategies) have sufficient items in common for there to be contradictory predictions made, that does not suffice to show inconsistency between the theories from which the predictions are drawn, since making predictions draws upon not only the theories but also a body of auxiliary hypotheses (Duhem-Quine). Secondly, where the competition between strategies is of type (1), it is still not possible to show inconsistency between theories developed under them using only the resources of the lexicons prescribed by the strategies – and the question of which theory best applies to the commonly described phenomena is answered only by determining which theory to accept (*SVF*, p. 229).

[7] I suggest that making this separation, and not grouping together under "methodological standards" both strategic prescriptions and cognitive values, enables us to transcend Doppelt's "moderate relativism" (Doppelt, this volume), while endorsing his account of Kuhn's legacy, "the promise of a more interdisciplinary model of scientific knowledge and the logic of its growth" (Doppelt, this volume, p. 159).

[8] Comparing quantitatively the actual numbers of data fit by the respective theories is *irrelevant*. Quite apart from the difficulties involved in individuating and therefore counting data, the numbers must reflect contingencies of the research processes.

[9] Cf., "In so far as Newtonian theory was ever a truly scientific [i.e., soundly accepted] theory supported by the evidence, it still is. Only extravagant claims for the theory – claims that were never properly parts of science [i.e., were not about phenomena of which the theory was soundly accepted] – can have been shown by Einstein to be wrong. Purged of these merely human extravagances, Newton's theory has never been challenged and cannot be" (Kuhn, 1970, p. 99).

[10] I have called them *impartial* (*SVF*, chapters 4, 10). Note that having the developed cognitive abilities to make such "universalistic" judgments, or being suitably located to be able to make them or having an interest in doing research on which they would be based, may depend (causally) upon adopting particular values. It is important to distinguish the grounds upon which such judgments are rationally based and the factors that explain that the conditions are available for them to be made (*SVF*, pp. 331–336). The latter, but not the former, may include social and cultural values. Note also that, in view of the second type of competition introduced above, there can be serious practical impediments to using MODEL 2 to make impartial judgments.

[11] The appeal of the notion that there be "universalistically" adoptable strategies, along with that of scientific knowledge as a public good (long associated with the view that science is *neutral* – *SVF*, chapter 4; Lacey, 1999c, forthcoming), is under strong challenge at the present time, especially in the light of the intertwining of biotechnology research with business interests that has been reinforced by extensions of patents to cover, e.g., genes. *Neutrality* may be rapidly disappearing as part of the self-understanding of the scientific community, as scientific research strategies progressively tend to become linked piecemeal with a variety of competing economic interests; then it becomes not so strange to think of contemporary scientific practices

as reflecting a particular culture (albeit linked with "globalization").

[12] For reasons of space I oversimplify and gloss over a range of complex, subtle and contested matters (but see *SVF*, chapters 6, 8).

[13] Cf.: "... professionals dealing with the ecological, social and medical parts of developmental aid have by now realized that the imposition of 'rational' or 'scientific' procedures, though occasionally beneficial (removal of some parasites and infectious diseases), can lead to serious material and spiritual problems. They did not abandon what they had learned in their universities, however; they combined this knowledge with local beliefs and customs and thereby established a much needed link with the problems of life that surround us everywhere ... " (Feyerabend, 1993, p. xiv).

[14] Cf. my discussion of "feminist strategies" (*SVF*, chapter 9).

[15] See *SVF*, chapter 8 and its extensive documentation (also Shiva, 1991; Lewontin, 1998). Seeds from which genetically modified plants are grown are *at one and the same time*: (a) Biological entities – under appropriate conditions they will grow into mature plants from which (e.g.) grain will be harvested; so they are subject to genetic, physiological, biochemical, cellular, developmental, etc analyses. (b) Parts of various ecological systems. (c) Human-produced entities – developed by scientists (mainly in corporate but also in university and other laboratories) by genetically modifying seeds that (for the most part) had previously been selected in the course of the practices by numerous farmers over many centuries (Kloppenburg 1987; 1991), and produced now (largely) by capital-intensive corporations; they cannot be understood (as seeds grown under different conditions can sometimes be) simply as the "natural" product of plants, simply (and sometimes not at all) as a part of the grain harvested, or as entities that regenerate themselves annually as a normal part of the crop (Shiva, 1991). (d) Commodities: bought and sold on the market; they are "property" whose users may not be their owners; they (sometimes) can be patented and otherwise regulated in accord with intellectual property rights. (e) Objects of scientific research under materialist strategies.

[16] I agree with Siegel (this volume, pp. 214–216) that there are plenty of items of knowledge that are both *local* and *universal* – produced under conditions that reflect particular social and cultural values *and* soundly accepted in virtue of high manifestation of the cognitive values. Not all of these items have been produced under materialist strategies. Being "universal," in this sense, means that the item has been properly and rigorously appraised favorably in light of the cognitive values; it may not be universally significant. I see no reason why the curriculum of science education should not be responsive to matters of significance. (Not every item of "universal" knowledge has a worthy place in the curriculum, or in all science curricula – in fact, I think the curriculum that Siegel (this volume, p. 215) endorses is itself responsive to significance for the values of modernization.) Should we teach (with respect to agriculture) biotechnology, agroecology, both, or one subordinate to the other?

[17] Cf.: "... being related to each other in lawful ways, [the entities unearthed by science] can be manipulated or predicted by using these laws. There can be new combinations of them and new entities may in this way arise at the phenomenological level. But these entities are important only if the resulting world is pleasant to live in, and if the gains of manipulation more than compensate for the losses entailed by the losses of unscientific layers. The objection that the entities and laws that connect them are 'real' and that we must adapt to them, no matter how dismal the consequences, has no weight" (Feyerabend, 1999, p. 12).

REFERENCES

Altieri, M. (1987). *Agroecology: The Scientific Basis of Alternative Agricultures*. Boulder: Westview Press.

Altieri, M. (1990). "Why Study Traditional Agriculture?" In C. Carroll, J. Vandemeer and P. Rosset, eds., *Agroecology*, pp. 551–564, New York: McGraw-Hill.

Altieri, M. (1999). "The Ecological Role of Biodiversity in Agroecosystems." *Agriculture, Ecosystems and Environment* **74**: 19–31.

Altieri, M., P. Rosset and C. Nicholls. (1997). "Biological Control and Agricultural Modernization: Towards Resolution of Some Contradictions." *Agriculture and Human Values* **14**: 27–58.

Anderson, E. (1995). "Knowledge, Human Interests, and Objectivity in Feminist Epistemology." *Philosophical Topics* **23**: 27–58.

Carrier, M. (this volume). "Changing Laws and Shifting Concepts: On the Nature and Impact of Incommensurability." pp. 65–90.

Doppelt, J. (this volume). "Incommensurabilty and the Normative Foundations of Scientific Knowledge." pp. 159–179.

Feyerabend, P. (1981). *Problems of Empiricism*. Cambridge: Cambridge University Press.

Feyerabend, P. (1993). *Against Method*. 3rd edition. London: Verso.

Feyerabend, P. (1999). *Conquest of Abundance: A Tale of Abstraction versus the Richness of Being*. Chicago: University of Chicago Press.

Hoyningen-Huene, P. (1993). *Reconstructing Scientific Revolutions: Thomas S. Kuhn's Philosophy of Science*, trans. A. Levine. Chicago: University of Chicago Press.

Kloppenburg, J., Jr. (1987). "The Plant Germplasm Controversy." *Bioscience* **37**: 190–198.

Kloppenburg, J., Jr. (1991). "Social Theory and the De/Reconstruction of Agricultural Science: Local Knowledge for an Alternative Agriculture." *Rural Sociology* **56**: 519–548.

Kuhn, T. (1970). *The Structure of Scientific Revolutions*. 2nd edition. Chicago: University of Chicago Press.

Kuhn, T. (1977). "Objectivity, Value Judgment and Theory Choice." In Kuhn, *The Essential Tension*. Chicago: University of Chicago Press.

Lacey, H. (1990). "Interpretation and Theory in the Natural and the Human Sciences: Comments on Kuhn and Taylor." *Journal for the Theory of Social Behavior* **20**: 197–212.

Lacey, H. (1999a). *Is Science Value Free? Values and Scientific Understanding*. London: Routledge. [Referred to throughout the text as "*SVF*"]

Lacey, H. (1999b). "On Cognitive and Social Values: A Reply to My Critics." *Science and Education* **8**: 89–103.

Lacey, H. (1999c). "Science and Values – 2." *Manuscrito* **22**: 165–203.

Lacey, H. (1999d). "Philosophically Reconstructing Kuhn." *Metascience* **8**: 188–192.

Lacey, H. (forthcoming) "The Ways in Which the Sciences Are and Are Not Value Free." In P. Gardenfors, K. Kijania-Placet and J. Wolenski, eds., *Proceedings of the 11th International Congress of Logic, Methodology and Philosophy of Science*. Dordrecht: Kluwer.

Lewontin, R. (1998). "The Maturing of Capitalist Agriculture: Farmer As Proletarian." *Monthly Review* **50**: 72–84.

Lewontin, R. and J. Berlan. (1990). "The Political Economy of Agricultural Research: The Case of Hybrid Corn." In C. Carroll, J. Vandemeer and P. Rosset, eds., *Agroecology*, pp. 613–628, New York: McGraw-Hill.

McMullin, E. (1999). "Materialist Categories?" *Science and Education* **8**: 36–41.

Nuffield Council on Bioethics (1999). *Genetically Modified Crops: The Social and Ethical Issues*. London: The Nuffield Foundation.

Rouse, J. (1987). *Knowledge and Power: Towards a Political Philosophy of Science*. Ithaca: Cornell University Press.

Sankey, H. (1997). *Rationality, Relativism and Incommensurability*. Aldershot: Ashgate.

Siegel, H. (this volume). "Incommensurability, Rationality and Relativism: In Science, Culture and Science Education." pp. 207–224.

Shapere, D. (this volume). "Reasons, Radical Change and Incommensurability in Science." pp. 181–206.

Shiva, V. (1991). *The Violence of the Green Revolution*. London: Zed.

Smith, J. (1996). *Biotechnology*. 3rd edition. Cambridge: Cambridge University Press.

Tilman, D. (1998). "The Greening of the Green Revolution." *Nature* **396**: 211–212.

PETER BARKER

INCOMMENSURABILITY AND CONCEPTUAL CHANGE DURING THE COPERNICAN REVOLUTION

Abstract. Kuhn's latest account locates incommensurability as a mismatch between taxonomies of natural kind terms. In collaboration with Hanne Andersen (University of Copenhagen) and Xiang Chen (California Lutheran University) I have developed a more general account along the same lines using the dynamic frame notation introduced by cognitive psychologist Lawrence Barsalou (Emory University, Atlanta). Here I apply a simplified version of this model to the conceptual systems of Ptolemaic, Copernican and Keplerian astronomy. I conclude that Copernicus's astronomy is only minimally incommensurable with Ptolemy's, but that Kepler's is strongly incommensurable with both.

1. KUHN AND THE COPERNICAN REVOLUTION

Thomas Kuhn's account of the Copernican revolution is problematic in several ways. Although this episode became the basis for Kuhn's first book-length study of scientific change (1957), its treatment in *The Structure of Scientific Revolutions* (1962; 1970) is uneven and not well integrated with the main theses of the book. There are specific problems in applying the concepts of anomaly, crisis and incommensurability. But, perhaps because of the attention it received from other historically oriented philosophers of science at about the same time, the Copernican revolution remains one of the central examples of incommensurability and conceptual change.

The historical picture that Kuhn presented in his book *Copernican Revolution* has proved durable, although we can now recognize that it was limited by the time and place where it was written. The changes in the historian's view of Copernicus's work and its aftermath have recently been examined in an excellent paper by Robert Westman (1994). My own list of the most important changes since Kuhn wrote his original historical account would begin with the identification by Bernard Goldstein in 1967 of that portion of the *Planetary Hypotheses* in which Claudius Ptolemy supplemented the mathematical models of the *Almagest* with a description of a set of physical shells (or, more properly for Ptolemy, 'rings') that carried individual planets and generated the deferent and epicycle motions by rotation (Goldstein, 1967). These models were appropriated and improved by Arabic astronomers and natural philosophers. But they were transmitted in very much their original form to the Latin West, where they became the basis for a genre of astronomical textbook called a *theorica* (plural: *theoricae*) and they rose to special prominence during the Renaissance (Van Helden, 1985). The models arising from the *Planetary Hypotheses* fill the heavens with spheres doing the work of deferents and epicycles. They show that astronomy was not a purely calculative or fictionalist discipline during the Middle Ages and Renaissance, at least in intent (Barker and Goldstein, 1999).

P. Hoyningen-Huene and H. Sankey (eds.). Incommensurability and Related Matters, 241–273.

Equally important additions to the historical information unavailable to Kuhn in the 1950s are the general recognition that Copernicus himself was very specifically indebted to one of the Arabic traditions that did most to improve Ptolemy (Swerdlow and Neugebauer, 1984; di Bono, 1995), and that the relationship between science and religion during the reception of Copernicus's work was not antagonism but mutual support. Lutherans, for example, played a special role in spreading Copernican astronomy, although they did not endorse Copernican cosmology (Westman, 1973; Barker, 2000; Barker and Goldstein, 2001). Kuhn's original account did not observe a clear boundary between astronomy and cosmology. But this boundary is clearly drawn in sixteenth century sources. Astronomy concerns only the calculation of the positions of celestial objects. And the position that is calculated is not a point in three-dimensional space but a direction that projects the position of the planet against the hypothetical sphere of the heavens. Cosmology, which specifies the actual configurations of the cosmos, and the sizes and distances of celestial objects, is a different science that operates from separate and more fundamental principles. Observing this boundary more clearly will lead to some interesting revisions in our account of the conceptual changes underlying the Copernican revolution.

According to Kuhn, and perhaps this is his most fundamental insight, each scientific community supports a relatively stable conceptual system, and the changes in these systems are the most important intellectual events in the history of science. The Copernican revolution remains one of the two or three most important such events.[1] For this reason, the changes in the historical narrative of the Copernican revolution have preserved the core of Kuhn's account. Astronomy from the time of Copernicus to the time of Newton was reconstructed on new foundations. The change is seemingly total: from a closed world to an infinite universe; from a finite, spherical and full Aristotelian cosmos with an astronomy based on Ptolemy to an infinite and largely empty mechanical universe of matter-in-motion, with an astronomy that, like everything else, is governed by Newton's laws. But we lack an account of conceptual systems which would allow us to describe or understand these changes in any detail. My aim later in this paper will be to suggest at least the beginnings of such an account.

If we turn our attention to the philosophical work begun by Kuhn in *Structure of Scientific Revolutions*, the Copernican revolution remains a conspicuous failure in his account of scientific change. The historical account of the Copernican revolution did not fit the philosophical model proposed in *Structure* when it was published, and it does not fit now. In *Structure* Kuhn proposed what Paul Hoyningen-Huene has elegantly described as a 'phase model' of scientific history (Hoyningen-Huene, 1993). The normal science tradition of one paradigm was supposed to generate one or more anomalies; the anomaly was supposed to generate a crisis during which a rival paradigm emerged, and the revolution was supposed to conclude the crisis. The new normal science tradition supported by the new paradigm resolved the anomaly that led the old paradigm to crisis, and for this and other reasons gained the allegiance of the scientific community and supplanted its predecessor.

In the case of the Copernican revolution it was easy enough to identify the rival normal science traditions. Before the revolution they were the tradition of Ptolemy in astronomy, and more generally the tradition of Aristotle in natural philosophy. After the revolution the normal science tradition was that of Newton, and more generally that of mechanical physics. But from the outset Kuhn found himself unable to specify the sequence of anomaly and crisis leading to the revolution that would connect these normal science traditions. The difficulty was especially acute for the work of Copernicus himself. The obvious candidate anomalies were discrepancies in the calendar. Agitation for calendar reform was prominent during Copernicus's life, and *De Revolutionibus* conspicuously redefined the astronomical year. But the final reform leading to the modern Gregorian calendar was carried through by the great Jesuit mathematician and astronomer Christopher Clavius a generation after Copernicus's death. And Clavius remained loyal to Ptolemaic astronomy, indeed he was its last great defender.[2] In Kuhn's original account rival paradigms were logically incompatible; the successor paradigm permitted the derivation of (an account of) the crisis causing anomaly, the predecessor did not. Conversely, if an effect could be accommodated by the predecessor paradigm, it could not be a crisis causing anomaly. Clavius saw no incompatibility between reforming the calendar and continued support for Ptolemaic astronomy, so calendar reform could not have been the anomaly that caused the Copernican revolution. Indeed, Kuhn acknowledged that the historical case for calendar reform as an anomaly was insufficient motivation for Copernicus, and made some remarks about the 'aesthetic' appeal of the Copernican system that were perhaps over-interpreted as offering an alternative reason for the change.

New historical work on the 'crisis state' also had a paradoxical effect. The widespread belief that Ptolemaic astronomy had adopted *ad hoc* accretions of epicycles by the time of Copernicus was shown to be unfounded (Gingerich, 1975). Sixteenth century astronomers continued to be satisfied with straightforward deferent plus epicycle systems. Even contemporary remarks like those of King Alfonso the Tenth of Castile allegedly referring to the crisis state were shown to have other meanings (Goldstein, 1991).[3]

All this aside, the Copernican revolution remained a favorite example of incommensurability. What could be a clearer example of the choice between incompatible paradigms competing to explain the same data than Kepler and Tycho Brahe standing on a hill watching the dawn or, if you prefer, Copernicus and Erasmus Reinhold? The first member of each pair sees a stationary sun revealed by the revolution of the earth; the second sees a moving sun appearing over the horizon of a stationary earth. At a more sophisticated level, the Copernican revolution offered a clear example of a key characteristic of the kind of revolutionary change that causes incommensurability: the silent amendment of meaning that makes historical sources hard to understand. Before the revolution all celestial bodies were referred to as 'stars'. The sun, moon, Mercury, Venus, Mars, Jupiter and Saturn were all referred to as planets. The earth was not a star or a planet. After the revolution, the sun was grouped with the fixed stars and was no longer a planet. The earth had become a planet and the moon had declined in status from a planet to a satellite. By almost any account of incommensurability these changes rendered the concepts after the revolution incommensurable with the concepts in use before.

But there are still some problems with the concept of incommensurability as it applies to the Copernican revolution. Although the main historical actors clearly did use the concept 'planet' in the way just outlined, none of the expected concomitants of incommensurability are evident in the historical record. There is no real evidence of mutual incomprehension between Ptolemaists and Copernicans, or failures of communication, or that actors were conceptually constrained from evaluating rival paradigms. Ptolemaic astronomers understood Copernicus's new concept of a planet well enough to be able to explain it in public. The problem was not that they failed to understand it – the problem was they did understand it and thought it was obviously wrong.

Many of the difficulties considered so far arise because we have been tempted (and in the past I was certainly tempted) to treat Copernican astronomy and its rivals as monolithic systems that had to be challenged or overturned by a single anomaly (or range of anomalies), and decisively rejected at a single moment in time. But the historical sequence leading from the general acceptance of Ptolemaic astronomy to the general acceptance of the Newtonian form of Copernicanism is a record of many small changes. Kuhn's later account of concepts allows the flexibility to accommodate such changes while retaining and refining the account of incommensurability, although Kuhn himself never returned to apply this account specifically to understand the Copernican revolution.

In what follows I want to use a particular model of concepts – the dynamic frame model – which incorporates the key features of Kuhn's mature account (see esp. Kuhn, 1990a; 1990b; 1991; 1993; Hoyningen-Huene, 1993) to try to show some of what I think really happened during the Copernican revolution. Although it is not my intention to present a detailed justification for the dynamic frame model, I hope the case I present will serve to stimulate work along similar lines and to support two general claims. First, for a long time philosophers have talked about conceptual schemes without being able to say anything very satisfactory about their structure. Attempts to model them as deductive structures, for example, have been a dismal failure. I suggest that the frame model now gives us a tool for representing (portions of) conceptual schemes in a concrete way. Second, the frame model allows us to pinpoint where incommensurability occurs in a conceptual structure and to assess its severity. I also hope to explain a number of otherwise puzzling features of the historical situation – for example the absence of the expected negative consequences of incommensurability for the immediate reception of Copernicus's work, and to better locate the historical moment when major incommensurability between Ptolemaic astronomy and Copernican astronomy first appears.

2. METHODOLOGICAL PRELIMINARIES

For a number of years I have collaborated with Hanne Andersen (University of Copenhagen) and Xiang Chen (California Lutheran University) on a project applying recent results in cognitive psychology and cognitive science to questions arising from Kuhn's mature philosophy of science. Among the main results achieved in this work, we demonstrated that the philosophically unpopular view of concepts developed by Wittgenstein and adopted by Kuhn was massively supported by empirical work in cognitive psychology (Andersen, Barker and Chen, 1996). The same work led us to adopt Barsalou's dynamic frame model as a notation for describing conceptual systems and their changes (Barsalou, 1992). We also showed that, when concepts were understood in this way, many features of Kuhn's phase model followed from the nature of such conceptual systems; that a detailed mechanism could be specified linking anomalies to conceptual change; and that incommensurability was a permanent and perhaps beneficial feature of scientific change (Andersen, 1996; Chen, Andersen and Barker, 1998; Chen & Barker, 2000). At the same time I have collaborated with the historian of science Bernard R. Goldstein (University of Pittsburgh) on a contextual study of Kepler that has uncovered much new information about astronomy and natural philosophy in the time from Copernicus to Kepler (Barker and Goldstein, 1988; 1994; 1999; 2001; Goldstein and Barker, 1995; also Barker 1990; 1993; 1997; 2000; forthcoming-a; forthcoming-b). In the present paper I find myself in the happy position of being able to bring these two research programs together, at least in a preliminary way, by applying the methods developed in the first to the historical information from the second.[4]

Kuhn's work began a tradition of philosophical reconstructions in the history of science (for a review see Gholson and Barker, 1985). Philosophers working in that tradition, for example Lakatos (1977), notoriously sought to substitute *a priori* philosophical reconstructions for the actual details of history. I want to make it clear at the outset that this is *not* what I am trying to do. The frame model is empirical, and hence vulnerable to empirical modification and supersession. I am not claiming it is unique. Any account with the same positive features might be substituted. Hence this is not an attempt to capture the immortal essence of the scientific method, and it does not place me in a position to criticize the actions of real scientists who lacked the philosophical insights I claim to possess. Rather the intent is to understand history as it actually happened – and especially the process of conceptual change that led from the conceptual system of Ptolemaic astronomy to the modern Copernican one.

Having expressed that reservation, let me say three things in defense of the frame model. First, it is based on very extensive, indeed literally global, research on the nature of human concepts by psychologists and cognitive scientists. So although the model is empirical, I believe I have strong grounds for claiming that it is a good empirical model (for more details see Andersen, Barker and Chen, 1996). Second, unlike earlier philosophical attempts to understand scientific change, this one applies to history without either falsifying the history or distorting the model. Hanne Andersen's recent work on the discovery of nuclear fission is one nice example (Andersen, 1996). Another – and really surprising one – was discovered by Xiang Chen recently. Andersen, Barker and Chen had been using Kuhn's well known story about a child learning the differences between ducks, geese and swans as a way of introducing the frame model (Kuhn, 1974). In the course of doing this we invented a series of examples using a real South American waterfowl called a Screamer – which combines a chicken's beak with duck's feet – as a way of illustrating the hypothetical dynamic of conceptual change (Chen, Andersen and Barker, 1998; Chen & Barker, 2000). When Xiang Chen checked the actual history of the discovery of this bird, he found that the real revisions made by real ornithologists in their classification schemes were those we had outlined in our hypothetical example. So, in contrast to Lakatos's methodology of scientific research programs and its alternatives, the dynamic frame model offers a method of reconstructing historical change in science that has (at least once) successfully predicted the historical facts before they were examined. I believe this success reflects the strength of the account's empirical base. The frame model captures something important about the way human beings use concepts, and so it can offer reliable expectations about historical actors – nineteenth century ornithologists or sixteenth century astronomers.

The third thing I would say in favor of the frame model of concepts and conceptual change is that it satisfies all the most stringent methodological requirements imposed on accounts of science by the sociologists who have supplanted the historical orientation philosophers of science like Lakatos. The frame model is clearly causal and reflexive, but most important it is impartial and symmetrical (cf. Bloor, 1991). The same factors are invoked to explain conceptual changes we currently accept and those we currently reject. But the frame model does this without losing the ability to talk about the content of science – indeed its object is to represent what following Kuhn is the main feature of science in any given era – its conceptual system.

3. A FRAME BASED MODEL OF INCOMMENSURABILITY

A frame is a pattern of nodes representing concepts (see *Figure 1*). It will be apparent that frames represent not single concepts but clusters or patterns of concepts. The connections between nodes are not necessarily all of the same kind; it will often be an advantage that frames can accommodate very heterogenous connections between concepts. Although the connections between nodes may be of many different kinds the overall structure is hierarchical. Reading from the left, as I have arranged my diagrams, we descend from superordinate concepts, through attributes of those concepts, and values of those attributes, to subordinate concepts. *Figure 1* shows two subordinate concepts, distinguished by patterns of values for attributes of the superordinate concept. *Figure 1a* shows an alternative method of displaying a single subordinate concept by shading nodes. Not all nodes in a frame need be in use at one time. Following an analogy from neural networks, when a particular node is brought into play it is said to be "activated". A subordinate concept may be represented as a pattern of activation within the hierarchy of nodes. For example, in *Figure 1a*, the attribute and value nodes have been activated to represent the concept 'moon'. Reading from left to right we see that a moon is an astronomical object which takes the value 'planet' for the attribute 'orbit center', takes the value 'elliptic' for the attribute 'orbit shape', takes the value 'near' for the attribute 'distance', takes the value 'reflective' for the attribute 'luminance', and takes the value 'small' for the attribute 'size'. This list of attributes and values is cumbersome to write down, but is quickly grasped from the frame diagram. More importantly, unlike the list of attributes and values, the frame specifies the alternative values that the attributes of 'astronomical object' might possess, or allows us to examine whether choosing one attribute in one way opens or closes possibilities for the values of other attributes.[5] In subsequent figures I will use both of the representational conventions illustrated here. Although my main concern in the present study will be with the overall structure and pattern of activation in a frame, it should also be apparent that the frame notation expresses compactly a great deal of information about individual concepts, including information about related concepts at the same level of generality and connected concepts at higher and lower levels of generality (Barsalou, 1992; Barsalou and Hale, 1993).

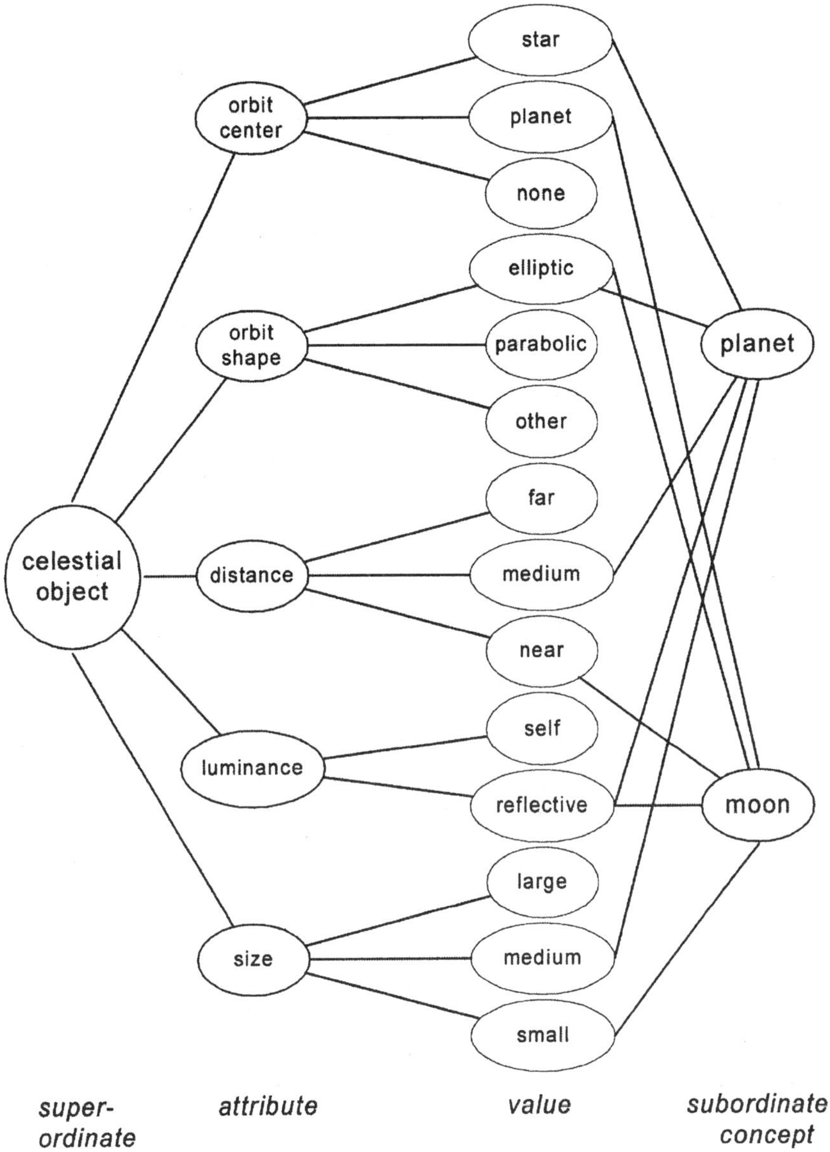

Figure 1: Partial frame for CELESTIAL OBJECT c. 1700

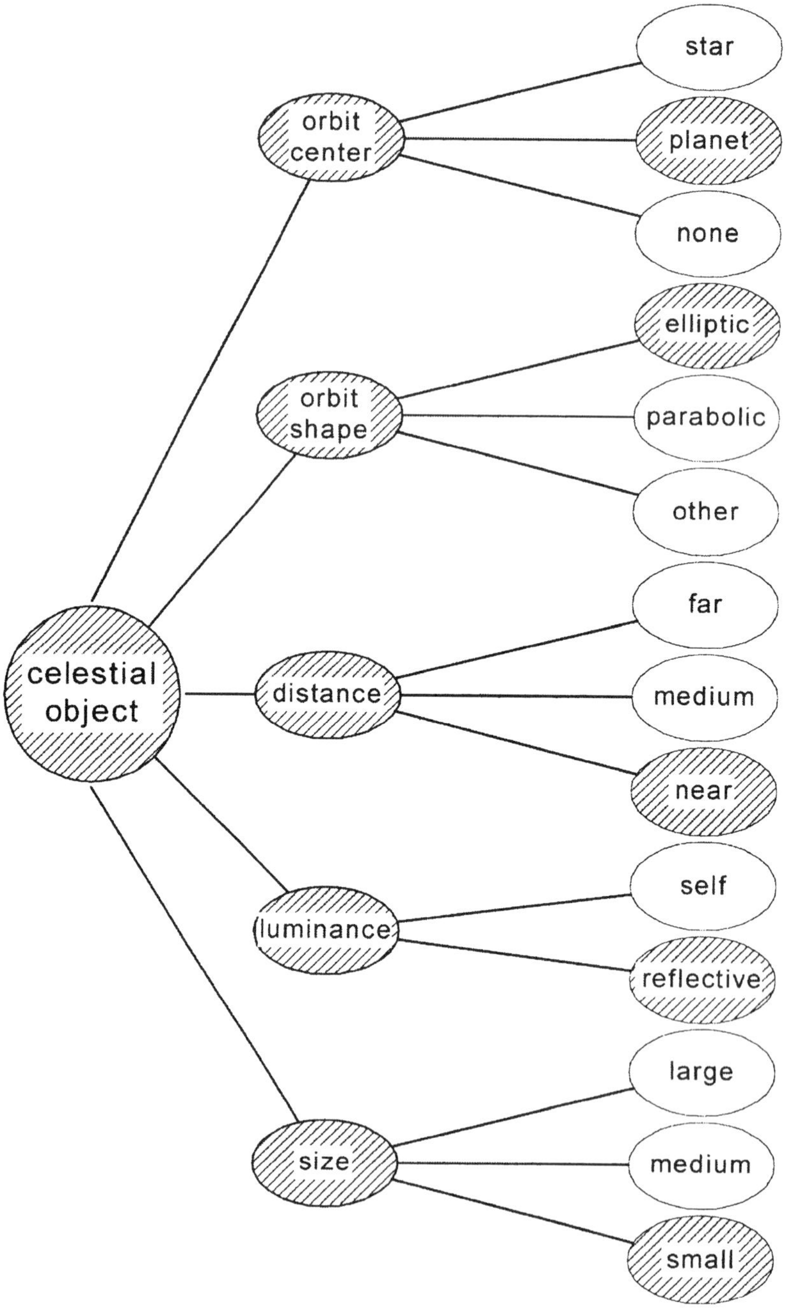

Figure 1a: Partial frame for CELESTIAL OBJECT c. 1700, with shaded nodes for subordinate concept 'moon'

The most important connection between Kuhn's mature philosophy of science and the frame model occurs at the lowest level of generality, which links clusters of concepts that are, first, attributes of a superordinate concept and, second, values of those attributes. The variation in which attributes and values are activated at this level corresponds to the similarity and dissimilarity classes introduced by Kuhn as the primary element in his account of concepts. The frame model can therefore capture all the important aspects of Kuhn's account, and particularly the family resemblance aspect of concepts, or to put it another way, the impossibility of giving necessary and sufficient condition analyses of important concepts.[6]

The frame model is more general than Kuhn's account in several respects. In his later work Kuhn placed special emphasis on taxonomies of natural kinds, especially in understanding incommensurability. Incommensurability occurs when two taxonomies proceed from apparently identical superordinate concepts to subordinate concepts represented by frames or parts of frames that cannot be mapped onto each other.[7] Thus, as we have already mentioned, parts of the frames for the subordinate concepts 'star' and 'planet' in the taxonomy of 'celestial object' after Newton cannot be mapped onto the concepts for the taxonomy of 'celestial object' in say, 1500. For Kuhn the lowest level in any taxonomy is a collection of (concepts representing classes of) entities constituted by similarity and dissimilarity classes of attributes. So although the attributes and their values play a central role in his account, he treats them separately from the taxonomic trees that he uses to explicate incommensurability. But frames are recursive – a property that will be important in our discussion later. Another way to put this is to say that *any* concept in a hierarchy could be selected for special attention and displayed as having attributes and values of its own. There is no essential difference between concepts that appear as superordinate concepts in a kind hierarchy, and the concepts that represent the attributes and values constituting the similarity and dissimilarity classes at the lowest level of the taxonomy. So frame diagrams collect in a single figure the information Kuhn provides through kind hierarchies on the one hand and specifications of similarity and dissimilarity classes on the other.

A second difference is that for Kuhn a taxonomy is a hierarchy of concepts built from only a single relation between the nodes, the relation "is a kind of". In frame diagrams nodes for (classes of) entities and nodes for attributes may all appear in a single frame. There is no requirement that relations between nodes be limited to "is a kind of" or indeed that they be homogeneous. Further freedom is gained by recognizing that the heterogeneity of node links can apply throughout the frame, allowing the representation of many other conceptual structures in addition to taxonomies. So while it is true that any of Kuhn's taxonomies can be represented by a frame diagram, the reverse is not the case. Not all frames are taxonomies (including some I will consider shortly). Consequently, when I use the terms 'superordinate' and 'subordinate' I do not necessarily indicate an "is a kind of" relation or a genus-species relation, but only that the subordinate concept (and any values it introduces) specify an attribute of the superordinate concept.

In the frame model incommensurability arises between conceptual structures, that is patterns of concepts, rather than individual concepts. In its simplest terms incommensurability is a mismatch between the nodes of two frames that contain what appear to be the same superordinate concepts. Reading from left to right we encounter the same series of subordinate concepts in descending levels of generality. But at some point we encounter structures in the two frames that do not map onto each other. Not just any mismatch will do. Division of subordinate concepts into further subclasses preserves the overall topology of the frame and will not generate incommensurability. The most serious problems will arise from the addition and deletion of nodes, or the replacement of superordinate nodes. Incommensurability occurs between two frames for the same superordinate concept when, in the attempt to read to the right of a particular subordinate node, we are confronted with two seemingly incompatible ways of decomposing that node into subordinate concepts. In the end the differences that matter – and generate incommensurability – are just those that create or reflect differences at the lowest level of attributes and values. These will be differences that correspond to differences in what Kuhn called similarity and dissimilarity classes, and hence in the fundamental objects that the conceptual structures represented by the frames allows us to talk about. Scientists who have been trained in one frame may find the claims made on the basis of the second incomprehensible or nonsensical. However there may still be structures that are common to both frames. So total failure of communication between communities with incommensurable conceptual structures is not something we should expect. Kuhn also denied that all incommensurability was total, although he was frequently and persistently misread as claiming total communication failure between proponents of rival conceptual systems. Incommensurability is also a matter of degree – the higher the level of the concept where the mismatch begins, the more severe will be the incommensurability. But there must be some connection between the two structures. Newtonian astronomy is incommensurable with Ptolemaic astronomy, but not with Galenic medicine. They are just two different fields.

An account as detailed and complex as Kuhn's, or the parallel account in terms of dynamic frames, offers many different ways of generating incommensurability. Kuhn did not explore all of these possibilities (subsequently Chen, 1997, and Andersen, 1998, have gone some way beyond his account). In the present paper I will be largely concerned with the simplest kind of incommensurability – mismatch between nodes of a subordinate concept. However, I will not limit my attention to hierarchies of kinds, but will consider quite generally the concepts needed to understand that portion of astronomy that deals with the planets during the sixteenth and seventeenth centuries.

4. THE CONCEPTUAL STRUCTURE OF PTOLEMAIC ASTRONOMY

If you ask a large modern audience "What shape is the path of the planet Mars?" you will get a few unorthodox answers but the overwhelming majority will respond that the path of Mars is an ellipse. Perhaps this is not surprising, since the planet Mars was Kepler's object of study in the book in which he introduced elliptical orbits into astronomy. But responding to the question in the modern way assumes that the right way to respond to a question about the *path* of a celestial object is by specifying the shape of its *orbit.*

The modern answer conceals an older stratum of science. In astronomy before Kepler the path of a planet was not its orbit but the pattern of its motion seen by an observer on a stationary earth against the hypothetical sphere of the heavens. It was recognized in antiquity that this pattern was not a real motion, but a complex outcome of the observer's viewpoint and a variety of circular motions that acted together. The task of astronomy was to define this pattern – to specify the path of the planet in its original sense. The real motion of the planet – it's track in what we would call three-dimensional space – was unknown, and possibly irrelevant. All other considerations – the causes of celestial motion, the actual dimensions of the heavens – were the business of a separate science, cosmology. It was well known that the goal of astronomy could be achieved, that is the path of a celestial object could be predicted, without making specific assumptions about its distance from the earth, once appropriate rates of rotation were introduced.

The motions treated in astronomy were not real motions (motions through three-dimensional space). Such motions might be specified in cosmology and as a matter of historical fact there was fairly general agreement on the cosmological constraints that led to definite values for the distances and sizes of planets. However the details of the planets' real motions remained controversial. Averroists insisted that the real motion must be concentric to the earth. Ptolemaic astronomers insisted that it must be at least eccentric, and modified the eccentric motion with an epicycle in all cases except the sun. Additional complications in Ptolemaic astronomy made it difficult to argue for the uniqueness of any given combination of eccentrics or epicycles. The shift from Ptolemy to Copernicus and Kepler therefore includes a conceptual shift from a concept of path that is not an orbit and does not specify a real motion to a concept that specifies an orbit and is a real motion. This is the kind of striking change Kuhn taught us to associate with the episodes called scientific revolutions.[8]

Returning to *Figure 1*, let us consider some differences between that frame and the corresponding frame for CELESTIAL OBJECT in, say, 1500 before Copernicus had published anything. (Here and elsewhere, I will use capitals to indicate nodes in frame diagrams.) In 1500 CELESTIAL OBJECT already has the attributes SIZE and LUMINANCE. The question of DISTANCE we will deal with shortly. Where there are two nodes for ORBIT CENTER and ORBIT SHAPE in the 1700 frame, the 1500 frame displays a single node, which I will call PATH.[9] Celestial objects may be differentiated by the varieties of motion they display (in this case without the further complication introduced by the values of each attribute node). Fixed stars have only a diurnal motion; the sun and moon have both a diurnal motion and a proper motion (or first anomaly); the remaining planets perform a diurnal motion, and a proper motion, and occasionally retrogress (that is they display both a first and a second anomaly). Rather than rewriting the whole frame for CELESTIAL OBJECT I will diagram only the attribute PATH and its values (see *Figure 2*).

As already indicated, PATH does not refer to the continuous motion of an object through three dimensional space. It refers rather to the pattern of an object's motion viewed against the sphere of the heavens. Kuhn expressed this well in *Copernican Revolution* when he referred to the 'two sphere' universe – the central earth surrounded by a hypothetical sphere of the heavens. It is the goal of astronomical calculation to calculate the successive positions of the planet along this path – including any retrogressions. As the size of the sphere of the heavens is arbitrary, you can see that the only important data here concern changes in angular position. It is not part of the main

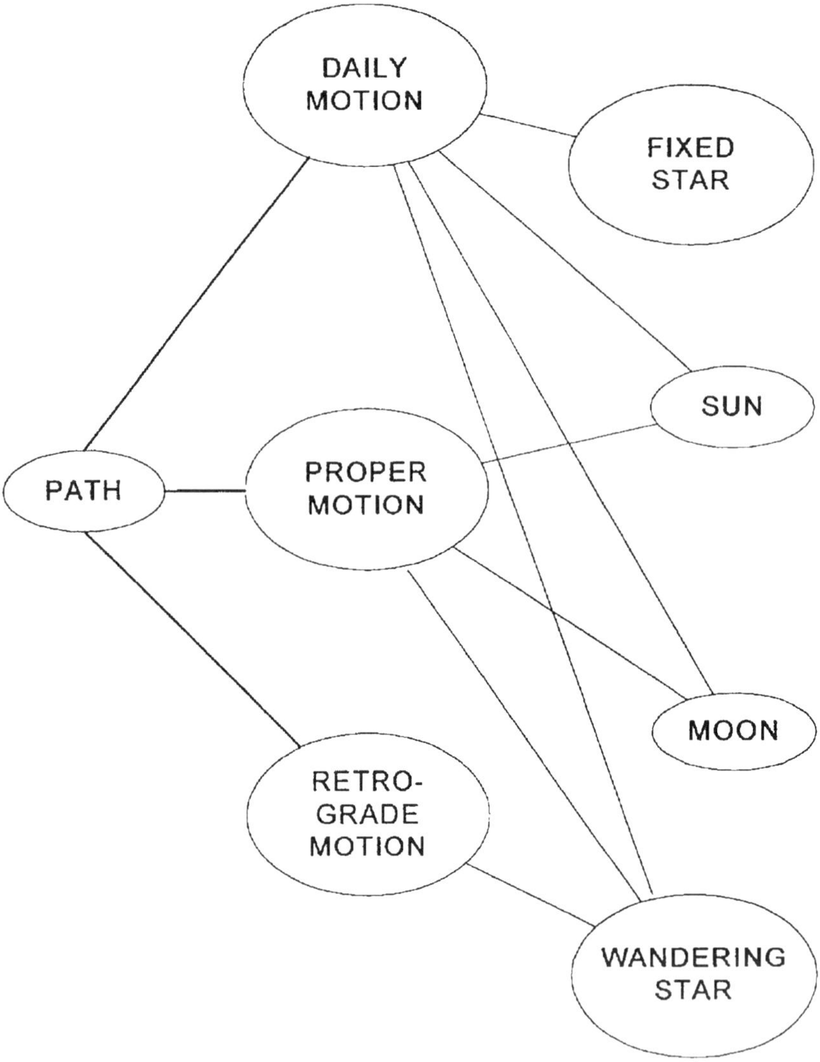

Figure 2: Partial frame for PATH, c. 1500

business of astronomy to also calculate planetary distances, although this may be done in an ancillary set of calculations that appeal to some premises outside astronomy.[10] It is, however, generally acknowledged that the moon is the closest celestial object, and that the fixed stars are the farthest. The irrelevance of distance in astronomy becomes clearer if we consider the frame for a related concept: CIRCULAR MOTION[11] (see *Figure 3*).

Circular motions have three attributes that need to be considered here: they have a center, a radius and a rate, which may be designated by an angular velocity or speed. As

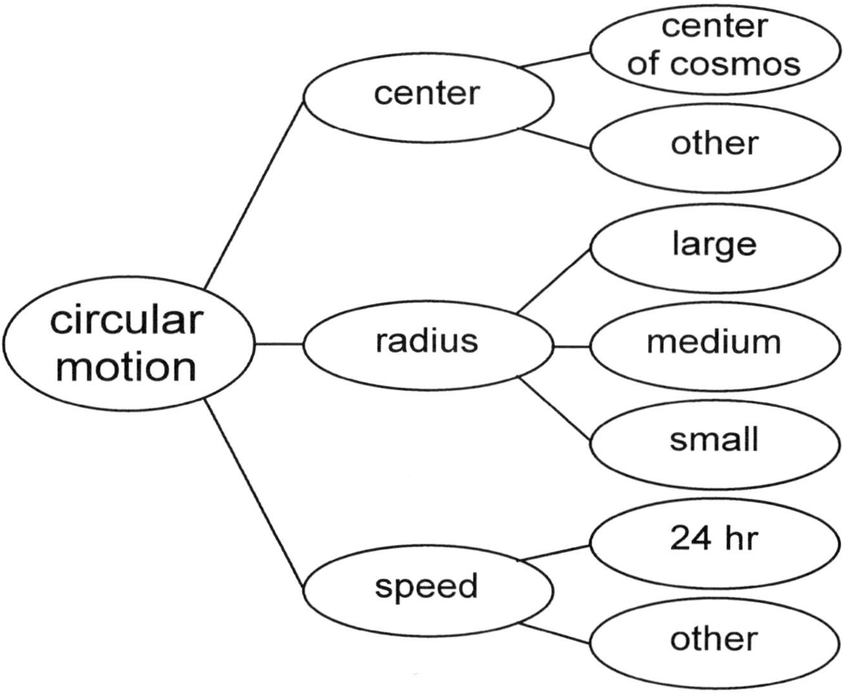

Figure 3: Partial frame for CIRCULAR MOTION, c. 1500

far as astronomy is concerned, the angular velocity can take any numerical value, so for constant angular velocities (uniform motions) the value nodes connected to this attribute should be an infinite array. For simplicity let us take special notice of just one value, that of making a single rotation in 24 hours, and group all the rest together as OTHER. In principle a circular motion must have a definite radius, but in practice this plays no role in astronomical calculations, so let us satisfy ourselves by grouping objects as near (like the moon), far (like the stars) or intermediate (for everything else). The possible values of the attribute CENTER are also indefinitely many. But again we may simplify matters by picking the single most important value (the center of the cosmos which is also the center of the earth) and grouping all the rest as OTHER. The main difference between the Ptolemaic tradition and its main rival during the sixteenth century may now be succinctly stated. Ptolemaic astronomers allow the activation of the OTHER value node for CENTER; their opponents the Averroists do not.

All celestial motions are circles traversed at constant speed. This fundamental tenet of sixteenth century astronomy can be displayed by combining the frames introduced so far (see *Figure 4*). PATH has three attribute nodes. Each of these is a separate mo-

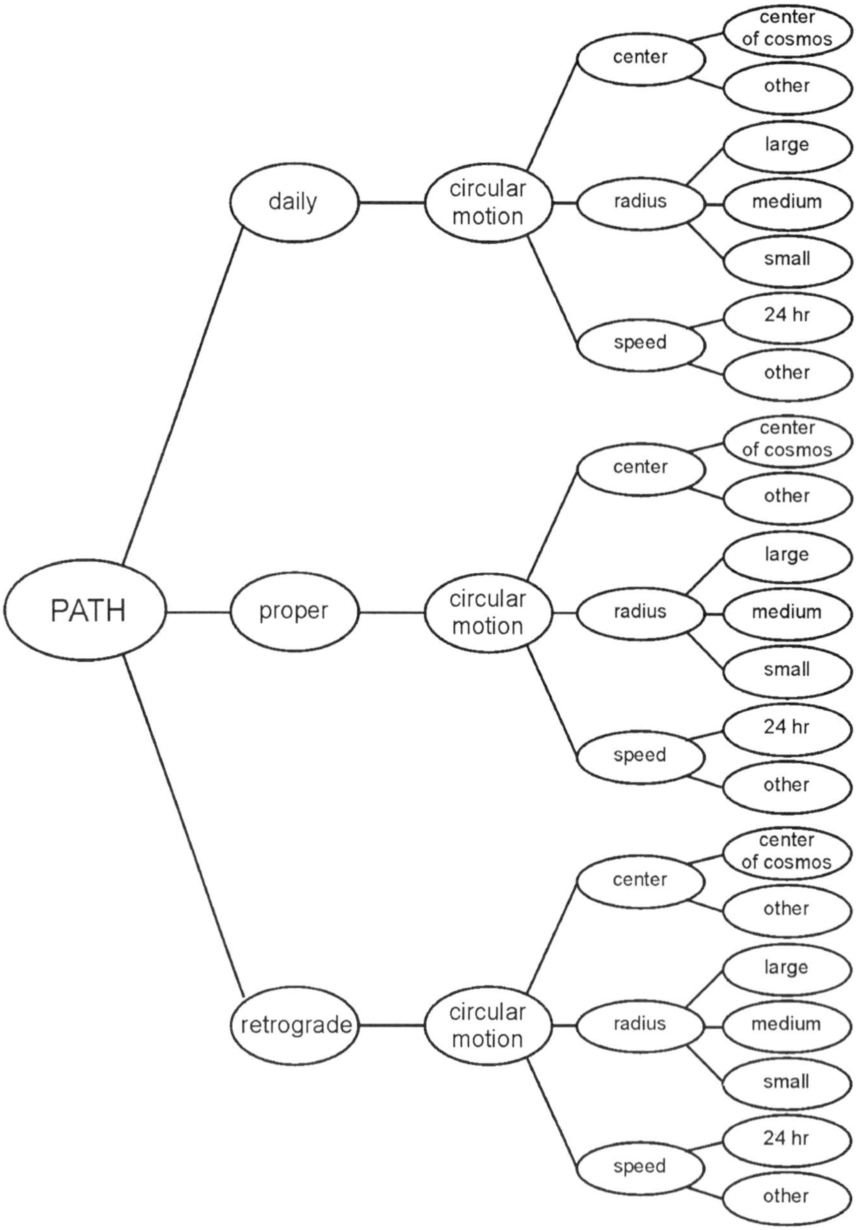

Figure 4: Partial frame for PATH and CIRCULAR MOTION, c. 1500

tion, hence each mus t display the attributes already introduced for CIRCULAR MOTION. By activating some of these nodes we can generate the conceptual structure of a Ptolemaic *theorica* for the planets Venus, Mars, Jupiter and Saturn, which were usually treated together (see *Figure 5*).[12] Note that the diurnal motion is earth centered, but that the proper motion and the retrograde motion have different centers, corresponding to the center of the eccentric deferent and the epicycle, respectively. I omit here the complication caused by the notorious equant point. The equant, in effect, separates the geometrical center of the deferent from its center of angular motion, and governs both the angular velocity of the deferent, and of the epicycle carried on it, so you may think of it as an ancillary technique used in calculations that begin from the conceptual structure shown in this frame and lead to calculations of actual positions. No-one is very happy with the equant. In a sense the main objection to it is the impossibility of representing it without modifying the frame we are considering.

Other *theoricae* are simple variations on this basic frame. The *theorica* for the sun lacks the lower cluster of nodes and has different values for the attributes under proper motion. That for the moon has a cluster of nodes representing a non-retrograding epicycle as its third element. Next let us consider how this frame changes to accommodate the astronomical models proposed by Copernicus in 1543.

5. THE CONCEPTUAL STRUCTURE OF COPERNICAN ASTRONOMY

Figure 6 is a partial frame for PATH and CIRCULAR MOTION c.1543, with activated nodes representing a Copernican *theorica* for Mars, Jupiter or Saturn. The top cluster of nodes now corresponds to the 24 hour motion of the earth. Copernicus's treatment of the diurnal motion uses practically the same attribute values as Ptolemy's. Only the radius of the motion changes from one of the largest allowed values to one of the smallest. The biggest change, of course, is that this whole motion is now regarded as apparent rather than real, however, when it comes to making astronomical calculations this makes no difference at all.

The proper motion of each planet is now understood primarily through the motion of a large circle eccentric to the mean sun. Notice that this change requires *no* major differences in the nodes activated, and no addition or deletion of nodes. Within the conceptual system of sixteenth century astronomy, the choice of the mean sun as the center of this circle is just the choice of a new excenter that differs from the center of the earth. But Ptolemaic astronomers were already using such a point.

Turning to the other two 'value' nodes: the distances involved remain intermediate. Although there is now a well known relation among these distances (changing one requires that you change the rest) this connection has no consequences for calculating planetary positions. The speed of these motions assumes different (but related) values from those in the Ptolemaic frame, but again, these are drawn from the existing set of allowed values. So in the cases of all three attributes we activate values that are already familiar from Ptolemaic astronomy. We do not need to add nodes and we do not need to delete nodes.[13]

What has just been said describes the mathematical models for calculating planetary positions presented in the body of *De Revolutionibus*, and not the cosmological sketch

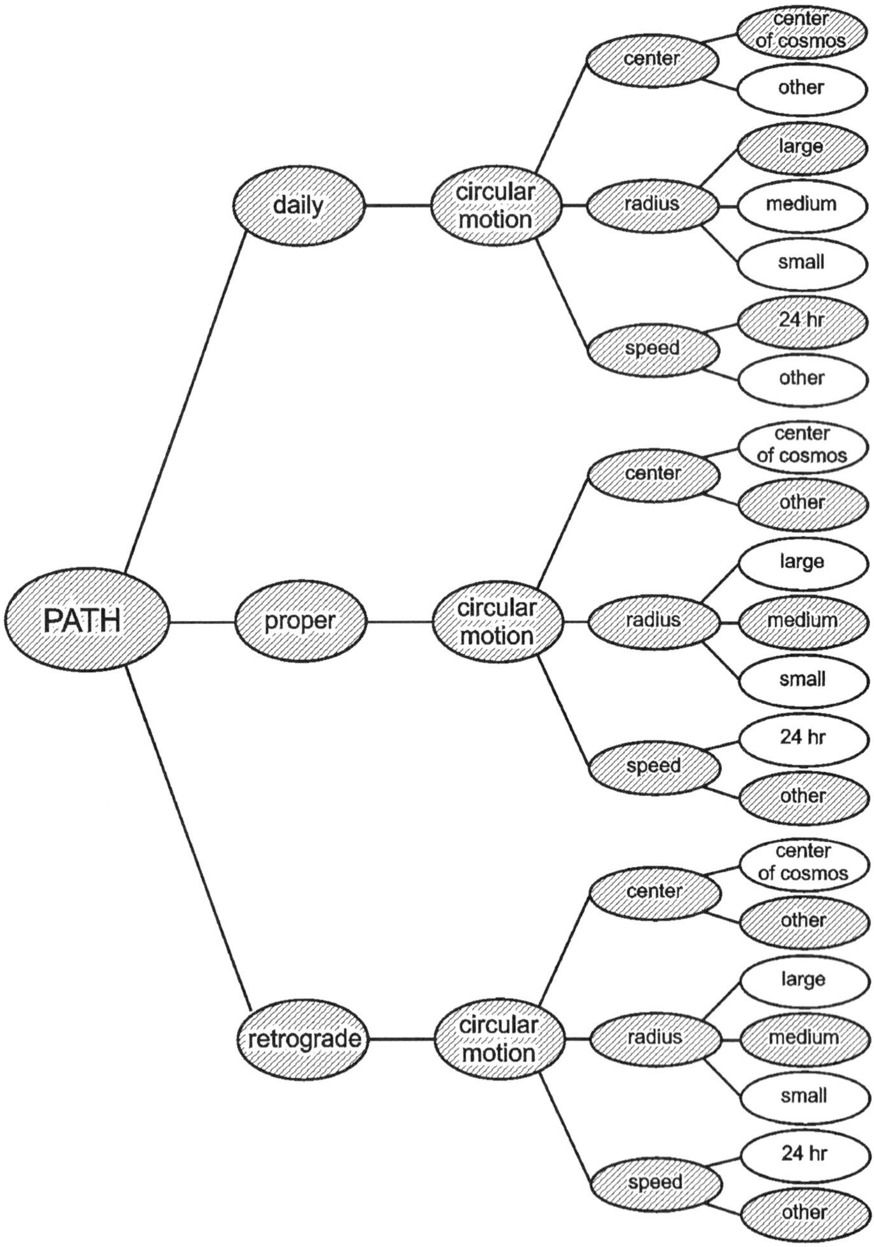

Figure 5: Partial frame for PATH and CIRCULAR MOTION, c. 1500, with activated nodes representing a Ptolemaic theorica *for Mars, Jupiter or Saturn*

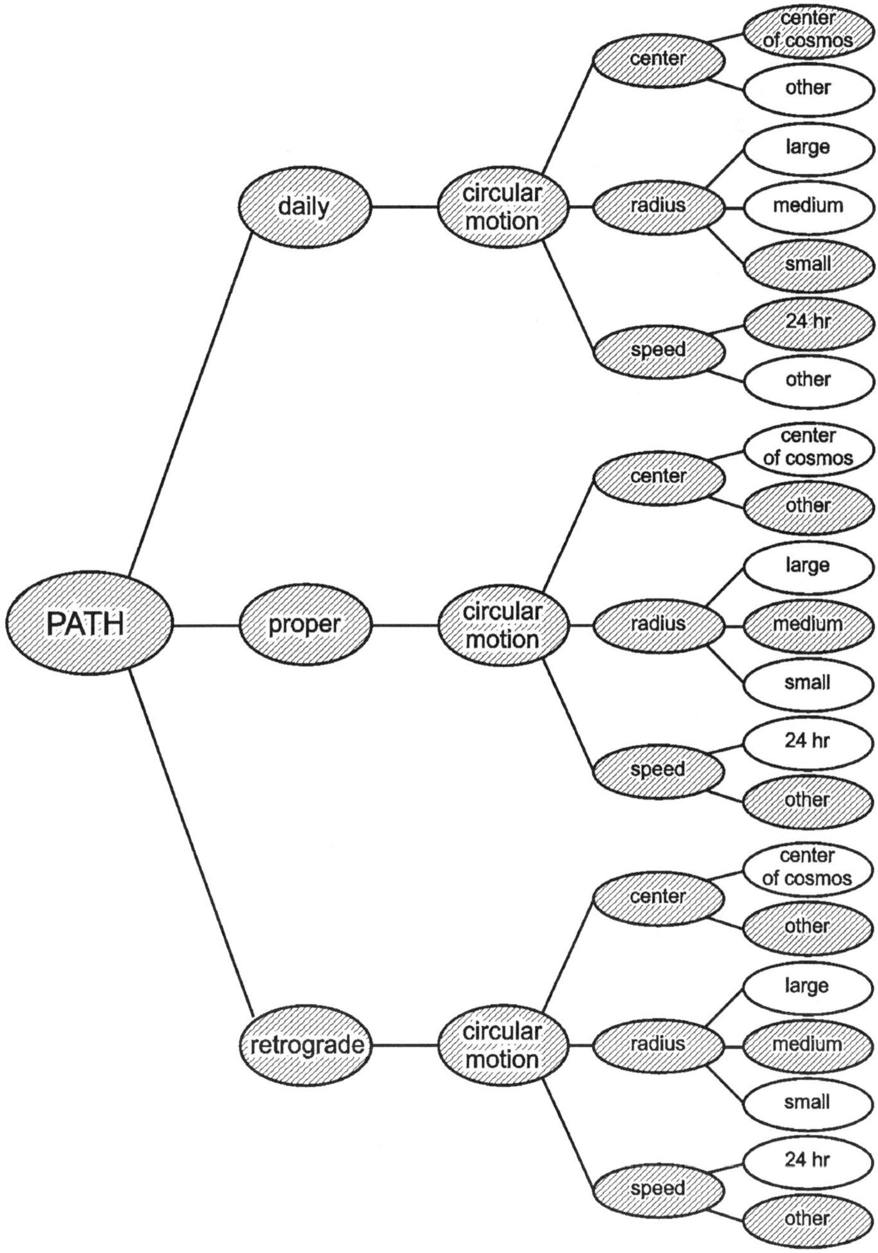

Figure 6: Partial frame for PATH and CIRCULAR MOTION, c. 1543, with activated nodes representing a Copernican theorica *for Mars, Jupiter or Saturn*

from Book I. In place of Ptolemy's equant Copernicus introduces a small epicycle carried by the main circle as an auxiliary calculating device (and also adjusts the eccentricity). From a mathematical viewpoint this device performs much the same services, but without dividing the geometrical center of rotation and the center of uniform motion. As I did not present the equant explicitly when considering Ptolemy, I will ignore this device now, for parallel reasons.

I will postpone for another discussion the question of whether Copernicus abolishes or merely conceals the equant, and whether these features of his models are incommensurable with Ptolemy's. Here I will merely suggest that we take Copernicus at his word. His intent is to restore an astronomy that uses only the attributes of circular motion that we have displayed in our frame. A Ptolemaic frame including the equant mechanism would be more complex. So Copernicus may be said to prune some nodes out of Ptolemy's frame. But his intent is to conserve an existing conceptual structure, and especially its superordinate nodes. So if incommensurability is judged by degree of mismatch between nodes, any incommensurability here will be confined to subordinate nodes, and consequently minor.

Once again, in the case of Copernicus's account of proper motion, the major difference from Ptolemy is obscured if we concentrate only on the basis for astronomical calculations. For Ptolemy, the main element in the proper motion is an eccentric circle. But exactly the same results would follow from a concentric circle carrying an epicycle. So judging only from its path (understood as defined in this paper) we cannot tell what the real motion of the planet is. For Copernicus, however, the eccentric circle that is the main contribution to the proper motion is a real motion.

Finally, let us examine the third cluster of nodes that in Ptolemaic astronomy correspond to the epicycle used to accommodate retrograde motion. For Copernicus, this motion is the result of the annual motion of the earth combined with the proper motion of the planets that has already been introduced. As is well known, Copernicus explains retrogression through change in the line-of-sight as a moving earth overtakes an outer planet (or is overtaken by an inner one). At the same time he can explain why outer planets retrogress while in opposition to the sun (and inner ones in conjunction), and why the retrogressions begin and end where they do.

To derive actual positions for retrogressions we need a theorica for the earth to replace the theorica for the sun. Copernicus provides this by giving the earth a purely circular path centered on the mean sun.[14] So, just like Ptolemy, he introduces a circular motion to explain retrogressions. The attributes and values of this circular motion are surprisingly familiar. Of course the center of this motion is the mean sun, that is a hypothetical point differing from the center of the earth. The speed of this motion is the speed attributed to the sun in the Ptolemaic *theorica*. And the radius of this motion is the distance attributed to the sun in the Ptolemaic *theorica*.

To summarize: Copernicus uses the same conceptual structure as Ptolemy for the key concept PATH which encompasses the positional data of astronomy (see *Figure 6a*). Not only does Copernicus employ the same attributes, but the values activated in his conceptual structure are almost the same pattern as the Ptolemaic ones (in contrast to those activated in an Averroist account) and the there is nothing objectionable in the particular values assigned to these attributes.[15] This includes the treatment of retro-

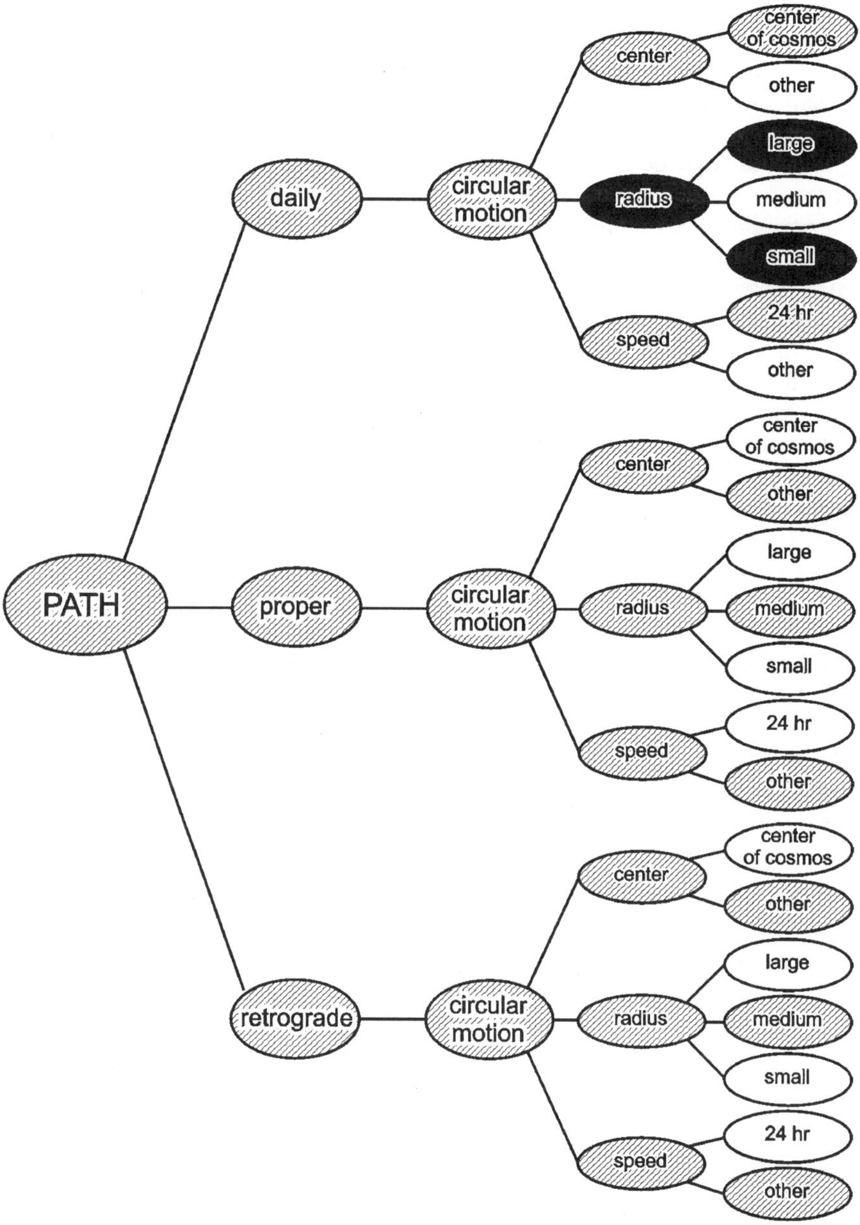

Figure 6a: Superimposition of Figures 5 and 6
with non-overlapping nodes highlighted, plus their first superordinate node

gression, where the proper motion of the earth serves the same function as the epicycle in a Ptolemaic *theorica*. Now remember that it is the goal of astronomy in the sixteenth century to calculate planetary positions against the sphere of the heavens as viewed from the earth. The astronomer reading *De Revolutionibus* is reading it with that goal in mind. And with that goal in mind the sixteenth century astronomer will find the conceptual structure underlying Copernican calculational techniques to be the same structure that appears in Ptolemaic astronomy. Copernicus's conceptual structure is not incommensurable with Ptolemy's – if anything it appears to be a variation on it. And this is exactly the way it was seen during the late sixteenth and early seventeenth centuries. Led by Erasmus Reinhold, who adopted Copernican calculations techniques to produce a new and improved set of astronomical tables (the *Prutenic Tables*), a whole host of Lutheran astronomers spread Copernican methods through Northern Europe. Further South, the Jesuit Christopher Clavius, who we have already had occasion to mention as the leader of the successful reform of the calendar, also counted Copernicus as an intellectual ally of Ptolemy in the common fight against the Averroists.

I have already mentioned that the frame diagrams displayed so far conceal an important divergence between Ptolemy and Copernicus on which motions are real and which apparent. For the first time, in Copernicus, the motion in longitude or proper motion of a planet (against the sphere of the heavens) arises from a real motion (around the Sun). Even if you do not specify the former motion in terms of direction *and* distance, if it is arises from a real motion it must *have* both a direction and a distance. This has made it practically irresistible for modern readers to begin talking in terms of *orbits*, even before the conceptual and theoretical apparatus needed to support the concept is in place. There were no orbits in the modern sense before Kepler gave the rules for computing a continuous series of angular positions with distances of Mars, thereby recovering the data gathered under the concept PATH in a new way.

6. FROM ORBS TO ORBITS

What then is real in sixteenth century astronomy? What are the fundamental entities and how do they produce the motions that we have been calling the PATH of a planet? Another way to put this is to ask, what are the real celestial objects? The answer is simple, but not as well known as it ought to be today (See *Figure 7*).

Before Copernicus it is agreed by almost everyone that beginning with the region of the moon, the cosmos consists of a series of concentric shells with a planet somehow confined to each one. If you are an Averroist you believe that the detailed motions of each planet can be recovered by dissecting each shell into a series of thinner concentric shells with offset axes and varying rates of rotation. Throughout Copernicus's lifetime, systems of this sort were regularly proposed by natural philosophers who objected to Ptolemy's use of more than one center of rotation (Barker, 1999). The most detailed efforts are due to Giovanni Battista Amico in 1536 and Girolamo Fracastoro in 1538 (di Bono, 1995). If you are a Ptolemaic astronomer, however, you believe that the interior of each celestial shell is divided in another way. Although the inner and outer surfaces remain concentric to the earth, interior surfaces may be eccentric, creating shells, or *orbs* as they are called in a *theorica*, that vary in thickness. These appear as crescent shapes when displayed in cross section (see *Figure 8*). Two of these

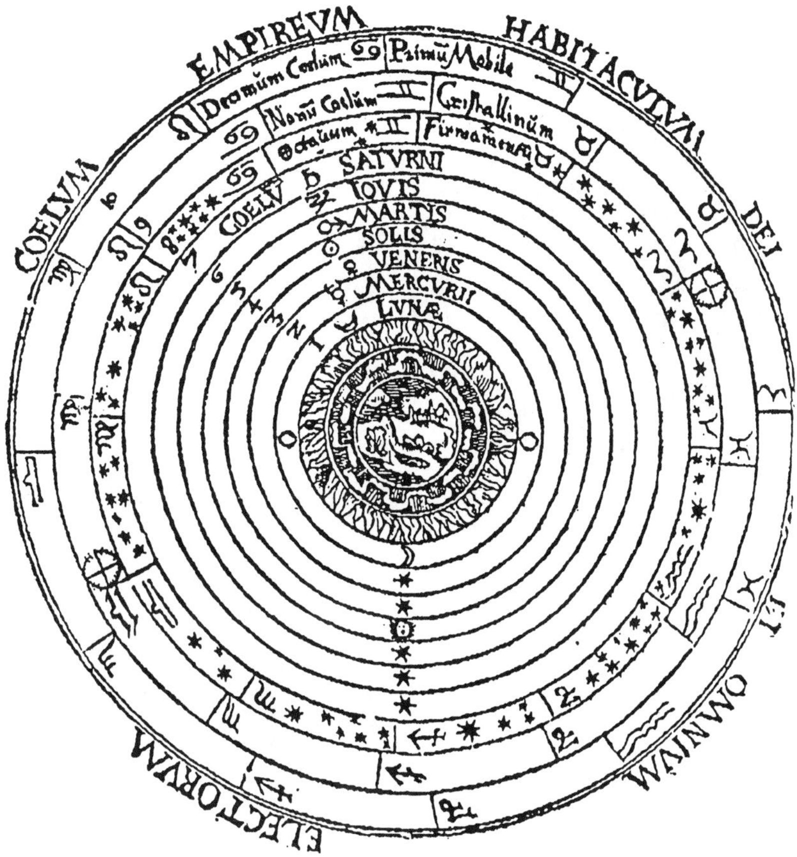

Figure 7: Geocentric system of the world
Peter Apian (1540) Cosmographia, *Antwerp, fol. 6 R*

shells can be arranged to sandwich a third which carries a small sphere corresponding to the planet's epicycle, or in the case of the sun, the spherical body of the sun itself. As they rotate, these shells generate motions equivalent to the eccentric and epicycle. Where the cluster of shells for one planet ends, the next begins (see *Figure 9*). By arranging systems of orbs inside each other, as illustrated here, a complete system of the world could be constructed, following the overall pattern of *Figure 7.* The planet is literally embedded in the epicycle sphere – indeed it is a commonplace that a planet is distinguished from the material of the orbs only by its density: "A planet is a denser part of its orb". So it might be said, in answer to the question with which this section began, that the real celestial objects are the invisible celestial spheres and the orbs into which they are divided.

Figure 8: Separate orb diagrams for the theoricae *of the sun and Venus Erasmus Reinhold (1542)* Theoricae novae planetarum, *Wittenberg, fol.s D iii R & I v V*

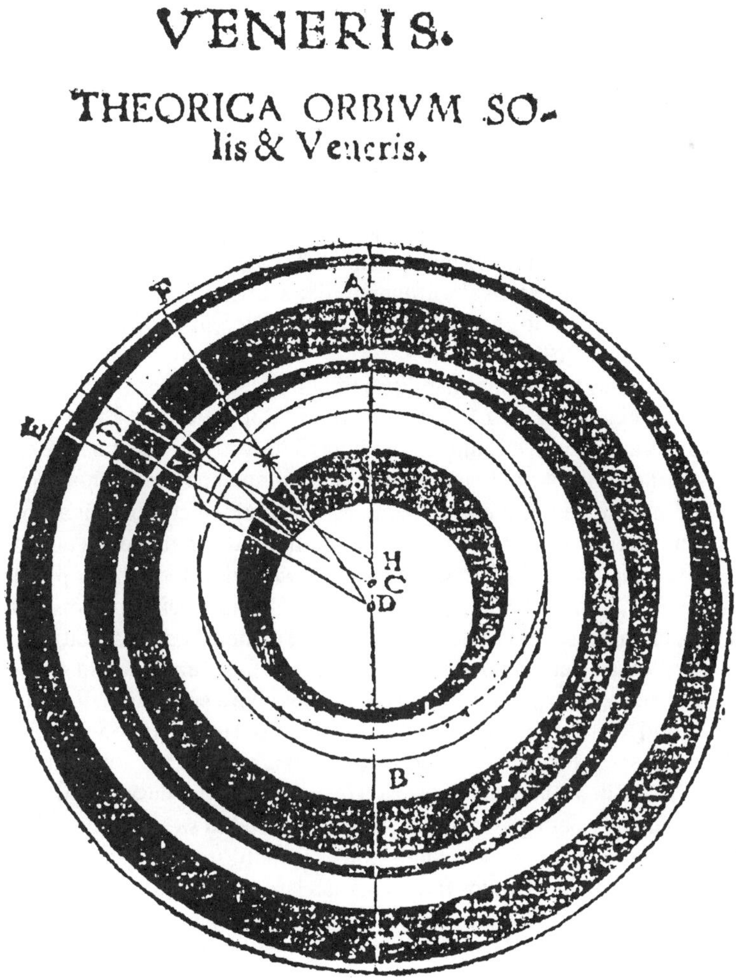

Figure 9: Combined orb diagrams for the theoricae *of Venus and the sun*
Erasmus Reinhold (1542) Theoricae novae planetarum, *Wittenberg, fol. M viii R.*
By arranging systems of orbs inside each other, as illustrated here,
a complete system of the world could be constructed.

The celestial ontology of the sixteenth century has its problems. Averroists have the problem that no system of homocentric spheres has ever been demonstrated to produce the observed path of a planet. The Ptolemaists have the problem that the equant motion cannot be accommodated as a uniform rotation within such a shell cluster, and the further problem that Apollonius's theorem about the equivalence of eccentrics and concentrics carrying epicycles makes it possible to generate the same path in several different ways, each corresponding to a different arrangement of orbs. So Ptolemaic astronomers are in the dubious position of being committed to orb clusters but not being able to tell you which particular set is up there in the heavens. Careful people like Phillip Melanchthon confine themselves to the claim that some such cluster of orbs is up there. Other defenders of mathematical astronomy, like Christopher Clavius, flatly insist on the existence of the specific clusters presented in his Ptolemaic *theoricae* (Lattis, 1994; Barker and Goldstein, 1998).[16]

Copernicus notes the general liabilities of both these positions in the letter to Pope Paul III at the beginning of *De Revolutionibus*. But little or nothing changes with Copernicus. As a perceptive Lutheran put it in 1576, you can generate the Copernican cosmos from the conventional one by swapping the sun and the earth-moon combination. Everything else stays in place. Copernicus's cosmological diagrams are not pictures of orbits, but patterns of orbs. There is both internal and external evidence for this: the significance of his drawings is quite clear once you look for orbs in place of orbits (Swerdlow, 1976; Barker, 1990). Copernicus continues to speak of planets being carried by their orbs or spheres. And Maestlin and Kepler, in their own presentations at the end of the century, clearly understood him as continuing to accept a nested shell cosmos.[17]

For Copernicus, each planet is again confined within a specific shell. How the shells are further divided is not specified – but it does not need to be. Anyone familiar with the orb models presented in a Ptolemaic *theorica* can construct equivalent patterns for Copernicus's detailed planetary models. To succeed at the business of astronomy – recovering planetary paths – Copernicus does not need to say anything new about the substance of the heavens, and he does not. And until the substance of the heavens changes, planets remain as minor flaws in much larger sets of divided spherical shells.

The arguments about the substance of the heavens and the demise of celestial spheres have several strands. One in particular concerns us here. Observers all across Europe described a comet that appeared in 1577. Many of them concluded that the comet was above the moon and thus a celestial object, contrary to Aristotle's teachings. Two observers were also unique in offering a new kind of data. They were Michael Maestlin, soon to become a professor at Tübingen and teacher of Johann Kepler, and Tycho Brahe, already beginning to establish himself on the island of Hven. Maestlin's account appeared immediately – Brahe's not for ten years.

Maestlin (1578) and Brahe (1588) published tables that gave the comet's position, that is its direction, every day over a period of months. But they also calculated the distance of the comet from the earth for each position they gave. This was the first time anyone had described the continuous track of a celestial object as it moved through the heavens – in effect delivering the information that Copernicus had implied should be available for every celestial object he treated. However, Maestlin and Brahe did not thereby acquire the concept of an orbit.

The track described by Maestlin and Brahe took the comet through several of the geocentric shells or spheres accepted by Ptolemy and Aristotle, which was supposed to be impossible. But neither observer concluded that there were no celestial spheres – at least not immediately. Maestlin continued to accept celestial spheres, but concluded that they must be centered on the sun. His most important conclusion was that the comet itself was confined to a sphere, just like one of the planets. Its sphere was located outside the sphere of Venus but inside that carrying the earth-moon system. The track of the comet showed only that there were no *geocentric* spheres. Brahe also assigned the comet to a heliocentric sphere just larger than that of Venus, but he was unable to accept Copernicanism because of physical objections to the motion of the earth. Consequently he preferred a geo-heliocentric arrangement for the planets (the earth is central, the moon and sun revolve around the earth, all the other planets and the 1577 comet revolve around the sun). In this arrangement the spherical shells for the sun and the planet Mars intersect, which is also supposed to be physically impossible if they are constituted as Aristotelians believed. In response to this problem, Tycho abandoned material shells entirely. By 1588 when he published his account of the comet he had adopted a fluid heavens in which the spheres of the planets became merely geometrical boundaries

Neither Maestlin nor Brahe changed the concept of PATH from the established pattern we have examined in Ptolemy and Copernicus. Both continued to decompose celestial motions into circles. The information they provided about distances for the comet of 1577 was used as negative evidence against Ptolemy, not as a positive contribution to astronomy. It was left to Kepler, as the intellectual heir of both men, to see the possibilities of replacing circles by ellipses, and spherical shells by orbits.

By the time Kepler began his career in the 1590s all the information we have just reviewed was readily available to him. In the *Mysterium Cosmographicum*, published in 1596, he opts for a fluid heavens through which planets move freely within the confines of heliocentric shells. These shells are no longer physical but purely geometrical structures defined by his ingenious Platonic solid construction. But the motions of these planets are still decomposed into circles, and the frame for the concept of PATH that Kepler is using here would be no different from the one we considered in the case of Copernicus.

The concept of an orbit receives its first book length treatment in the *Astronomia Nova* of 1609. Although the book is presented as a narrative of Kepler's discoveries, and the ellipse which is the orbit of Mars does not appear until right at the end, Kepler clearly has the concept of an orbit in view from the very beginning of the book (see *Figure 10*).[18] On page 4 of the *Astronomia Nova* Kepler presents a picture of the orbit of Mars in Ptolemaic astronomy – that is he draws the track of the planet as a continuous curve in two dimensions. Kepler intends the reader to contrast the complexity of this figure with the comparative simplicity of the ellipse he will introduce. Examining the frame diagram for PATH in Kepler's *Astronomia Nova* will show us several interesting things about how concepts change and also allow us to pinpoint the first major incommensurability in astronomical concepts due to Copernicanism.

Although Kepler ultimately subverts the whole structure of the PATH frame, at the outset the work that does this fits entirely within the established pattern (see *Figure 11*). Let us deal first with the top and bottom of the figure. Kepler's treatment of diurnal

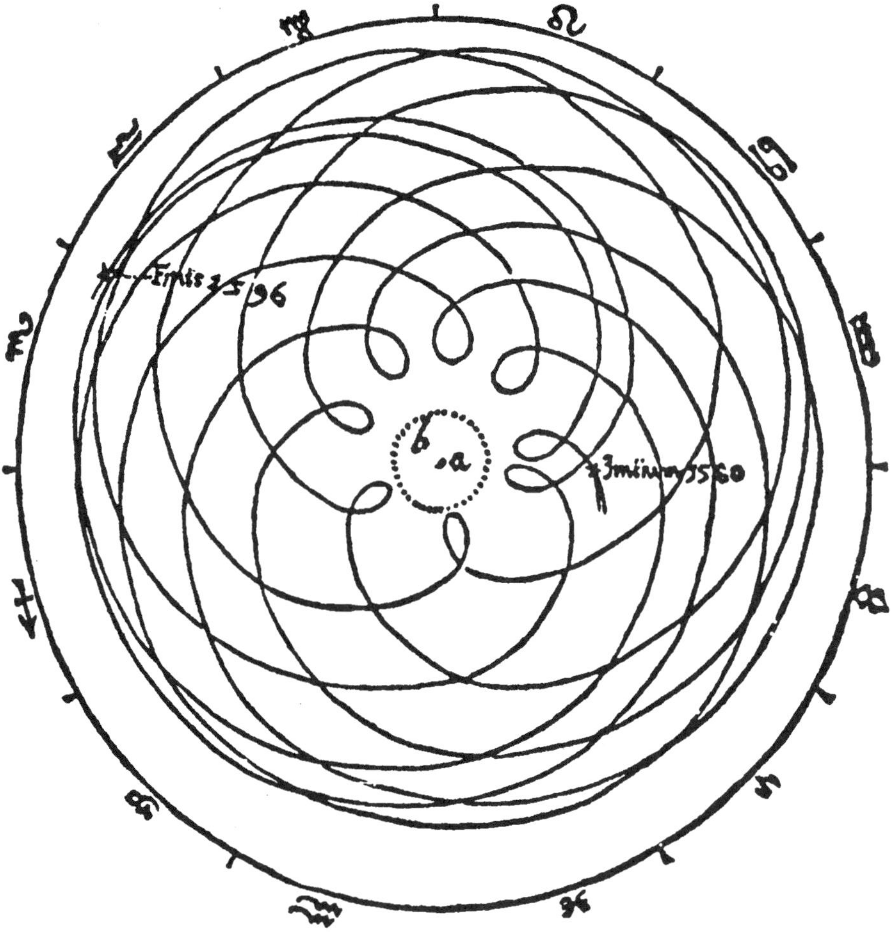

Figure 10: Kepler's illustration of the geocentric trajectory of Mars 1580–1596
Ioannes Kepler (1609) Astronomia Nova, *Prague, p. 4*

motion will be exactly the same as Copernicus's. His account of retrogression will also be the same as Copernicus's in general terms – although it differs in one important specific. Copernicus had attributed a single circular motion to the earth, while all the other planets moved on eccentric circles modified by small epicycles. He thus gave a different account of the motions of the planets and the earth. Kepler insisted from the beginning of the *Astronomia Nova* that all planets including the earth should be treated in the same way. Ultimately this would mean giving each one an elliptical orbit. So the set of nodes subordinate to RETROGRESSION must differ from Copernicus's, which specify a circular motion, and be replaced by whatever Kepler wants to say about the elliptical orbit of the Earth.

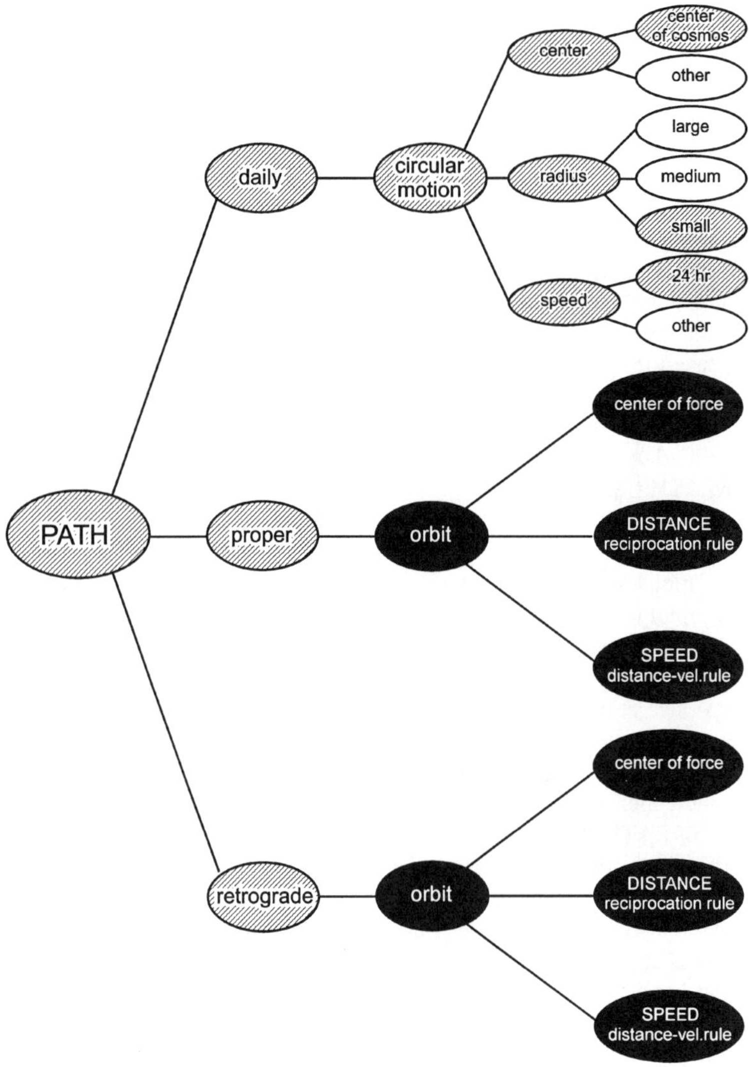

Figure 11: Frame diagram for PATH showing modifications introduced by Kepler In the Astronomia Nova *(1609). New nodes are solid black*

The most important changes occur in the account of proper motion. First, the concept of center of motion is replaced by the concept of a center of force. As soon as the material spheres of Aristotle and Ptolemy were abandoned the question of what moved the planets became acute. Kepler still accepted a version of the Aristotelian concept of inertia – objects move only while a force acts on them, and the motion is in proportion to the intensity of the force. In common with contemporaries including Bruno and Galileo, Kepler assumed that a power or force located in the physical sun was responsible for moving the planets. He believed this force diminished with distance (in our terms an inverse first power rather than an inverse square law). As the solar force diminished with distance it would propel a planet more slowly when further from the sun and more quickly when closer to it. If the circle on which a planet moved was eccentric to the sun, then, the planet would move more slowly near aphelion and more quickly near perihelion. Kepler used this relation between distance and velocity (understood now as the velocity of the planet *along its track*) to calculate the angular position of the planet, viewed from the sun, and ultimately the earth. This set of techniques (later – and in a modified form – to be labeled Kepler's second law) replaces the SPEED node in the earlier PATH frame.

The initial presentation of the distance-velocity relation uses an eccentric circle for the planet's track and does not agree completely with observations of positions for Mars or calculations of its distance. To accommodate the distance data Kepler introduces a second principle that governs the motion of the planet towards or away from the sun along the radial line defined by the distance-velocity relation. The planet is made to reciprocate on a diameter of a hypothetical epicycle. The motion is real – the epicycle is only a calculating device, it no longer carries the planet. This set of techniques allows Kepler to define the heliocentric distance to the planet at all points along its track and replaces the DISTANCE node in the earlier frame of PATH.

By these means Kepler is able to recover all the positional data corresponding to the concept of PATH. He shows that the actual track of the planet is an ellipse inside the eccentric circle introduced earlier. Thus, the superordinate node between the value nodes and PROPER MOTION is no longer CIRCULAR MOTION but something new. It is not merely the concept of an elliptical motion, but also the concept of a continuous, real track in space, which we may now correctly call an ORBIT.

What happened to the frame of PATH? Here I will offer a brief reconstruction summarizing a complex set of conceptual changes. In the work of Kepler and the astronomical tradition descended from him, the new concept ORBIT assumes a central place in all astronomical calculations that lead to comparisons with observations of position, which I originally gathered under the concept of PATH. ORBIT replaces the central attribute node under PROPER MOTION. What became of the other two attributes? The Copernican account of diurnal motion was universally accepted and those nodes were eliminated as corresponding to apparent rather than real motions. Next the account of retrogression was rewritten following Copernicus's proposals as modified by Kepler. Those nodes also changed status permanently when it was universally accepted that this motion too was apparent. The real motions that generated the retrogressions were those specified under ORBIT – so in effect the RETROGRESSION node is retained but repositioned as an attribute of ORBIT. Notice that these changes prune all the subordinate nodes out of the frame except those leading directly from PATH to ORBIT. Lacking branch points that would lead to alternative subordinate concepts, later astronomers began to refer to the orbital motion as the proper motion of a planet, and to the orbit as the path of the planet. And that, I think, is why we today find it natural to call to mind the concept of an orbit when someone speaks of the path of a planet, and only with some effort remind ourselves that path originally meant something completely different.

7. CONCLUSION

I emphasized at the outset that there are many ways of generating incommensurability in a theory of concepts as rich as Kuhn's mature view, or the frame model. The conclusions reached here are only preliminary, showing one particularly obvious kind of incommensurability generated by differences between nodes in frames with the same superordinate concepts. Judged by these, admittedly limited, standards we have reached an interesting result: in astronomy proper we see that the conceptual system of Copernican astronomy is minimally incommensurable with the conceptual system of Ptolemaic astronomy, but the conceptual system of Kepler is incommensurable with both. If degree of incommensurability is judged by the level of the superordinate node where the incommensurability begins, then this is a severe incommensurability. These results from the theory of concepts go some way towards explaining the historical facts. First, as already noted, Copernican astronomy, but not Copernican cosmology, was adopted and spread by many Northern Europeans. This would not be expected if Copernican astronomy was incommensurable with the existing Ptolemaic tradition. The tenacity of the older conceptual structure is also shown by the way that Kepler's astronomy developed. Initially, the two components of the solar force introduced by Kepler match the existing node structure for DISTANCE and SPEED, preserving the conceptual structure of the PATH frame. But the introduction of orbits ultimately subverts the whole frame. It is also worth noting that Kepler's astronomy was very slow to be accepted, and its reception included prominent efforts to remove those features that were incommensurable with the previous tradition. In our terms, these would be attempts to resist the revisions to the frame of PATH. An example is Bullialdus, who attempted to resolve the ellipse into collections of circles, preserving the superordinate node replaced by Kepler. The real break with the Ptolemaic tradition comes with Kepler. The analysis presented here shows that the main incommensurability between the new astronomy and the old astronomy appears only with Kepler's treatment of planetary motion in terms of orbits.

The episode we have considered is a classic example of the sort of change in concept that occurs when one incommensurable conceptual structure replaces another, and which makes the history of science so hard to understand. The replacement of PATH by ORBIT is so complete that it is difficult to understand the state of astronomy before ORBIT was in everyday use. However, this kind of cognitive obstacle occurs only after there is no longer ready access to the older conceptual system in texts, research reports or scientific practice. As the case of Bullialdus suggests, there was also a problem at the time, but this must have been of a different sort. Kepler just did not do a very good job of persuading readers of the *Astronomia Nova* to accept his new conceptual framework.

Despite the considerable rhetorical and dialectical skill employed in writing the book, it was hard to understand then and it is hard to understand now. Seventeenth century readers did not have frame diagrams to help them disentangle the components of Kepler's solar force.

A subsidiary issue of some importance is the fate of the celestial spheres. Major scientific revolutions and episodes that generate incommensurability are marked by the elimination and replacement of fundamental entities. In the case of the Copernican revolution it was not clear what was eliminated and replaced – unless you count the whole cosmos. The conceptual change in astronomy that allowed the introduction of ORBIT required the elimination of the celestial orbs, or shells as I have called them. This is an important change in fundamental entities that deserves more attention.

In this paper I have confined my attention to astronomy, with only occasional reference to cosmology or physics. The relations between these disciplines is itself a complex matter requiring some historical finesse during the period we have examined. It is clear that these relations are changing at the same time, and partly because of, the changes in astronomy that I have attempted to describe here. Astronomy used to be regarded as the engine driving scientific change during the Scientific Revolution as a whole. We have seen that this is at best partially true: changes in astronomy depended upon changes elsewhere (for example the elimination of celestial spheres in cosmology).

There is much more that needs to be said here about the conceptual systems of the different cosmological schemes. But judging from the results of the present investigation we should resist the temptation to conclude that the rival schemes are incommensurable until we have examined their detailed structure by means of a method like the frame model. The revolution in astronomy did lead to incommensurability between the older conceptual system and the new one, but not where we have been lead to expect it. If the present analysis is correct, Kepler, not Copernicus created the real break with ancient astronomy. The same analysis also suggests that the changes that cause severe incommensurabilities need not occur globally and all at once, but may depend upon the accumulation over time of many small changes, each of which may be quite minor. I am much more confident about the historical account I have presented than my reconstruction of the concepts in terms of dynamic frames. However, I believe that incommensurability is an undeniable feature of the historical changes in the concepts considered here, and that the introduction of the concept of an orbit locates one of the most acute incommensurabilities between the old and new conceptual systems during the Copernican revolution.

The University of Oklahoma

ACKNOWLEDGEMENTS

Earlier versions of this paper was presented at the conference on "Incommensurability (and related matters)" at the University of Hannover in June 1999, and at the conference "Kuhn Reconsidered" at Virginia Tech in March 2000. I would like to thank Paul Hoyningen-Huene and Howard Sankey for the occasion to present this work, and both of them together with Hanne Andersen, Xiang Chen, Bernard Goldstein and Steven Wagner for helpful comments. See also note 4. *Figures 7–10* are reproduced here by permission of the History of Science Collections at the University of Oklahoma, Marilyn B. Ogilvie, Curator.

NOTES

[1] Other events of comparable importance include the revolution in geology and the life sciences centering on the work of Lyell and Darwin, and the transition from Newtonian macroscopic theories to quantum microscopic theories in physics, centering on the work of Rutherford and Bohr.

[2] Although Clavius and his followers remained loyal to Ptolemaic astronomy, they were quite flexible in matters of natural philosophy connected with the cosmology. For example, they endorsed Galileo's telescopic discoveries and adopted a fluid substance for the heavens (Lattis, 1994).

[3] It is perhaps surprising that Kuhn did not consider Copernicus's main, announced objection to Ptolemy's models, the device known as the equant, but this issue may have been difficult to deal with for several reasons. First, Kuhn's concept of an anomaly in *Structure* was limited to observational or experimental failures, that is empirical problems. The equant however is a conceptual problem, and independent of the theory's success or failure in matching observations. Second, the connection between the equant problem and the physical models in Ptolemaic astronomy was obscured by the general denial that Copernicus himself subscribed to celestial spheres (Swerdlow, 1976; Barker, 1990). More on the status of physical models in section 6 below.

[4] I hereby absolve all my collaborators from any follies in the sequel.

[5] As an example of the latter sort of constraint, consider that 'orbit shape : elliptic' cannot be paired with 'orbit center : none'. Constraints of this sort are introduced as an explicit feature of some frame diagrams (see, for example Barsalou 1992 and Andersen et al., 1996). For simplicity I will not be considering constraints of this sort in the present paper.

[6] On these aspects of Kuhn's mature philosophy, see Hoyningen-Huene (1993). On the comparison between Kuhn's account and the frame model see Andersen, Barker and Chen, (1996) and Chen, Andersen and Barker (1998).

[7] For the development of Kuhn's view see Kuhn (1990a; 1990b; 1991; 1993).

[8] The basic data of astronomy from antiquity to the sixteenth century – the explananda or, if you prefer, the 'phenomena' that needed to be 'saved' – were recorded observations of planetary positions. Sixteenth century astronomy texts devoted most of their attention to motion in longitude. Motion in latitude was usually handled by a brief section at the end of the book and after the main business had been completed. It is also worth noting that the data to be explained were extremely sparse: before the programs of systematic observation initiated at Kassel by the Landgrave Wilhelm IV and at Hven by Tycho Brahe, most observations of celestial positions were made when celestial objects were doing something unusual, such as retrogressing or passing close by another object.

[9] My reasons for choosing this term are connected with the very limited Latin vocabulary available for celestial motions – basically just three terms. The most common, *cursus,* translates directly as 'path'. The least common, *orbita*, is adopted by Kepler and becomes the root of our term 'orbit'.

[10] For details see Barker and Goldstein (1994).

[11] From here on the frames I consider are no longer taxonomies in any simple sense.

[12] The term *theorica* identifies both to a genre of astronomical textbooks, and the most important contents of those books, which are mathematical models for the motions of particular celestial objects. I use the term in the latter sense from here on. The standard order of presentation was a *theorica* for the sun, followed by one for the moon, followed by a *theorica* that could be used for all the outer planets and Venus, and last a *theorica* for Mercury, which required special treatment. For details see Barker and Goldstein (1998; and Barker, 2000).

[13] It might be objected that the appearance of different values for the deferent center (for example) requires the activation of new nodes. But these values are, by hypothesis, part of the array of values already admitted.

The important point is not that these values differ for Ptolemy and Copernicus but that this diversity can be accommodated within the existing conceptual structure and without addition or deletion of nodes.

[14] Here I ignore Copernicus's account of precession and other long term changes.

[15] That is, no objection based on the *structure* of the system of concepts.

[16] It is rather difficult to say what kind of substance these spheres are made of. Although they are referred to as 'crystalline' this should be understood to indicate their transparency, and not their hardness. Hardness may have been attributed to them only later, and most conspicuously by people like Tycho Brahe who opposed spheres in any form in favor of fluid heavens. For a discussion see Goldstein and Barker (1995).

[17] On all of these points see Barker (2000).

[18] It is now generally accepted that the order of presentation in the *Astronomia Nova* is not the order in which Kepler made his discoveries but a literary device. See especially Stevenson (1994) and Donahue (1988; 1993).

REFERENCES

Primary Sources

Amico, Giovanni Battista. (1536). *De motibus corporum coelestium*, Venice.
Apianus, Petrus. (1540). *Cosmographia*, Antwerp.
Brahe, Tycho (1588) *De mundi aetherii recentioribus phaenomenis*, Uraniborg.
Copernicus, Nicolaus. (1543). *De Revolutionibus Orbium Coelestium*, Nuremberg.
Fracastoro, Girolamo. (1538). *Homocentrica*, Venice.
Kepler, Ioannes. (1596). *Mysterium Cosmographicum*, Tübingen.
Kepler, Ioannes. (1609). *Astronomia Nova*, Prague.
Maestlin, Michael. (1578). *Observatio et demonstratio cometae aetherei, qui in anno 1577 et 1578 constitutus in sphaera Veneris, apparuit*, Tübingen.
Reinhold, Erasmus. (1542). *Theoricae novae planetarum*. Wittenberg.

Secondary Sources

Andersen, H. (1996). "Categorization, Anomalies and the Discovery of Nuclear Fission." *Studies in the History and Philosophy of Modern Physics* **27**: 463-492.
Andersen, H., X. Chen and P. Barker. (1996). "Kuhn's Mature Philosophy of Science and Cognitive Psychology." *Philosophical Psychology* **9**: 347–363.
Barker, P. (1990). "Copernicus, the Orbs and the Equant." In R. Ariew and P. Barker, eds., *Pierre Duhem: Historian and Philosopher of Science*, Synthese **83**: 317-323.
Barker, P. (1993). "The Optical Theory of Comets from Apian to Kepler." Physis **30**: 1–25.
Barker, P. (1997). "Kepler's Epistemology." In C. Methuen, D. Liscia and E. Kessler, eds., *Method and Order in Renaissance Natural Philosophy*, pp. 355–68, New York: Kluwer.
Barker, P. (1999). "Copernicus and the Critics of Ptolemy." *Journal for the History of Astronomy* **30**: 343–358.
Barker, P. (2000). "The Role of Religion in the Lutheran Response to Copernicus." In M. Osler, ed., *Rethinking the Scientific Revolution*, pp. 59–88, Cambridge: Cambridge University Press.
Barker, P. (forthcoming-a). "Brahe and Rothmann on Atmospheric Refraction." In R. Halleux, ed., *Proceedings of the XXth International Congress of History of Science*. Turnhout: Brepols (1998–2000).
Barker, P. (forthcoming-b). "Stoic Alternatives to Aristotelian Cosmology: Pena and Rothmann." *Revue d'Histoire des Sciences*.
Barker, P. and B. Goldstein. (1988). "The Role of Comets in the Copernican Revolution." *Studies in History and Philosophy of Science* **19**: 299-319.
Barker, P. and B. Goldstein. (1994). "Distance and velocity in Kepler's astronomy." *Annals of Science* **51**: 59–73.
Barker, P. and B. Goldstein, R. Bernard (1998). "Realism and Instrumentalism in Sixteenth Century Astronomy: A Reappraisal." *Perspectives on Science* **6**: 232–258.
Barker, P. and Goldstein, R. Bernard R. (2001). "Theological Foundations of Kepler's Astronomy." In J. Brooke, M. Osler and J. van der Meer, eds., *Osiris* **16**: Forthcoming.

Barsalou, L. (1992). "Frames, Concepts, and Conceptual Fields." In A. Lehrer and E. Kittay, eds., *Frames, Fields and Contrasts: New essays in semantical and lexical organization,* pp. 21–74, Hillsdale: Erlbaum.

Barsalou, L. and Hale, C. (1993). "Components of Conceptual Representation: From Feature-Lists to Recursive Frames." In I. Mechelen, J. Hampton, R. Michalski, and P. Theuns, eds., *Categories and Concepts: Theoretical views and inductive data analysis*, pp. 97–144, New York: Academic Press.

Bloor, D. (1991). *Knowledge and Social Imagery*. Chicago: University of Chicago Press.

di Bono, M. (1995). "Copernicus, Amico, Fracastoro and Tusi's device: Observations on the use and transmission of a model." *Journal for the History of Astronomy* **26**: 133–154.

Chen, X., H. Andersen and P. Barker. (1998). "Kuhn's Theory of Scientific Revolutions and Cognitive Psychology." *Philosophical Psychology* **11**: 5–28.

Chen, X., H. Andersen and P. Barker. (2000). "Continuity Through Revolutions: A frame-based account of conceptual change." *Philosophy of Science (Proceedings)* **67:** forthcoming.

Donahue, W. (1988). "Kepler's Fabricated Figures: Covering Up the Mess in the *New Astronomy*." *Journal for the History of Astronomy* **19**: 217-237.

Gingerich, O. (1975). "'Crisis' versus Aesthetic in the Copernican Revolution." In A. Beer, ed., *Vistas in Astronomy, vol. 17*, pp. 85–94, Oxford: Pergamon Press.

Gholson, B., P. Barker. (1985). "Kuhn, Lakatos and Laudan: Applications to the History of Physics and Psychology." *American Psychologist* **40**: 755-769.

Goldstein, B. (1967). *The Arabic Version of Ptolemy's Planetary Hypotheses. Transactions of the American Philosophical Society* **57**: Pt. 4.

Goldstein, B. (1991). "The Blasphemy of Alfonso X: History or Myth?" In P. Barker and R. Ariew, eds., *Revolution and Continuity: Essays in the History and Philosophy of Early Modern Science*, pp. 143–153, Washington: Catholic University of America Press.

Goldstein, Bernard R. and Barker, P. (1995). "The Role of Rothmann in the Dissolution of the Celestial Spheres." *British Journal for the History of Science* **28**: 385–403.

Hoyningen-Huene, P. (1993). *Reconstructing Scientific Revolutions: Thomas S. Kuhn's Philosophy of Science*. Chicago: University of Chicago Press.

Kuhn, T. (1957). *The Copernican Revolution: Planetary Astronomy in Western Thought*. Harvard: Harvard University Press.

Kuhn, T. (1962). *The Structure of Scientific Revolutions*. Chicago: Chicago University Press. 2nd edition with Postscript, 1970.

Kuhn, T. (1990a). "An Historian's Theory of Meaning." Talk to Cognitive Science Colloquium. UCLA (unpublished manuscript).

Kuhn, T. (1990b). "Dubbing and Redubbing: The Vulnerability of Rigid Designation." In C. Savage, ed., *Scientific Theory*, pp.298–318, Minneapolis: University of Minnesota Press.

Kuhn, T. (1991). "The road Since Structure." In A. Fine, M. Forbes, and L. Wessels, eds., *PSA 1990*, Volume 2, pp. 3–13, East Lansing: Philosophy of Science Association.

Kuhn, T. (1993). "Afterwords." In P. Horwich, ed., *World Changes*, pp. 311–341, Cambridge: MIT Press.

Lakatos, I. (1977). *The Methodology of Scientific Research Programmes*. Cambridge: Cambridge University Press.

Lattis, J. (1994). *Between Copernicus and Galileo: Christopher Clavius and the Collapse of Ptolemaic Astronomy*. Chicago: University of Chicago Press.

Stevenson, B. (1994). *Kepler's Physical Astronomy*. Princeton: Princeton University Press.

Swerdlow, N. (1976). "Pseudodoxia Copernicana." *Archives Internationales d'Histoire des Sciences* **26**: 105–158.

Swerdlow, N. and O. Neugebauer. (1984). *Mathematical Astronomy in Copernicus's De Revolutionibus*, 2 volumes. Berlin: Springer.

Van Helden, A. (1985). *Measuring the Universe*. Chicago: Chicago University Press.

Westman, R. (1974). "The Melanchthon Circle, Rheticus and the Wittenberg Interpretation of the Copernican Theory." *Isis* **85**: 79-115.

Westman, R. (1994). "Two Cultures or One? A Second Look at Kuhn's *The Copernican Revolution*." *Isis* **85**: 79-115.

NANCY J. NERSESSIAN

CONCEPT FORMATION AND COMMENSURABILITY

Abstract. The paper addresses the issue of how the processes of concept formation and change in science can be brought to bear on the problem of incommensurability. It argues that the problem arises out of a methodological approach that identifies the conceptual structure of a science with a language and transfers what is thought to be known about languages to science. Employing a cognitive-historical method that shifts the focus to the representational and reasoning practices of scientists in constructing new concepts provides a way of uncovering the nature of the *commensurability* relations between successive representations of a domain.

1. INTRODUCTION

A central feature of the "problem of incommensurability" is the contention that it is not possible to compare the conceptual structures of comprehensive scientific theories. The problem arises becauseofwhat Feyerabend (Feyerabend, 1962; 1970) called the "meaning variance thesis": although successive theories of a domain may use the same words, many or all of their concepts have substantially different meanings. As many, including Kuhn (Kuhn, 1970, Postscript), have pointed out, "incommensurability" was not the best choice of expression, since although complete comparison is not possible, various bases for at least partial comparison can be found. Further, historians then and now seem able to locate and discuss similarities and differences between conceptual structures such as those of Newtonian mechanics and relativistic mechanics. I have been arguing that although the introduction of the problem into the philosophy of science has led to significant new insights about the nature of meaning change in science, incommensurability is not a problem in and of itself. In earlier work I traced the origins of the problem as it relates to meaning change to the positivist reductionist theory of meaning and the classical view of concepts (Nersessian, 1982; 1984; 1985). Here I first will sketch that analysis and then discuss the fruits seeking of what I called for then: a way of understanding nature of the commensurability relations between conceptual structures.

How the concepts of science are related to its empirical basis was a significant concern of logical positivism. Their solution to the problem became known as "the reductionist theory of meaning." Briefly, this theory held that 1) conceptual structures are languages and 2) all the terms – and later, sentences – of a theory reduce to a theory-neutral observational language. The observation language makes for continuity, and thus for comparisons between the conceptual structures of earlier and later theories of a domain, since each theoretical language can be reduced to the same observation language. Reduction was held first as taking place through definition and later through

P. Hoyningen-Huene and H. Sankey (eds.). Incommensurability and Related Matters, 275–301.

the looser connection of meaning postulates. Although Feyerabend (1962) and Kuhn (Kuhn, 1962) mounted a challenge to this theory based largely on the contention that the history of science does not support the continuous and cumulative nature of science that the theory leads to, cracks in its foundation had already appeared in the critiques of several of the distinctions upon which the theory depends. These distinctions are those between the conceptual and the given aspects of experience, the theoretical and the observational languages, a conceptual structure and a theory, analytic and synthetic knowledge, and pragmatic and factual dimensions of verification (see Nersessian, 1979; 1982; 1984 for an extended discussion). The most influential critiques where those of Lewis (Lewis, 1956), Carnap (Carnap, 1956b; 1956a), Duhem (Duhem, 1914), and Quine (Quine, 1960; 1963).

In place of the reductionist theory, Quine and Feyerabend proposed a "network" or "holistic" theory of meaning. The theory maintained that the sentences of a theory form an intricately woven network, ranging from highly theoretical to highly observational sentences with so sharp boundary between them. Since meaning is completely relative to a network, sentences are meaningless intertheoretically and incommensurability is a *fait accompli*. Feyerabend accepted this, but Quine attempted to salvage some form of network-independent meaning to serve as a basis for comparison by developing a notion of "stimulus meaning" which rested on soon-to-be discredited behaviorist notions of language learning. Kuhn did not immediately supply an alternative to the reductionist theory, instead over the next thirty years he attempted to mitigate the impact of incommensurability by developing a "family resemblance" account of concepts, especially of natural kinds. On the classical view, the representation of a concept is a set of conditions each of which is necessary and all of which are jointly sufficient to define it. Discussions of this view go back to Plato and Aristotle and its influence on contemporary thinking has come through the work of Frege and Russell. We can readily see how this notion of concept representation contributes to the problem of incommensurability. If differing sets of necessary and sufficient conditions are ascribed, e.g., to the concept 'field' a in theory and to the concept 'field' in a later theory, they are different concepts even though they are named by the same word. The notion of conceptual *change* makes no sense on this view; there is only conceptual *replacement*. In *Structure* Kuhn had employed the critique of the classical view of concepts by the later Wittgenstein to argue that the concepts and methods of a paradigm are not learned by assimilating rules but by abstracting family resemblances from exemplary problem solutions. In later work he sought to develop an account of concepts based on the notions of similarity and dissimilarity rather than rules and definitions.

In this work Kuhn saw himself as shifting the focus of philosophical analysis from a static view to the dynamical perspective first opened by examining the history of scientific practices. However, this shift ultimately led him to abandon history as a source for building a theory of scientific change. As he stated in the Rothschild Lecture at Harvard in 1992:

> Given what I shall call the historical perspective, one may reach many of the central conclusions we drew with scarcely a glance at the historical record itself. That historical perspective was, of course, initially foreign to all of us. The questions which led us to examine the historical records were products of a philosophical tradition that took science as a static body of knowledge and asked what rational warrant there was for taking one or

> another of its component beliefs to be true. Only gradually, as a by-product of our study of historical "facts" did we learn to replace that static image with a dynamical one, an image that made science an ever developing enterprise or practice. And it is taking longer still to realize that, with that perspective achieved, many of the most central conclusions we drew from the historical record can be derived instead from first principles. Approaching them in that way reduces their apparent contingency, making them harder to dismiss as a product of muckraking investigations by those hostile to science (Kuhn, 1992, p. 10; see also Kuhn, 1990, p. 6)

The question is from where do the "first principles" arise? Here I see Kuhn as reverting to the strategy that the numerous other approaches to the problem of incommensurability have in common with that of logical positivism; namely, placing too much analytical focus on scientific conceptual structure as languages and so transferring to science what might be said of languages generally. Clearly scientific conceptual structures can be represented linguistically. But this does not mean we can learn about the nature of conceptual change in science simply by investigating the nature of languages. One important difference between ordinary languages and the language of a science is that the former do not change as drastically as latter can within a short span of time. Theories ordinary of languages largely consider them as static, i.e, they consider how a language is acquired and transmitted within a community, not the dynamics of how languages are constructed and change. Kuhn is right that the history of science has shown us that the language of a science is dynamic, continually undergoing a process of construction and refinement. I believe the earlier Kuhn's insight that a theory of conceptual change would have to be grounded in an examination of the history of scientific practices has not yet fully been exploited. The representational and reasoning practices employed by scientists in creating scientific conceptual structures need to be examined and figured into the account. Clearly also a way needs to be devised to go beyond the "apparent contingency" and particulars of case studies to a more general account of the nature of concept formation and change in science. My approach, which will be discussed in the next section, has been to employ what I call a "cognitive-historical" method of analysis. The following sections will address the problem of how to countenance the continuity between successive representations that is observed in the historical record. This requires addressing the issues of what the meaning of a scientific concept comprises and how to represent meaning diachronically as well as synchronically ("representational dimension") and of how new concepts arise ("generative dimension").

2. COGNITIVE-HISTORICAL ANALYSIS OF CONCEPT FORMATION

One of the most important aspects of the critiques leading to the problem of incommensurability was the methodological challenge posed by Quine, Kuhn, and Feyerabend. This challenge has led to a burgeoning of naturalism in philosophical analysis where history, psychology, and sociology are all resources to be employed in philosophical investigations. As I have argued elsewhere, in shifting the focus from the conceptual structures themselves to the representational and reasoning practices employed by scientists, both the historical and the cognitive dimensions of the micro-processes of conceptual change come into play (Nersessian, 1985; 1992a; 1995).

To investigate these dimensions requires employing what I have called a "cognitive-historical" method of analysis. Cognitive-historical analysis presupposes that the cognitive practices scientists have invented and developed over course of the history of science are sophisticated outgrowths of ordinary reasoning and representational practices. Thus, understanding how science develops and changes requires knowledge both of the actual practices of scientists and of how human cognitive abilities and limitations produce and constrain these practices. Neither of these can be determined *a priori*. The cognitive-historical method, on the one hand, conducts fine-structure historical inquires about cognitive practices, such as concept formation and change, employed in science. Numerous historical investigations establish conceptual change to be a problem-solving process, extended in time and embedded in social contexts. On the other hand, the method draws on scientific investigations of how humans reason and represent more generally to help interpret these practices. The cognitive sciences offer analyses and techniques which, if used with proper respect for their scope and their limitations, can help us develop and test models of how conceptual change takes place in science

Research in cognitive science provides an additional basis for taking seriously the practices exhibited in the historical cases as generative of problem solutions, even though these practices do not fit traditional philosophical notions of reasoning as deductive or inductive argument. Further, it provides a basis for generalization. Scientific cognition and ordinary cognition are not different in kind but lie along a continuum. Those reasoning and representational practices that are employed productively by humans generally are more likely to be generative in specific endeavors such as science. But, that ordinary and scientific representational and reasoning practices lie on a continuum does not rule out the possibility of significant differences in interpretation from what current cognitive theories claim of ordinary processes. At present cognitive theories are largely uninformed by data arising from examination of scientific practices, making the fit between cognitive theories and scientific practices something that still needs to be determined. The process of cognitive-historical analysis is reflexive in nature. It is a dialectical process that makes a contribution both to philosophical analysis and cognitive theorizing.

As discussed above, a significant aspect of the problem of incommensurability is: Given that scientific theories change, what is the nature of the relationship between successive representations of a domain? This paper is addressing the issue from the perspective of how representational structures emerge, where in many cases the relationship is one of continuous but not simply cumulative development. Issues pertaining to the developmental aspect of conceptual change lie along two dimensions; these are the metatheoretical question of the form of representation of concepts and the generative question of the nature of the constructive practices that create concepts, which here will be limited to considering reasoning practices.

3. THE REPRESENTATIONAL DIMENSION

From a cognitive-historical perspective, a scientific conceptual structure is one kind of representational system. Examining the representational practices of scientists leads to the following insights. Some earlier concepts *disappear*, such as 'phlogiston' from chemistry and totally new concepts are *created*, such as 'spin' in quantum mechanics. However, often many concepts in a new system are *descendants* of earlier ones, such as 'mass' (descendant of 'mass' in Newtonian mechanics) and 'field' (descendant of 'field' in electrodynamics) in the special theory of relativity. Also, some, such as 'aether', though appearing to be eliminated, have significant aspects *absorbed* by other concepts, in this case 'field' and 'space-time'. In many cases of conceptual change in the history of science all four phenomena occur with those of descendance and absorption predominating. This situation led to my characterizing such conceptual change as "continuous but not simply cumulative" (Nersessian, 1984). Another way of characterizing the idea would be as commensurability without cumulativity. Initial discussions of the problem of incommensurability characterized conceptual change as wholesale replacement of one structure by another (creation and disappearance). This characterization stems in part from the traditional view that concepts are represented by necessary and sufficient defining conditions. If this were the case, then something either does or does not fit the specified conditions. On this view of representation, changing the conditions does not change the meaning of an existing concept, rather it creates a new concept. Such a characterization cannot accommodate the phenomena of descendants and absorption. If we couple this problem with the widespread fact that it is not possible to specify necessary and sufficient conditions for scientific concepts, generally, we are led to the conclusion that a different form of representation is needed to handle change.

3.1 Frame-Based Representations

From a cognitive-historical perspective, psychological research on human categorization in general should be brought to bear on the issue. This research substantiates that in many instances people do not represent concepts by means of sets of necessary and sufficient conditions (Rosch and Mervis, 1975; Rosch and Lloyd, 1978; Smith and Medin, 1981). Rather, people represent concepts by prototypical exemplars with category membership determined by similarities and differences (family resemblances) to the features of the prototype. Additionally, concepts show graded structure with better and worse instances. Further, categories possess a hierarchical structure – basic level, subordinate level, and superordinate level – showing that concepts form taxonomies. The two main psychological accounts of the format of representation for taxonomic concepts are feature lists and frames. Feature list representation are based on the notion of similarity and have been criticized for not accommodating well-known effects of categorization, such as context dependence and goals. Additionally, empirical evidence strongly suggests that features are organized into complex structures, possibly on the basis of intuitive theories of the instances (see, e.g., Armstrong, Gleitman and Gleitman, 1983; Medin, 1989). For example, building nests and laying eggs are typical features of 'bird', but these features bear complex relations to one a other as well as to other features such as having feathers. Frame accounts have three important aspects: 1) they structure features hierarchically in attribute-value relations, where features at the value level are instantiations of the more general attribute; 2) frames capture "structural invariants", i.e., stable relations among attributes, called structural invariants; and

3) frames "capture constraints", i.e., correlations between values instantiating different attributes.

In various papers Hanne Andersen, Peter Barker, and Xiang Chen (Andersen, Chen, and Barker, 1996; Andersen, 1996; Chen, 1995; Chen, Andersen, and Barker, 1998) have attacked the representation problem by making use of the "dynamic frame" form of representation developed by the cognitive psychologist Lawrence Barsalou (Barsalou, 1992). *Figure 1* provides an illustration of how dynamic frames might be developed for the taxonomic concepts 'goose' and 'celestial body'.

Their research has established how this form of representation enables tracking the commensurability relations between successive representations of concepts. The paper by Barker in this volume deals exclusively with dynamic frame representations and since there is much that I agree with in it, I will focus my attention here on what I see as limitations in the taxonomic approach and the use of Barsalou's version of dynamic frames. The main problem is that only a limited range of scientific concepts can be ostended individually and thus form contrast sets (family resemblance on the superordinate level), such as 'duck', 'goose', and 'swan' or 'planet', 'comet', and 'asteroid'. Instances of theoretical concepts such as 'force' and 'electromagnetic field' are not pointed out individually to reveal similarities between instances or differences from contrasting concepts. Rather, what are usually pointed out are instances of the application of a natural law, such as Newton's second law $\mathbf{F} = m\mathbf{a}$, in which the concepts 'force', 'mass' and 'acceleration' are involved simultaneously. Thus for these concepts another form of representation needs to be developed.

3.2 The "Meaning Schema" Representation

In my 1984 analysis I introduced a form of representation called a 'meaning schema' that was designed to capture not only the phenomena of descendance and absorption but also the fact that scientific concepts are embedded in complex problem situations and are intertwined with theory. Recent collaboration with Hanne Andersen (Andersen and Nersessian, 2000; Nersessian and Andersen, 1998) on the limitations discussed above have led to our attempting to connect this form of representation to an extension of dynamic frame representation to problem situations rather than individual concepts.[1] This work is best illustrated by example.

The summary of the pertinent results from my study of the electromagnetic field case (Nersessian, 1984) are as follows. First, there are no necessary and sufficient conditions for various individual historical usages of the field concept and none that would encompass all. Second there did seem to be a core of meaning that carried across and allowed for identification of the various historical instances as 'field': physical processes take place in the region surrounding charges and bodies. Third, the meaning of the concept and theoretical principles were intertwined. These results had some parallels to findings about concepts and categorization in the cognitive science literature, specifically research on graded structures and prototypes (see, e.g., Rosch and Mervis,1975; Rosch and Lloyd 1978) and that pointing out that 1) some concepts exhibit a common core and 2) that concepts are intertwined with intuitive theories in that they are learned together with certain principles (see, e.g., Armstrong, Gleitman and Gleitman, 1983; Medin, 1989; Smith and Medin, 1981). However, in 1984 there was no form of rep-

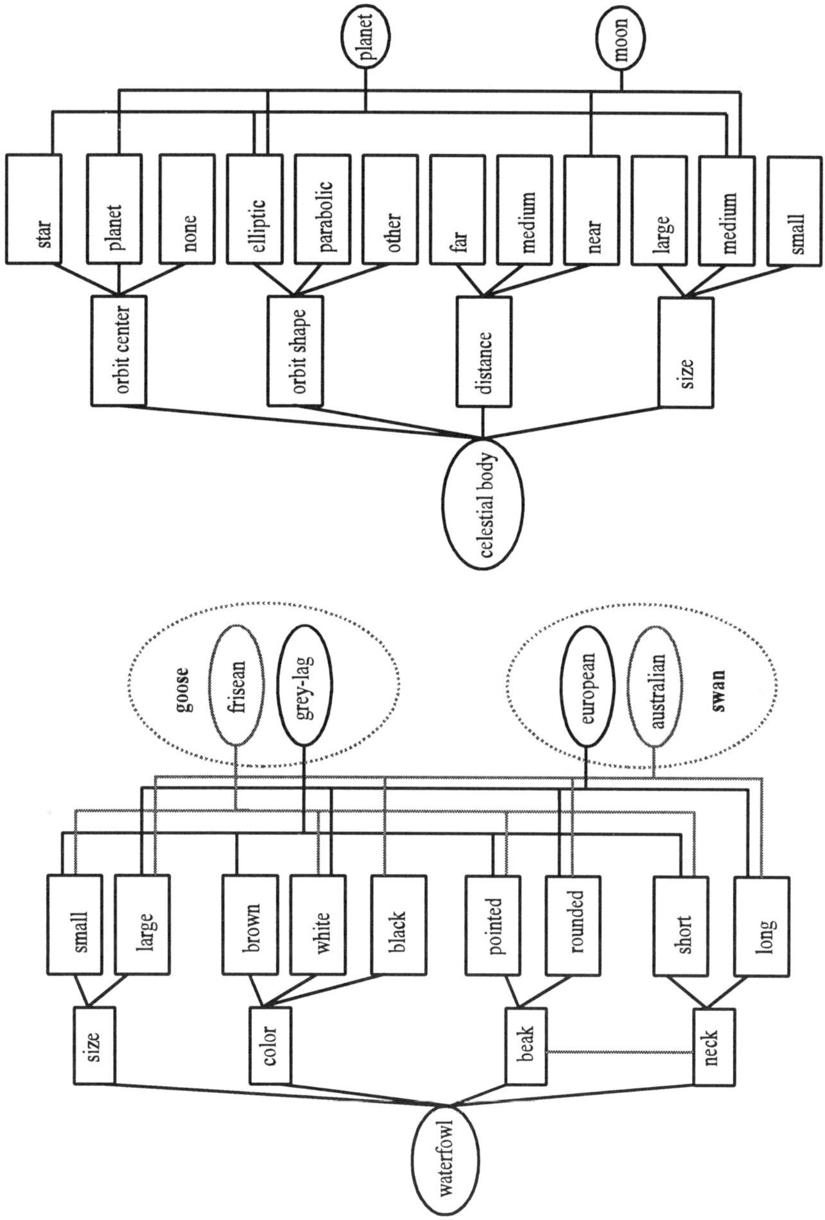

Figure 1: Partial frames for taxonomic concepts such as 'goose', 'swan', 'planet' or 'moon'

resentation of a concept proposed in cognitive science that accommodated these findings. The meaning schema was my attempt to construct a form of representation that could capture these cognitive-historical findings plus my claim that concepts are embedded in problem situations and connected through reasoning. It is a frame-like structure that captures the salient components of a scientific concept central to its descriptive and explanatory functions within a problem situation. The meaning schema encompasses theoretical principles and exhibits the family resemblances over their various historical instantiations (token). An instantiation of concept at a given time is represented by four components: 'ontological status', 'function', 'mathematical structure', and 'causal power'. The components of successive concepts are connected via "chains of reasoning" along both the vertical and horizontal dimensions of the schema.

Figure 2 provides an outline of a meaning schema for four historical tokens of the concept 'electromagnetic field'.

The *causal power* feature of a concept marks out the problem situations in which the referent of a concept comes into use in order to explain the situation (i.e. the situations that the concept is used to explain). Note that this dimension captures the cumulative nature of experimental situations Theodore Arabatzis (Arabatzis, 1999) has argued determines whether there is referential stability or not. The *mathematical structure* corresponds to the laws associated with a group of similar problem situations. Once Maxwell had constructed the field equations, they remained relatively stable in the form reformulated by Hertz, but their interpretation changed significantly. The *function* of a concept marks out a specific part of problem situation in which a concept plays an explanatory role and clarifies that role. Here the function is new for physics – nothing had played the role of transmitting action in space where there is no "ordinary" matter before. The *ontological status* of a concept represents a belief about what kind of "stuff" is responsible for this particular function. For example, corresponding to the function of transmitting the electric action is a state of a mechanical aether in the Maxwellian representation and a state of space in the Einsteinian. In the Lorentzian and Einsteinian representations, charge corresponds to a particle (electron). However, in the Faradayan and Maxwellian representations charge has the ontological status of a state of stress in the medium. For Faraday, it is a state of stress in the lines of force; for Maxwell, a state of stress in the mechanical aether.

Each component of the meaning schema is connected with subsequent forms through the "chains of reasoning" (COR connections) (see Shapere, this volume) leading to that change. This provides a dimension beyond that of family resemblances for determining whether there is continuous but non-cumulative development in a specific case. So, for example, the preferred ontological status of the electromagnetic field in the Faradayan representation is a substance (force in a region of space). In the Maxwellian representation, forces in regions of space are produced in an underlying mechanical medium, the aether. But Faraday's lines of force become specific configurations of Maxwell's mechanical aether, so there is continuity between the concept representations. Lorentz's goal of incorporating charged particles into the laws of the electromagnetic field constrains the aether to be non-mechanical, which in turn requires the introduction of the Lorentz force for field-particle interactions into the Maxwellian mathematical structure. Finally, the ontological change introduced by Einstein stems from reasoning that there is no

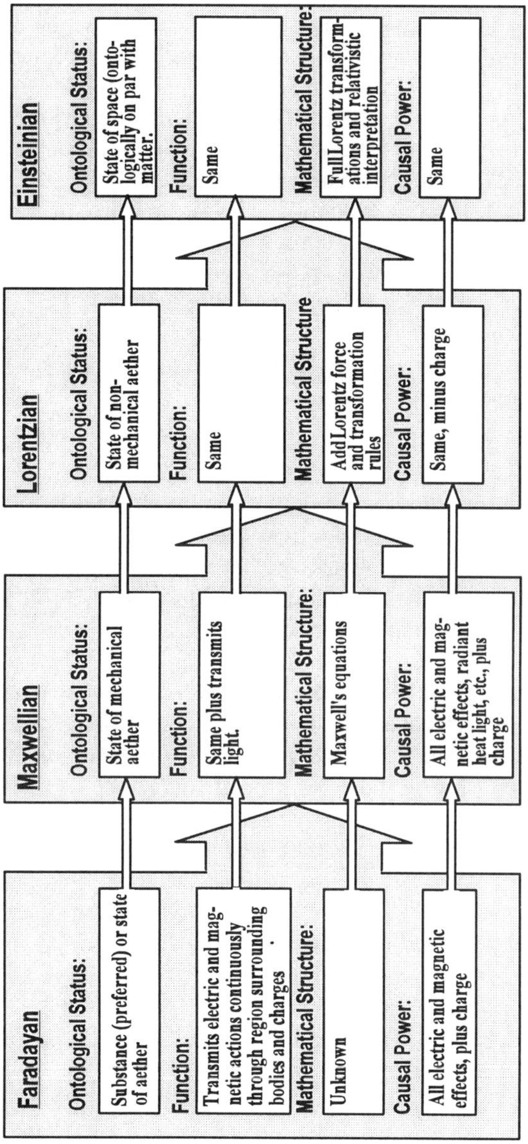

Arrows indicate chain of reasoning connections

Figure 2: Partial meaning schema of historical instantiations of 'electromagnetic field'

physical role for the Lorentzian aether in production of the field, so it is eliminable. However, the function and causal power components remain the same, as does the form of mathematical structure (with stream-lined transformation rules), but not the interpretation.

I am not wedded to the meaning schema form of representation as a means of individuating concepts, but that something like it is needed to represent scientific concepts underscores the essential historical dimension of meaning change and the role concept formation plays in the problem of incommensurability. With it we can represent the nature of the *commensurability* connections that allow us to call all the instantiations 'electromagnetic field'. Without the cognitive-historical analysis, the case would be seen to be paradigmatic of incommensurability. It is even worse than customary exemplar case of 'mass' because there is no possibility of reverting to a limiting-case to provide commensurability. The Maxwellian electromagnetic field is a state of a mechanical aether, where charge and current arise from the field. The Lorentzian electromagnetic field is a state of a non-mechanical aether, where the field arises from charge and current. The Einsteinian electromagnetic field is a state of space, where the field arises from charge and current.

3.3 Non-Taxonomic Concepts

The meaning schema has also proved useful in research conducted jointly with Hanne Andersen on the problem of representing non-taxonomic concepts (Andersen and Nersessian 2000). In addressing this problem, we have generalized the dynamical frame representation to capture the family resemblance structure among problem situations. *Figure 3* provides a partial representation of problem situations pertaining to electrostatic action.

These problem situations all involve the attributes **charge distribution**, **electric field, electric action** and **electrostatic potential**. The attributes of the problem situation can take different values, for example, **charge distribution** takes – among others – the values **point charge**, **line charge** and **surface charge**. On this representation of 'electrostatic action', the different forms of the electrostatic equations are associated with different instantiations of the frame, that is, with different patterns of values of the attributes. For example, the various forms of Gauss' equation are associated with instantiating specific values of **charge distributions** and **electric field**. Thus, instantiating the values **point charge** (**charge distribution**) and **spherical** (**electric field**) is associated with the equation

$$\boldsymbol{E} = Q/(4\pi\epsilon_0 r^2)\ \boldsymbol{a}_r \tag{1}$$

Likewise, instantiating the values **line charge** and **radial** is associated with the equation

$$\boldsymbol{E} = \rho/(2\pi\epsilon_0 r^2)\ \boldsymbol{a}_r \tag{2}$$

Here, the various situations to which the equation applies as well as the specific forms of the equation are related by family resemblance and the individual attributes in the situation, here the **charge distribution** and the **electric field,** are related via a scientific law.

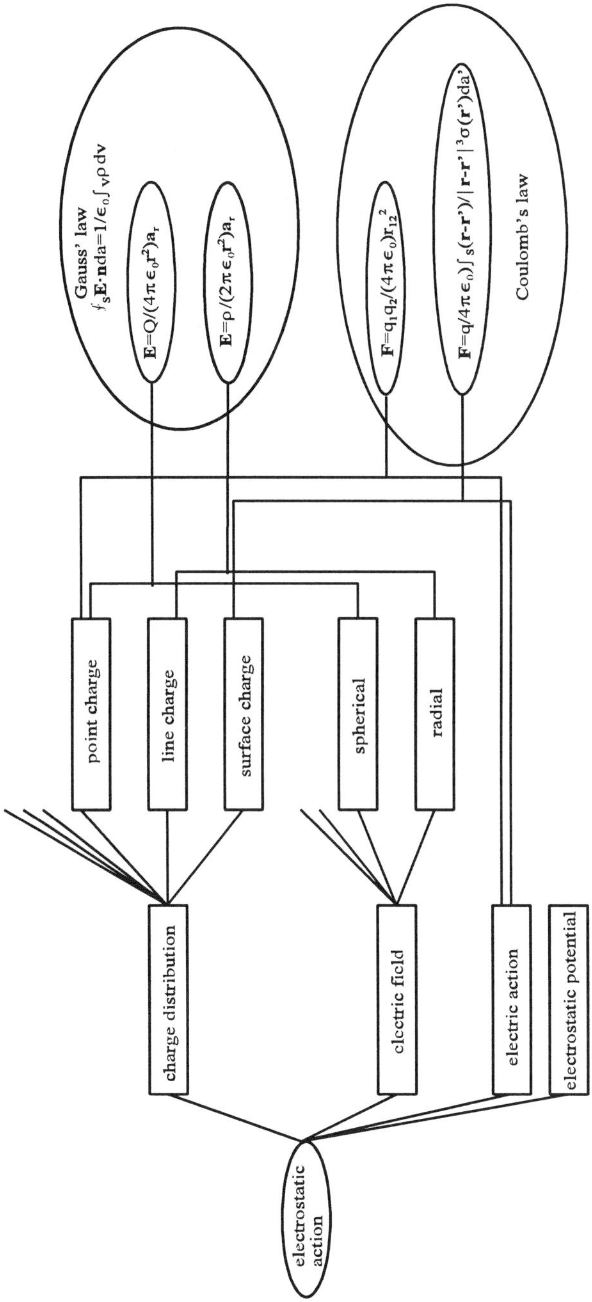

Figure 3: Partial frame for non-taxonomic concept 'electrostatic action'

Another instantiation pattern is associated with the similarity class of problem situations that can be described by Coulomb's law. The instantiations of the values **point charge** and **electric action** are associated with the equation

$$\boldsymbol{F}=q_1 q_2/(4\pi\epsilon_0)\boldsymbol{r}_{12}^{\,2} \tag{3}$$

while those of **surface charge** and **electric action** are associated with the equation

$$\boldsymbol{F}=q/(4\pi\epsilon_0)\int_S(\boldsymbol{r}-\boldsymbol{r}')/|\boldsymbol{r}-\boldsymbol{r}'|^3\sigma(\boldsymbol{r}')da' \tag{4}$$

Again, the various situations to which the equation applies as well as the specific forms of the equation are related by family resemblance and the relation between the individual attributes in the situation, here the **charge distribution** and the **electric action**, is a scientific law.

The modified frame of the problem situation, however, does not in itself enable individuating concepts in a problem situation. To resolve this problem we link the component of the meaning schema for the historical instantiation of the concept. In this case, the contemporary (relativistic) concept of 'electromagnetic field' is linked to partial frame for the problem situations pertaining to electrostatic action, as shown in *Figure 4*.

The *causal power* marks out the problem situations in which the concept comes into use in order to explain the situation and thus can be linked to the frame representing a similarity class of problem situations. Likewise, the *mathematical structure* corresponds to the laws associated with this similarity class of problem situations. The additional components of the meaning schema – 'function' and 'ontological status' – are the components which serve to distinguish individual concepts within the complex situation.

The *function* of a concept marks out a specific part of the problem situation in which it plays an explanatory role. For the two kinds of problem situations dealing with electrostatic action introduced above, situations dealing with the electric field arising from a charge distribution (i.e., *causal power*) and described by Gauss' equation, with *mathematical structure*:

$$\oint_S \boldsymbol{E}\cdot\boldsymbol{n}da=(1/\epsilon_0)q_i \tag{5}$$

or situations dealing with the electric force exerted on a charged body due to the presence of other electric charges (*causal power*) and described by the Coulomb equation, with *mathematical structure*:

$$\boldsymbol{F}_i=q_i\Sigma(q_j\boldsymbol{r}_{ij})/(4\pi\epsilon_0 r_{ij}^{\,3}) \tag{6}$$

the various quantities contained in these equations play different roles in the explanation of why a given problem situation develops as it does. For example, the electric field intensity (**E**) transmits the electric action (**F**) that is exerted on a charged body due to the presence of other electric charges (q). To the various functions corresponds an *ontological status*, that is, a belief about what kind of "stuff" is responsible for this particular function. Here charge is a property of a particle (electron) and the electromagnetic field is a state of space.

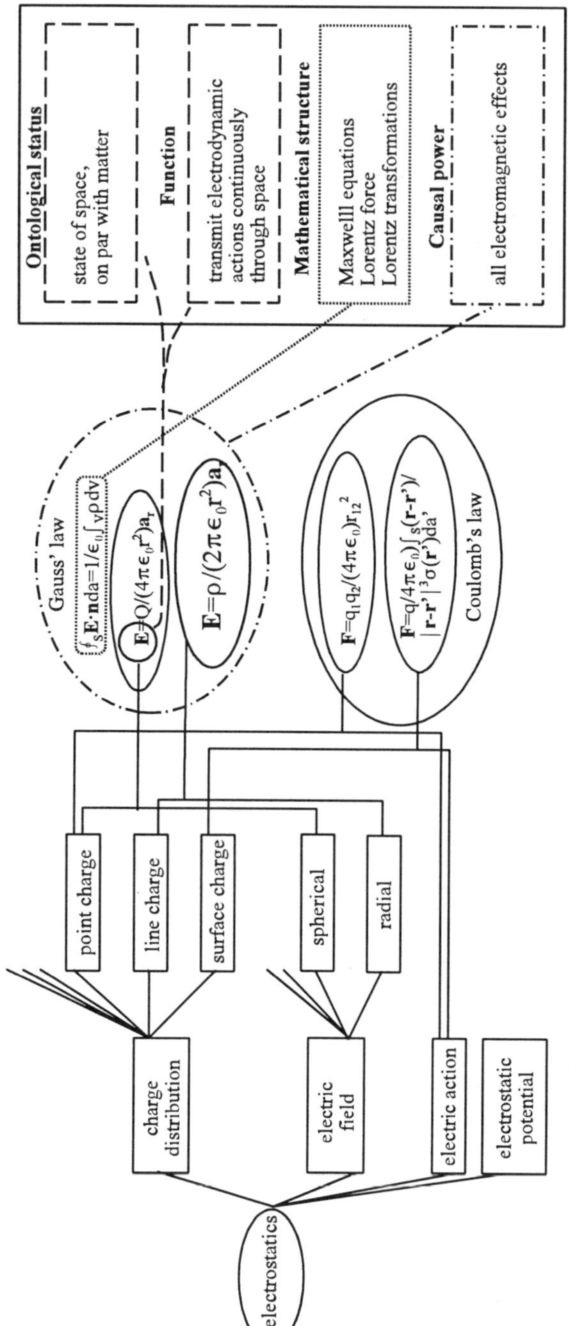

Figure 4: Meaning schema for individuating 'electromagnetic field' in electrostatic problem situations

4. THE GENERATIVE DIMENSION

The chains-of-reasoning aspect of the meaning schema captures the idea that the specifics of the nature of any commensurability that might exist between earlier and later tokens of a concept can only be determined through unpacking the fine structure of the chains of reasoning leading from one instantiation to the next. However, there is an a broader question of the general nature of the reasoning practices that are generative of new conceptual structures and how that kind of reasoning might establish continuity among various tokens. Although the notion of 'chains of reasoning' was drawn from the work of Shapere (Shapere, 1966), my focus since the 1984 book has been on understanding the general nature of reasoning practices rather than on unpacking the specific reasons for various change (see, e.g., Nersessian, 1992b; 1992a; 1999). In this research cognitive-historical analysis yields further means of understanding the nature of the commensurability between successive representations of a domain. Given the frequency of reasoning through analogy and visual modeling and thought experimenting in cases of conceptual change across the sciences, I have been trying to understand how these modeling practices could be generative of new concepts. The interpretation I have constructed has implications for the problem of incommensurability in that it establishes how that kind of reasoning can lead to continuity between successive representations of a domain.

From the perspective of the most radical interpretation of incommensurability, that of complete discontinuity, there would seem to be no role existing representations could play in the construction of new conceptual structures. Feyerabend's theory of conceptual change called for proliferating competing theories, but offered no insights into where the new theories come from. The replacement conceptual structure is the one that emerges victorious from a largely non-rational competition process. For Kuhn, conceptual change follows a pattern of first there are anomalies, then crises, and then a new theory arises, but that process is also left mysterious. However, the radical discontinuity view is decidedly unhistorical, and Kuhn, having carried out many historical analyses, could not subscribe to it. Instead, he maintained that "since the new paradigms are born from old ones, they ordinarily incorporate much of the vocabulary and apparatus, both conceptual and manipulative" of the old paradigm (Kuhn, 1970, p. 149). My contention is answering the question of how they are "born from new ones" reveals the nature of the commensurability relations between successive representations of a domain.

My cognitive-historical analysis of the field case suggests a dimension of continuity between Newtonian mechanics and relativistic mechanics, if they are viewed from the generative perspective of the processes that included the development of the electromagnetic field concept. The chief point I will be making here is that a major source of the continuity between the mechanical and the relativistic representational structures lies in how Maxwell used analogy to create generic abstractions in formulating the mathematical structure of the electromagnetic field concept. This use of analogy differs from a formal analogy, such as used by Thompson in the analogy between heat and electrostatic, where one takes the equations from one domain and substitutes in the variables corresponding to the target domain. Instead, Maxwell's use of analogy involved taking constraints from a target domain to create a base analogy and then abstracting the mathematical relations common to both domains, creating a dimension of commensurability between them. I call the process Maxwell employed, "abstraction via generic modeling" (Nersessian, 1996; 1999; 2000a; 2000b). What I mean by this will be illustrated with an analysis of how Maxwell used generic abstraction in representing electromagnetic induction and in representing energy. But first we need some general discussion of the process of generic abstraction.

4.1 Generic Abstraction

To understand the role of model-based reasoning in concept formation and conceptual change in science requires a fundamental revision of the understandings of concepts, conceptual structures, conceptual change, and reasoning customarily employed explicitly in philosophy and at least tacitly in the other science studies fields. It is not possible to provide all the necessary details and arguments in the confines of this paper. Only an outline of my account will be developed here. A basic ingredient of the revision is to view the representation of a concept as providing sets of constraints for generating members of classes of models. Concept formation and change is a process of generating new, and modifying existing constraints. When model-based reasoning is employed, this is accomplished through iteratively constructing models embodying specific constraints until a model of the *same type* with respect to the salient constraints of the phenomena under investigation, the "target" phenomena, is achieved. My hypothesis is that the prevalence of analogies, visual representations, and thought experiments in periods of radical conceptual change indicates that model-based reasoning is a highly effective means of examining, revising, and abstracting constraints of existing representational systems and, in light of constraints provided by the target problem, effective means of generating new sets of constraints that the new representational structures come to embody. I make no claims that they are the only methods employed in concept formation. However, they are ubiquitous in cases across the sciences (see, e.g. Darden; 1980; 1991; Giere, 1988; 1992; 1994; Gooding, 1990; Gentner *et al.*, 1997; Griesemer and Wimsatt, 1989; Griesemer, 1991b; 1991a; Holmes, 1981; 1985; Latour, 1987; 1986; Lynch and Woolgar, 1990; Rudwick, 1976; Shelley, 1996; Thagard, 1991; Trumpler, 1997; Tweney, 1992).

In my full analysis of the various forms of reasoning noted above, I treat them in a unified fashion as model-based reasoning because they are often employed together in reasoning episodes. For example, in the case that will be discussed in section 4.2, the idle wheel - vortex model employed by Maxwell in his derivation of the electromagnetic field equations and illustrated by him in *Figure 5* exemplifies why a unified account is needed.

On my interpretation this is a *visual* representation of an *analogical* model that is accompanied with instructions for *simulating* it correctly in thought:

> Let the current from left to right commence in AB. The row of vortices *gh* above AB will be set in motion in the opposite direction to a watch... We shall suppose the row of vortices *kl* still at rest, then the layer of particles between these rows will be acted on by the row *gh* on their lower sides and will be at rest above. If they are free to move, they will rotate in

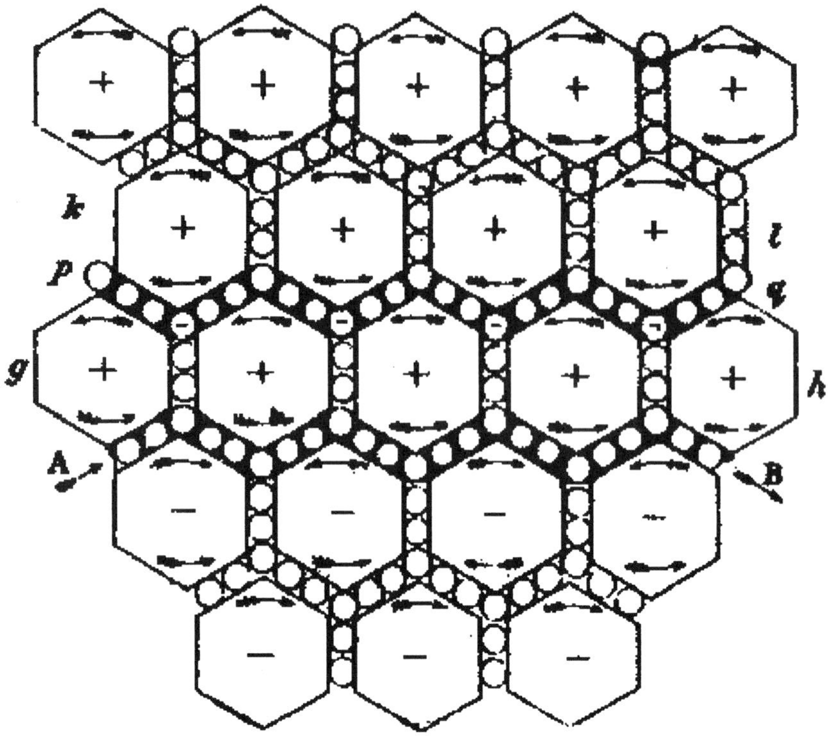

Figure 5. Maxwell's drawing of the vortex-idle wheel medium (Maxwell 1890, Volume I, Plate VII)

> the negative direction, and will at the same time move from right to left, or in the opposite direction from the current, and so form an *induced* electric current.

(Maxwell, 1890, Volume 1, p. 477, italics in original). However, here I am only going to discuss analogical modeling since that will be sufficient to understand how model-based reasoning creates commensurability relations between representations.

To engage in analogical modeling one calls on knowledge of the generative principles and constraints for models in a known "source" domain. These constraints and principles can be represented in different informational formats and knowledge structures that act as explicit or tacit assumptions employed in constructing and transforming models during problem solving. Inter- or intra-domain models can be retrieved directly from the source domain and applied with suitable adaptation, but often, and especially in cases of conceptual change, no direct analogy exists and construction of an initial model itself is required. In these cases the source domain(s) provides constraints that are used together with those provided by the target problem to create the initial as well as subsequent models (Nersessian, 1992a; 1999; 2000a). Evaluation of the analogical modeling process is in terms of how well the salient constraints of a model fit the salient constraints of a target problem, with key differences playing a significant role for further model generation (Griffith, Nersessian and Goel, 1996; Griffith, 1999; Griffith, Nersessian and Goel, 2000). There is an extensive cognitive science literature on analogy with much empirical evidence that substantiates the claim that it is generative in instances of conceptual change. This literature provides theories of the processes of retrieval, mapping, transfer, elaboration, and learning employed in analogy and the syntactic, semantic, and pragmatic constraints operating on these processes (see, e.g., Gentner, 1983; 1989; Gentner *et al.*, 1997; Gick and Holyoak, 1980; Gick and Holyoak, 1983; Holyoak and Thagard, 1989; Holyoak and Thagard, 1996; Thagard *et al.*, 1990). Although no current cognitive theory is able to handle the complexity of the Maxwell case, the literature does agree with my analysis in that analogies are not "merely" guides to reasoning but form the creative heart of the reasoning processes in which they are employed. There is also widespread agreement on criteria for good analogical reasoning drawn from psychological studies of productive and non-productive use of analogy and formulated by Gentner (Gentner, 1983; Gentner, 1989): 1. "structural focus": preserves relational systems, 2. "structural consistency": isomorphic mapping of objects and relations, and 3. "systematicity": maps systems of interconnected relationships, especially causal and mathematical relationships.

Constraints in both the target and source domains are domain-specific and need to be understood in the reasoning process at a sufficient level of abstraction for retrieval, transfer, and integration to occur. I call this level of abstraction "generic". That is, the various representations employed have to function with some of their features considered as unspecified. In model-based reasoning processes, a central objective is to create a model that is of the *same kind* with respect to salient dimensions of the target phenomena one is trying to represent. Thus, although an instance of a model is specific, inferences made with it in a reasoning process are generic. In viewing a model generically, one takes it as representing features, such as structure and behaviors, common to members of a class of phenomena. The relation between the generic model and the specific instantiation is similar to the type-token distinction used in logic. Generality in representation is achieved by interpreting the components of the representation as referring to object, property, relation, or behavior types rather than tokens of these. One cannot draw or imagine a "triangle in general" but only some specific instance of a triangle. However, in considering what it has in common with all triangles, humans have the ability to view the specific triangle as lacking specificity in its angles and sides. In considering the behavior of a physical system such as a spring, again one often draws or imagines a specific representation. However, to consider what it has in common with all springs, one needs to reason as though it as lacked specificity in length and width and number of coils; to consider what it has in common with all simple harmonic oscillators, one needs to reason as though it lacked specificity in structure and aspects of behavior. That is, the reasoning context demands that the interpretation of the specific spring be as generic.

The kind of creative reasoning employed in conceptual innovation involves not only applying generic abstractions but creating and transforming them during the reasoning process. The process of abstracting to the generic level is a significant reasoning process in analogical modeling in conceptual change which often requires recognition of potential similarities across disparate domains, and abstraction and integration of information from these. There are many significant examples of generic abstraction in conceptual change in science. In the domain of classical mechanics, for example, Newton can be interpreted as employing generic abstraction in reasoning about the commonalities among the motions of planets and of projectiles, which enabled his formulating a unified mathematical representation of their motions. The models he employed, understood generically, represent what is common among the members of specific classes of physical systems, viewed with respect to a problem context. Newton's inverse-square law of gravitation abstracts what a projectile and a planet have in common in the context of determining motion, for example, that within the context of determining motion, planets and projectiles can both be represented as point masses. After Newton, the inverse-square-law model of gravitational force served as a generic model of action-at-a-distance forces for those who tried to bring all forces into the scope of Newtonian mechanics.

One key feature of Maxwell's mechanical analogical models is that they are meant to be understood as lacking specificity as to the mechanisms creating the stresses. A concrete mechanism is supplied to aid in reasoning, but it is meant to represent generic processes. When one reasons about a generic object, situation, or process one often draws a concrete representation or imagines one, but from our reasoning context one understands it as being without specificity of salient dimensions. Thus, the same concrete representation can be generic or specific depending on the context. So, although the visual representation of the analogical model provided by Maxwell in *Figure 5* supplies a concrete representation of specific mechanisms, we know from the context that Maxwell is considering them generically. That is, these mechanisms are treated in the way that the spring is treated generically when it is taken to represent the class of simple harmonic oscillators.

4.2 Abstraction via Generic Modeling in Maxwell's Construction of the Field Concept

In deriving the mathematical representation of the electromagnetic field concept Maxwell constructed a series of analogical models drawn from continuum mechanics (the mechanics of fluids, elastic bodies, etc.) and machine mechanics (Maxwell, 1861–1862). Details of these analogies and his complete modeling process can be found in several of my previous publications (Nersessian, 1984; 2000a; 2000b). Here only a sketch of the overall process will be provided, with attention focused on how continuities between the domains of mechanics and electrodynamics were created. Maxwell hypothesized that the causes of electromagnetic phenomena are stresses in a mechanical continuum, the electromagnetic aether, which transmits electromagnetic actions continuously through the space surrounding bodies and charges. Given the hypothesis one could assume a "resemblance in form" between the dynamical relations that hold in the domains of continuum mechanics, such as fluids and elastic solids, and those that hold in the domain of electromagnetism. Thus continuum mechanics could serve as a source domain for constructing models. Maxwell began with a set of constraints from the target domain of electromagnetism and constructed an analogical model of a fluid-dynamical system under stress. To account for the specific nature of the stresses he filled the fluid system with infinitesimal vortices. In deriving the equations pertaining to magnetic phenomena, such as magnetic induction, he was able to focus on a single vortex. *Figure 6a* is my rendering of a vortex drawn from Maxwell's description.

The system of infinitesimal vortices does not correspond to any known physical system. Maxwell constructed it to serve as the basis for deriving mappings between known dynamical relations in continuum mechanics and those thought to produce electromagnetic phenomena. The mathematical expressions for the magnetic phenomena are derived from the mathematical formula for the stresses in the vortex-fluid model by substitution. The vortex-fluid model is "generic" in that it is to be understood as satisfying constraints that apply to the *types* of entities and processes that can be considered as constituting either domain. The model represents the class of phenomena in each domain that are capable of producing the specific configurations of stresses. The modeling process Maxwell used throughout the analysis went as follows. First he constructed a model representing a specific mechanism. Then he treated the dynamical properties and relations generically by abstracting features common to the mechanical and the electromagnetic classes of phenomena. He proceeded to formulate the mathematical equations of the generic model and substituted in the electromagnetic variables.

A mechanical inconsistency in the vortex-fluid model led Maxwell to a means of representing the causal relationships between magnetism and electricity. He began this part of the analysis by stating that his purpose was to inquire into the connection between the magnetic vortices and current. Thus he could no longer simply consider a single generic vortex in his analysis. He admitted a serious problem with the model in that he "found great difficulty in conceiving of the existence of vortices in a medium, side by side, revolving in the same direction." (p. 468) *Figure 6b* is my drawing of a cross section of the vortex-fluid model as described by Maxwell. By imagining the motion of the vortices in this figure, it becomes evident that direct contact between consecutive vortices poses a problem in that there will be friction and, thus, jamming. Further, since they are all going in the same direction and, thus, at points of contact they would be going in opposite directions, in the case where they are revolving at the same rate, the whole mechanism should stop. Maxwell noted that in machine mechanics this kind of problem is solved by the introduction of "idle wheels." On that basis he proposed to enhance his imaginary model by supposing that "a layer of particles, acting as idle wheels is interposed between each vortex and the next" (p. 468). *Figure 6c* is Maxwell's rendering of the vortex-idle wheel model. The diagram shows a cross section of the medium. The vortex cross sections are represented by hexagons rather than circles, presumably to provide a better representation of how the particles are packed around the vortices, with the 3-dimensional dodecahedra approximating to spheres in the limit.

Again, it does not matter whether the mechanical system of the models could or could not exist in nature; all that matters is that they supplied mechanisms belonging to the class of phenomena with dynamical structure common to mechanics and electromagnetism. Here we see again how Maxwell abstracted from the specific mechanism to find the mathematical form of that class of mechanism, i.e., of the generic dynamical structure common to both continuum mechanics and electromagnetism. The vortex-idle

Figure 6a: A single vortex segment

Figure 6b: A cross section of the initial vortex medium

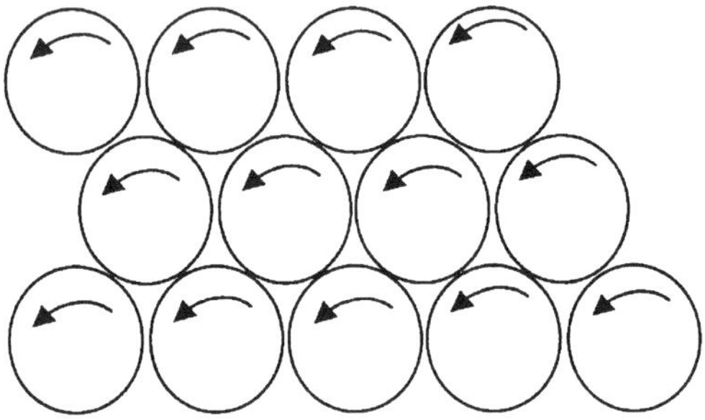

Figure 6c: Maxwell's drawing of idle-wheel-vortex model

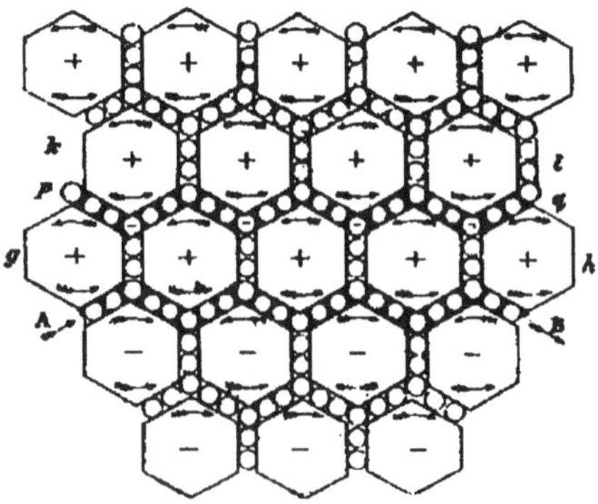

Figure 6: Maxwell's models

wheel system is understood as representing the class of such dynamical systems and that class includes electromagnetic actions on the assumptions of Maxwell's analysis. The causal structure is to be viewed as separated from the specific physical systems by means of which it has been made concrete. Thus, in using the analogical model in his analysis of electromagnetic induction, causal structure is maintained but not specific causal mechanisms. The vortex-idle wheel mechanism is not the cause of electromagnetic induction; it represents the *generic causal structure* common to such mechanical systems and electromagnetic systems.

The vortex-idle wheel model is a hybrid constructed from two source domains: fluid dynamics and machine mechanics. As discussed earlier, to combine salient entities and processes from two disparate domains requires abstraction of these to a sufficient level. My explanation of how generic abstraction could have led to the introduction of the idle-wheel particles is illustrated in *Figure 7.* The problem was to find a way of connecting the vortices so that they would spin without touching one another.

First Maxwell abstracted the generic model of spinning wheels from the initial vortex model (A). Then the generic model reminded him of specific mechanical systems containing machine gears (B). This provides an analogy between the vortices and the machine gears (C). Next, from the model of machine gears he abstracted the generic model of fly wheels (D) as a mode of connection and further abstracted that into a generic model of a dynamically smooth connector (E). Finally, he instantiated this generic mode of connection in the vortex model as idle wheel particles (F), where the instantiation is guided by both the analogous case of fly wheels (G) and the constraints of the continuum mechanical system.

With the model he derived the equations for electromagnetic induction. He solved the final problem of incorporating electrostatic induction into the account by endowing the vortices in the vortex-idle wheel model with elasticity and calculated the velocity of propagation in the medium to be nearly identical with the velocity of propagation of light in the light aether. However he did not conclude here that the two aethers are the same and that light is an electromagnetic phenomenon. We can interpret Maxwell's reticence to draw the inference in this analysis as due, among other reasons, to the fact that there were no grounds on which to assume vortex motion in the light aether. On the then prevailing view light is a transverse wave in an elastic medium and this does not belong to the same class of mechanism as that provided by the model for propagating electromagnetic actions. In his next paper Maxwell (Maxwell, 1864) treated the aether as a generic elastic medium whose constraints could be satisfied by many specific mechanical instantiations and thus saw no reason for multiplying aethers.

To summarize, on my cognitive-historical interpretation, Maxwell first formulated the mathematical laws of the electromagnetic field by abstracting from the models what continuum-mechanical systems, certain machine mechanisms, and electromagnetic systems have in common. In their mathematical treatment, these common dynamical properties and relationships are separated from the specific systems by means of which they had been made concrete. Once he had abstracted these properties and relationships, he was in a position to reconstruct the mathematical representation using generalized dynamics, which is what he did in the 1864 paper. That analysis assumes only that the electromagnetic medium is a "connected system", possessing elasticity and thus having

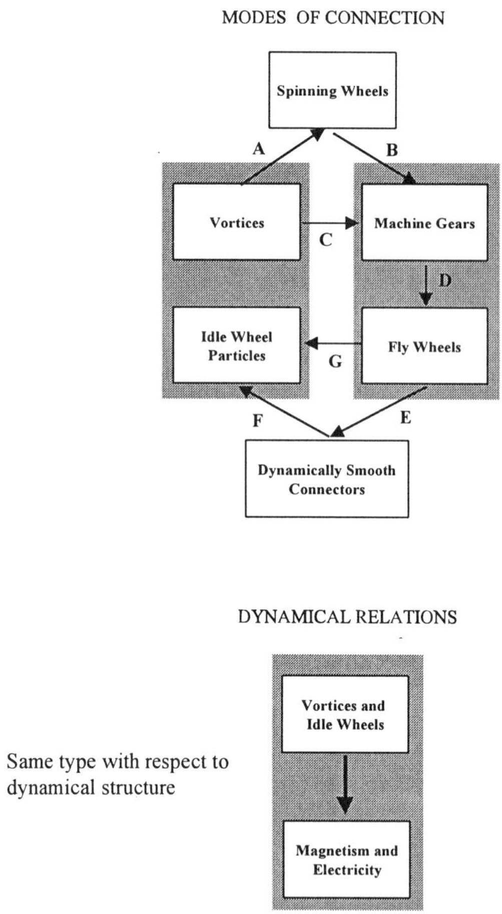

Figure 7: Introducing idle wheels via generic modeling

energy. The connected system needs to be elastic to provide for the time delay. Elastic systems can receive and store energy. The energy of such a system has two forms: "energy of motion", or kinetic energy, and "energy of tension", or potential energy. To apply the formalism of generalized dynamics to the system one needs to know how to represent its potential and kinetic energy. Maxwell identified kinetic energy with magnetic polarization and potential energy with electric polarization. Here we can see how the 1861–1862 analysis enabled him to do this through generic abstraction from the analogical mechanical model. *Figure 8* illustrates schematically that kinetic energy, which in the earlier analysis is associated the rotation of vortices, considered generically

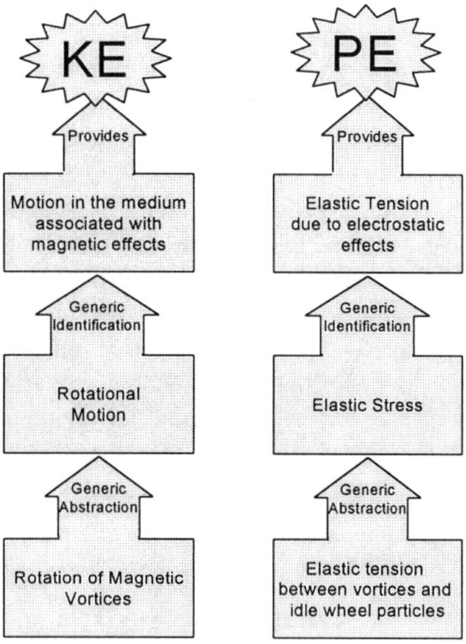

Figure 8: Identifying energy components via generic modeling

becomes rotation, which in the later analysis is just motion in the medium associated with magnetic effects. Potential energy is associated with elastic tension between the vortices and the particles, which, generically, becomes elastic stress, which here is elastic tension due to electrostatic effects.

What we know that Maxwell did not is that many different kinds of dynamical systems can be formulated in generalized dynamics. Electrodynamical systems are not the same kind of dynamical system as Newtonian systems. What he abstracted through the generic modeling process is a representation of the general dynamical properties and relationships for electromagnetism. From the interpretive perspective of the special theory of relativity, the abstract laws, when applied to the class of electromagnetic systems yield the laws of a dynamical system that is non-mechanical; that is, one that cannot be mapped back onto the mechanical domains used in Maxwell's construction of the laws. It is through the process of abstraction via generic modeling that the new conceptual system of electrodynamics is "born out of" mechanics. Ultimately, the generic abstractions created in the reasoning process and captured in the general dynamical representation provide a basis for commensurability between the Newtonian and the special relativistic representational structures. To establish more details of the nature of the commensurability relations would require examining how generic abstraction figured in Lorentz' reasoning when combining the ostensively inconsistent particle conception of charge and field representation of force and Einstein's abstractive process in creating an consistent representational structure for mechanics and electromagnetism in the special theory of relativity.

4. CONCLUSION

I have argued that the problem of incommensurability of meaning arises from approaching problems pertaining to science as though they were problems pertaining to languages alone. Kuhn and Feyerabend followed out the implications of abandoning the reductionist theory of meaning, but not the implications of their methodological challenge. One cannot address problems of conceptual change without examining the representational and reasoning practices employed by scientists in constructing new conceptual structures. But we need to move beyond just describing those practices to developing an explanatory account that utilizes them in addressing the representational problem of nature the format for scientific concepts that can capture commensurability between earlier and later forms of concepts and the generative problem of how the reasoning employed in concept formation and change can establish commensurability among successive representations of a domain. To do this demands following the naturalistic path opened by their challenge, in the case at hand, this means employing the resources of cognitive-historical analysis.

Georgia Institute of Technology

ACKNOWLEDGEMENTS

I appreciate the comments by Andreas Bartels on the version of the paper delivered at the Hannover conference and those of the editors of the volume. This work was supported by a grant from the National Science Foundation SBE9810913. I also appreciate the support of the Dibner Institute for the History of Science and Technology, where I was a Senior Fellow during the writing of this paper.

NOTE

[1] Hanne Andersen and I (Andersen & Nersessian 2000) argue that the problems I was grappling with in 1984 are similar to what Kuhn was trying to address in his latest work in 1993 where he introduced distinction between 'normic' and 'nomic' concepts. Normic concepts acquired in contrast sets (these are the ones with which he mostly concerned himself) and nomic concepts are learned together in complex problem situations exemplifying laws of nature. In making this distinction Kuhn had realized the limitations of his analysis of concepts in terms of family resemblances: many scientific concepts cannot be ostended because they occur in complex problem situations and are intertwined with theory.

REFERENCES

Andersen, H. (1996). "Categorization, Anomalies and the Discovery of Nuclear Fission." *Studies in the History and Philosophy of Modern Physics* **27**: 463–492.

Andersen, H., X. Chen and P. Barker. (1996). "Kuhn's Mature Philosophy of Science and Cognitive Psychology." *Philosophical Psychology* **9**: 347–363.

Andersen, H., and N. Nersessian. (2000). "Nomic Concepts, Frames and Conceptual Change." *Philosophy of Science* **67** (Proceedings): in press.

Arabatzis, T. (1999). "The Historicity of Scientific Realism: Reflections on Meaning Variance." Unpublished lecture, Hannover Conference on Incommensurability (and related matters).

Armstrong, S., L. Gleitman and H. Gleitman. (1983). "What Some Concepts Might Not Be." *Cognition* **13**: 263–308.

Barsalou, L. (1992). "Frames, Concepts and Conceptual Fields." In A. Lehrer and E. Kittay, eds., *Frames, Fields and Contrasts: New Essays in Semantical and Lexical Organization*, pp. 325–340, Hillsdale: Erlbaum.

Carnap, R. (1956a). "Empiricism, Semantics and Ontology." In R. Carnap, ed., *Meaning and Necessity*. Chicago: University of Chicago Press.

Carnap, R. (1956b). "The Methodological Character of Theoretical Concepts." In H. Feigl and M. Scriven, eds., *Minnesota Studies in the Philosophy of Science*. Minnesota: University of Minnesota Press.

Chen, X. (1995). "Taxonomic Changes and the Particle-Wave Debate in Early Nineteenth-Century Britain." *Studies in the History and Philosophy of Science* **26**: 251–271.

Chen, X., H. Andersen and P. Barker. (1998). "Kuhn's Theory of Scientific Revolutions and Cognitive Psychology." *Philosophical Psychology* **11**: 5–28.

Darden, L. (1980). "Theory Construction in Genetics." In T. Nickles, ed., *Scientific Discovery: Case Studies*, pp. 151–170, Dordrecht: Reidel.

Darden, L. (1991). *Theory Change in Science: Strategies from Mendelian Genetics*. New York: Oxford University.

Duhem, P. (1914). *The Aim and Structure of Physical Theory*, trans. P. Weiner. New York: Atheneum.

Feyerabend, P. (1962). "Explanation, Reduction and Empiricism." In H. Feigl, G. Maxwell, and M. Scriven, eds., *Minnesota Studies in the Philosophy of Science: Volume 3*, pp. 231–272, Minneapolis: University of Minnesota Press.

Feyerabend, P. (1970). "Against Method: An Outline of an Anarchistic Theory of Knowledge." In H. Feigl and G. Maxwell, eds., *Minnesota Studies in the Philosophy of Science*. Minnesota: University of Minnesota Press.

Gentner, D. (1983). "Structure-Mapping: A Theoretical Framework for Analogy. *Cognitive Science.*" **7**: 155–170.

Gentner, D. (1989). "The Mechanisms of Analogical Learning." In S. Vosniadou and A. Ortony, eds., *Similarity and Analogical Reasoning*, pp. 200–241, New York: Cambridge University Press.

Gentner, D., S. Brem, R. Ferguson, A. Markman, B. Levidow, P. Wolff and K. Forbus. (1997). "Analogical Reasoning and Conceptual Change: A Case Study of Johannes Kepler." *The Journal of the Learning Sciences* **6**: 3–40.

Gick, M. and K. Holyoak. (1980). "Analogical Problem Solving." *Cognitive Psychology* **12**: 306–355.

Gick, M. and K. Holyoak. (1983). "Schema Induction and Analogical Transfer." *Cognitive Psychology* **15**: 1–38.

Giere, R. (1988). *Explaining Science: A Cognitive Approach*. Chicago: University of Chicago Press.

Giere, R. (1992). *Cognitive Models of Science. Minnesota Studies in the Philosophy of Science, Volmen 15*. Minneapolis: University of Minnesota Press.

Giere, R. (1994). "The Cognitive Structure of Scientific Theories." *Philosophy of Science* **61**: 276–296.

Gooding, D. (1990). *Experiment and the Making of Meaning: Human Agency in Scientific Observation and Experiment*. Dordrecht: Kluwer.

Griesemer, J. (1991a). "Material Models in Biology." In A. Fine, M. Forbes and L. Wessels, eds., *PSA 1990*, pp. 79–94, East Lansing: Philosophy of Science Association.

Griesemer, J. (1991b). "Must Scientific Diagrams Be Eliminable? The Case of Path Analysis." *Biology and Philosophy* **6**: 177–202.

Griesemer, J. and W. Wimsatt. (1989). "Picturing Weismannism: A Case Study of Conceptual Evolution." In M. Ruse, ed., *What the Philosophy of Biology Is, Essays for David Hull*, pp. 75–137, Dordrecht: Kluwer.

Griffith, T. (1999). "A Computational Theory of Generative Modeling in Scientific Reasoning." Ph.D. Dissertation, College of Computing, Georgia Institute of Technology, Atlanta.

Griffith, T., N. Nersessian and A. Goel. (1996). "The Role of Generic Models in Conceptual Change." In G. Cattrell, ed., *Proceedings of the Cognitive Science Society 18*, pp. 312–317, Hillsdale: Lawrence Erlbaum.

Griffith, T., N. Nersessian and A. Goel. (2000). "Function-Follows-Form Transformations in Scientific Problem Solving." In L. Gleitman and A. Joshi, eds., *Proceedings of the Cognitive Science Society 22*, pp. 196–201, Hillsdale: Lawrence Erlbaum.

Holmes, F. (1981). "The Fine Structure of Scientific Creativity." *History of Science* **19**: 60–70.

Holmes, F. (1985). *Lavoisier and the Chemistry of Life: An Exploration of Scientific Creativity*. Madison: University of Wisconsin Press.

Holyoak, K. and P. Thagard. (1989). "Analogical Mapping by Constraint Satisfaction: A Computational Theory." *Cognitive Science* **13**: 295–356.

Holyoak, K. and P. Thagard. (1996). *Mental Leaps: Analogy in Creative Thought*. Cambridge: MIT Press.

Kuhn, T. (1962). *The Structure of Scientific Revolutions: International Encyclopedia of Unified Science. Volume 2.* Chicago: University of Chicago Press.
Kuhn, T. (1970). *The Structure of Scientific Revolutions.* 2nd edition. Chicago: University of Chicago Press.
Kuhn, T. (1990). The Road Since Structure. In A. Fine, M. Forbes and L. Wessels, eds., *PSA 1990, Volume 2*, pp. 3–13.
Kuhn, T. (1992). "The Trouble with the Historical Philosophy of Science." Unpublished.
Latour, B. (1986). "Visualisation and Cognition: Thinking with Eyes and Hands." *Knowledge and Society* **6**: 1–40.
Latour, B. (1987). *Science in Action.* Cambridge: Harvard University Press.
Lewis, C. (1956). *Mind and the Work Order.* New York: Dover Publications.
Lynch, M. and S. Woolgar. (1990). *Representation in Scientific Practice.* Cambridge: MIT Press.
Maxwell, J. (1861–1862). "On Physical Lines of Force." In W. Niven, ed., *Scientific Papers*, pp. 451–513, Cambridge: Cambridge University.
Maxwell, J. (1864). "A Dynamical Theory of the Electromagnetic Field." In W. Niven, ed., *Scientific Papers*, pp. 526–597.
Maxwell, J. (1890). *The Scientific Papers of James Clerk Maxwell, Volumes 1 and 2*, W. Niven, ed., Cambridge: Cambridge University.
Medin, D. (1989). "Concepts and Conceptual Structure." *American Psychologist* **44**: 1469–1481.
Nersessian, N. (1979). "The Roots of Epistemological 'Anarchy'." *Inquiry* **22**: 423–440.
Nersessian, N. (1982). "Why Is 'Incommensurability' a Problem?" *Acta Biotheoretica* **31**:205–218.
Nersessian, N. (1984). *Faraday to Einstein: Constructing Meaning in Scientific Theories.* Dordrecht: Martinus Nijhoff/Kluwer Academic Publishers.
Nersessian, N. (1985). "Faraday's Field Concept." In D. Gooding and F. James, eds., *Faraday Rediscovered: Essays on the Life & Work of Michael Faraday*, pp. 377–406, London: Macmillan.
Nersessian, N. (1992a). "How Do Scientists Think? Capturing the Dynamics of Conceptual Change in Science." In R. Giere, ed., *Minnesota Studies in the Philosophy of Science, Volume 15*, pp. 3–45, Minneapolis: University of Minnesota Press.
Nersessian, N. (1992b). "In the Theoretician's Laboratory: Thought Experimenting as Mental Modeling." In D. Hull, M. Forbes and K. Okruhlik, eds., *PSA 1992*, pp. 291–301, East Lansing: Philosophy of Science Association.
Nersessian, N. (1995). "Opening the Black Box: Cognitive Science and the History of Science."In A. Thackray, ed., *Osiris* **10**:194–211.
Nersessian, N. (1996). "Should Physicists Preach What They Practice? Constructive Modeling in Doing and Learning Physics." In C. Bernardini, C. Tarsitani and M. Vincentini, eds., *Thinking Science for Teaching*, pp. 77–96, New York: Plenum.
Nersessian, N. (1999). "Model-Based Reasoning in Conceptual Change." In L. Magnani, N. Nersessian and P. Thagard, eds., *Model-Based Reasoning in Scientific Discovery.* New York: Kluwer Academic/ Plenum Publishers.
Nersessian, N. (2000a)." Abstraction Via Generic Modeling in Concept Formation in Science." In M. Jones and N. Cartwright, eds., *Correcting the Model: Abstraction and Idealization in Science.* Amsterdam: Rodopi.
Nersessian, N. (2000b). "Maxwell and the 'Method of Physical Analogy'". In D. Malamet, ed., *Essays in Honor of Howard Stein*, in press. Lasalle: Open Court.
Nersessian, N. and H. Andersen. (1998). "Conceptual Change and Incommensurability: A Cognitive-Historical View." *Danish Yearbook of Philosophy* **32**/1997:111–151.
Quine, W. (1960). *Word and Object.* Cambridge, Mass.: MIT Press.
Quine, W. (1963). "Two Dogmas of Empiricism." In W. Quine, ed., *From a Logical Point of View.* New York: Harper and Row.
Rosch, E. and B. Lloyd. (1978). *Cognition and Categorization.* Hillsdale: Lawrence Erlbaum.
Rosch, E. and C. Mervis. (1975). "Family Resemblance Studies in the Internal Structure of Categories." *Cognitive Psychology* **7**: 573–605.
Rudwick, M. (1976). "The Emergence of a Visual Language for Geological Science." *History of Science* **14**:149–195.
Shapere, D. (1966). "Meaning and Scientific Change." In R. Colodny, ed., *Mind and Cosmos*, pp. 41–85, Pittsburg: The University of Pittsburgh Press.
Shelley, C. (1996). "Visual Abductive Reasoning in Archeology." *Philosophy of Science* **63**: 278–301.

Smith, E. and D. Medin. (1981). *Concepts and Categories*. Cambridge, Mass.: Harvard University Press.

Thagard, P. (1991). *Conceptual Revolutions*. Princeton: Princeton University Press.

Thagard, P., K. Holyoak, G. Nelson and D. Gochfield. (1990). "Analog Retrieval by Constraint Satisfaction." *Artificial Intelligence* **46**: 259–310.

Trumpler, M. (1997). "Converging Images: Techniques of Intervention and Forms of Representation of Sodium-Channel Proteins in Nerve Cell Membranes." *Journal of the History of Biology* **20**:55–89.

Tweney, R. (1992). "Stopping Time: Faraday and the Scientific Creation of Perceptual Order." *Physis* **29**:149–164.

INCOMMENSURABILITY BIBLIOGRAPHY

Achinstein, P. (1964). "On the Meaning of Scientific Terms." *Journal of Philosophy* **61**: 497–509.
Agassi, J. (1994). "Minimal Criteria for Intellectual Progress." *Iyyun* **43**: 61–83.
Agazzi, E. (1985). "Commensurability, Incommensurability, and Cumulativity in Scientific Knowledge." *Erkenntnis* **22**: 51–77.
Agazzi, E. (1885). "The Historical Dimensions of Science and its Philosophy." *Diogenes* **132**: 60–79.
Agazzi, E. (1988). *Die Objektivität in den verschiedenen Wissenschaften.* Fribourg: Editions Universitaires.
Agazzi, E. (1990). *Die Vergleichbarkeit wissenschaftlicher Theorien.* Freiburg: Universitätsverlag.
Allchin, D. (1994). "The Super Bowl and The Ox Phos Controversy: "Winner Take All" Competition in Philosophy Of Science." *Proceedings of the Biennial Meetings of the Philosophy of Science Association* **1**: 22–33.
Allchin, D. (1996). "Points East And West: Acupuncture And Comparative Philosophy Of Science." *Proceedings Of The Biennial Meetings Of The Philosophy Of Science Association* **3**: 107–115.
Ambrus, V. (1999). "Is Putnam's Causal Theory of Meaning Compatible with Internal Realism?" *Journal for General Philosophy of Science* **30**: 1–16.
Andersen, H. (1996). "Categorization, Anomalies and the Discovery of Nuclear Fission." *Studies in the History and Philosophy of Science* **27**: 463–492.
Andersen, H., P. Barker and X. Chen. (1996). "Kuhn's mature philosophy of science and cognitive psychology." *Philosophical Psychology* **9**: 347–363.
Andersson, G. (1988). *Kritik und Wissenschaftsgeschichte. Kuhns, Lakatos und Feyerabends Kritik des kritischen Rationalismus.* Tübingen: Mohr.
Andersson, G. (1988). "Die Wissenschaft als objektive Erkenntnis." In E. Agazzi, ed., *Die Objektivität in den verschiedenen Wissenschaften*, pp. 27–39, Fribourg: Editions Universitaires.
Aronson, J.L. (1989). "Testing for Convergent Realism." *British Journal for the Philosophy of Science* **40**: 255–259.
Arregui, J. (1997). "Incommensurabilidad y relativismo: el reconocimiento de lo humano." *Contrastes* **2**: 27–51.
Austin, W. (1972). "Paradigms, Rationality, and Partial Communication." *Zeitschrift für allgemeine Wissenschaftstheorie* **3**: 203–218.
Baigrie, B. (1988). "Commentary to Margolis's "In Defense of Relativism"." *Social Epistemology* **2**: 228–232.
Bailey, A. (1997). "Split Level Contexts, Intersubjective Facts and Semi-Naive Empiricism: A Semi-Naive Fairy Story." *Prima Philosophia* **10**: 457–471.
Balsiger, F. and A. Burri. (1990). "Sind die klassische Mechanik und die spezielle Relativitätstheorie kommensurabel?" *Journal for General Philosophy of Science* **21**: 157–162.
Balzer, W. (1978). "Incommensurability and Reduction." *Acta Philosophica Fennica* **30**: 313–335.
Balzer, W. (1985). "Was ist Inkommensurabilität?" *Kant-Studien* **76**: 196–213.
Balzer, W. (1985). "Incommensurability, Reduction, and Translation." *Erkenntnis* **23**: 255–267.
Balzer, W. (1989). "On Incommensurability." In K. Gavroglu, Y. Goudaroulis and P. Nicolacopoulos, eds., *Imre Lakatos and Theories of Scientific Change*, Dordrecht: Kluwer.
Balzer, W., C. Moulines and J. Sneed. (1987). *An Architectonic for Science. TheStructuralist Program.* Dordrecht: Reidel.
Barker, P., H. Andersen and X. Chen. (1996). "Kuhn's Mature Philosophy of Science and Cognitive Psychology." *Philosophical Psychology* **9**: 347–363.

P. Hoyningen-Huene and H. Sankey (eds.). Incommensurability and Related Matters, 303–316.

Barnes, B. and D. Bloor. (1982). "Relativism, Rationalism and the Sociology of Knowledge." In M. Hollis and S. Lukes, eds., *Rationality and Relativism*, Cambridge Mass.: M.I.T. Press.

Barnhart, M. (1994). "'Sunyata', Textualism, and Incommensurability." *Philosophy East and West* **44**: 647–658.

Barrotta, P. (1988). "Teoria della Argomentazione e Filosofia della la Scienza." *Epistemologia* **11**: 255–276.

Barrotta, P. (1998). "Contemporary Philosophy of Science in Italy: An Overview." *Journal for General Philosophy of Science* **29**: 327–345.

Bartels, A. (1990). "Review of P. Hoyningen-Huene: Die Wissenschaftsphilosophie Thomas S. Kuhns." *Conceptus* **24**: 104–109.

Bartels, A. (1995). "Chains of Meaning: A Model For Concept Formation in Contemporary Physics Theories." *Synthese* **105**: 347–379.

Batens, D. (1983). "The Relevance of Theory-Ladenness and Incommensurability, and a Survey of the Contributions to this Issue." *Philosophica* **31**: 3–6.

Batens, D. (1983). "Incommensurability is not a Threat to the Rationality of Science or to the Anti-Dogmatic Tradition." *Philosophica* **32**: 117–132.

Bayley, C. (1995). "Our World Views (May Be) Incommensurable: Now What?" *Journal of Medicine and Philosophy* **20**: 271–283.

Bellone, E. (1978). "Thomas Kuhn e Il Libro della Natura." *Scientia* **113**: 675–682.

Bernstein, R. (1983). *Beyond Objectivism and Relativism: Science, Hermeneutics, and Praxis*. Philadelphia: University of Pennsylvania Press.

Bernstein, R. (1991). "Incommensurability and Otherness Revisited." In E. Deutsch, ed., *Culture and Modernity. East-West Philosophic Perspectives*, pp. 85–103, Honolulu: University of Hawaii Press.

Bhaskar, R. (1986). *Scientific Realism and Human Emancipation*. London NewYork: Verso.

Biagioli, M. (1990). "The Anthropology of Incommensurability." *Studies in History and Philosophy of Science* **21**: 183–209.

Bishop, M. (1991). "Why the Semantic Incommensurability Thesis is Self-Defeating." *Philosophical Studies* **63**: 343–356.

Bishop, M. and S. Stich. (1998). "The Flight to Reference, or How Not to Make Progress in the Philosophy of Science." *Philosophy of Science* **65**.

Bohm, D. (1974). "Science as Perception-Communication." In F. Suppe, ed., *The Structure of Scientific Theories*, pp. 374–391, Urbana: University of Illinois Press.

Bombardi, R. (1988). "Davidson in Flatland." *Australasian Journal of Philosophy* **66**: 66–74.

Boniolo, G. (1992). "Theory and Experiment: The Case of Eotvos' Experiments." *British Journal for the Philosophy of Science* **43**: 459–486.

Boros, J. (1990). "Probleme der Inkommensurabilität." In E. Agazzi, ed., *Die Vergleichbarkeit wissenschaftlicher Theorien*, pp. 125–132, Fribourg: Editions Universitaires.

Boyd, R. (1979). "Metaphor and Theory Change: What is 'Metaphor' a Metaphor for?" In A. Ortony, ed., *Metaphor and Thought*, pp. 356–408, Cambridge: Cambridge University Press.

Boyd, R. (1984). "The Current Status of Scientific Realism." In J. Leplin, ed., *Scientific Realism*, pp. 41–82, Berkeley: University of California Press.

Boyd, R. (1992). "Constructivism, Realism, and Philosophical Method." In J. Earman, ed., *Inference, Explanation, and other Frustrations. Essays in the Philosophy of Science*, pp. 131–198, Berkeley: University of California Press.

Broccard, N. (1982). "Going on with Systematology." *Metaphilosophy* **13**: 263–276.

Bromberger, S. (1990). "L'Incommensurabilité en Science et La Possibilité d'Argumenter." In M. Meyer and A. Lempereur, eds., *Figures at Conflits Rhetoriques*, pp. 241–253, Bruxelles: Editions de l'Université de Bruxelles.

Brown, H. (1977). *Perception, Theory and Commitment. The New Philosophy of Science*. Chicago: University of Chicago Press.

Brown, H. (1983). "Incommensurability." *Inquiry* **26**: 3–29.

Brown, H. (1983). "Response to Siegel." *Synthese* **56**: 91–103.

Brown, H. (1990). "Prospective Realism." *Studies in History and Philosophy of Science* **21**: 211–242.

Burian, R. (1984). "Scientific Realism and Incommensurability: Some Criticisms of Kuhn and Feyerabend." In R. Cohen and M. Wartofsky, eds., *Methodology, Metaphysics and the History of Science; in Memory of Benjamin Nelson, Boston Studies in the Philosophy of Science*, pp. 1–31, Dordrecht: Reidel.

Buzzoni, M. (1998). "Scienza, tecnica e conoscenza comune in Bachelard." *Epistemologia* **21**: 343–354.

Carey, S. (1988). "Conceptual differences between children and adults." *Mind & Language* **3**: 167–181.

Carey, S. (1991). "Knowledge acquisition: Enrichment or conceptual change?" In S. Carey and R. Gelman, eds., *The Epigenesis of Mind: Essays on Biology and Cognition*, pp. 257–291, Hillsdale: Erlbaum.

Carey, S. and R. Gelman. (1991). *The Epigenesis of Mind: Essays on Biology and Cognition*. Hillsdale: Erlbaum.

Carnap, R. (1966). *Philosophical Foundations of Physics*. New York: Basic Books.

Causey, R. (1974). "Professor Bohm's View of the Structure and Development of Theories." In F. Suppe, ed., *The Structure of Scientific Theories*, pp. 392–401, Urbana: University of Illinois Press.

Chalmers, A. (1999) *What is this Thing called Science?* 3rd edition. Hackett Pub. Co.: Indianapolis.

Chen, X. (1990). "Local Incommensurability and Communicability." In A. Fine, M. Forbes and L. Wessels, eds., *PSA 1990: Proceedings of the 1990 Biennial Meeting of Philosophy of Science Association*, pp. 67–76, East Lansing: Philosophy of Science Association.

Chen, X. (1994). "How Do Scientists Have Disagreements about Experiments? Incommensurability in the Use of Goal-Derived Categories." *Perspectives on Science* **2**: 275–301.

Chen, X. (1997). "Thomas Kuhn's Latest Notion of Incommensurability." *Journal for General Philosophy of Science* **28**: 257–273.

Chen, X., H. Andersen and P. Barker. (1998). "Kuhn's Theory Of Scientific Revolutions And Cognitive Psychology." *Philosophical Psychology* **11**: 5–28.

Churchill, J. (1989). "'If a Lion Could Talk'." *Philosophical Investigations* **12**: 308–324.

Churchill, J. (1990). "The Bellman's Map: Does Antifoundationalism Entail Incommensurability and Relativism?" *Southern Journal of Philosophy* **28**: 469–484.

Cohen, J. (1999). "Deliberation, Tradition, and the Problem of Incommensurability: Philosophical Reflections on Curriculum Decision Making." *Educational Theory* **49**: 71–89.

Colapietro, V. (1996). "Marking Distinctions and Making Differences: Being "As" Dialectic." *Journal Of Speculative Philosophy* **10**: 1–18.

Collier, J. (1984). "Pragmatic Incommensurability." In P. Asquith and P. Kitcher, eds., *PSA 1984. Proceedings of the 1984 Biennial Meeting of the Philosophy of Science Association*, pp. 146–153, East Lansing: Philosophy of Science Association.

Collins, H. and T. Pinch. (1982). *Frames of Meaning: The Social Construction of Extraordinary Science*. Boston: Routledge.

Corry, L. (1993). "Kuhnian Issues, Scientific Revolutions and the History of Mathematics." *Studies in History and Philosophy of Science* **24**: 95–117.

Cortois, P. (1995). "Farewell to a Brilliant Dish Cleaner: Paul Feyerabend and the Merry Philosophy of Science." *Tijdschrift Voor Filosofie* **57**: 91–110.

Couvalis, G. (1989). *Feyerabend's Critique of Foundationalism*. Brookfield: Gower.

Crofts, A. (1979). "Peter Mitchell and the Chemiosmotic Hypothesis." *Biochemical Society Bulletin* **1**: 4–7.

Crowder, G. (1998). "John Gray's Pluralist Critique of Liberalism." *Journal of Applied Philosophy* **15**: 287–298.

Danneberg, L. (1989). *Struktur, Aufbau und Evaluation*. Berlin: Duncker & Humblot.

Davidson, D. (1974). "The Very Idea of a Conceptual Scheme." *Proceedings & Addresses of the American Philosophical Association* **47**: 5–20.

Davidson, D. (1982). "The Very Idea of a Conceptual Scheme." In J. Meiland and M. Krausz, eds., *Relativism. Cognitive and Moral*. Notre Dame: University of Notre Dame Press.

Devitt, M. (1979). "Against Incommensurability." *Australasian Journal of Philosophy* **57**: 29–50.

Devitt, M. (1984). *Realism and Truth*. Oxford: Basil Blackwell.

Devitt, M. and K. Sterelny. (1987). *Language and Reality*. Cambridge Mass.: MIT Press.

Diederich, W. (1989). "The Development Of Structuralism, A Re-Evaluation on the Occasion of W. Stegmuller's *Theorie Und Erfahrung*." *Erkenntnis* **30**: 363–386.

Doppelt, G. (1978). "Kuhn's Epistemological Relativism: An Interpretation and Defense." *Inquiry* **21**: 33–86.

Doppelt, G. (1980). "A Reply to Siegel on Kuhnian Relativism." *Inquiry* **23**: 117–123.

Doppelt, G. (1981). "Laudan's Pragmatic Alternative to Positivism and Historicism." *Inquiry* **24**: 253–271.

Doppelt, G. (1982). "Kuhn's Epistemological Relativism: An Interpretation and Defense." In J. Meiland and M. Krausz, eds., *Relativism. Cognitive and Moral*, pp. 113–146, Notre Dame: University of Notre Dame Press.

Doppelt, G. (1983). "Relativism and Recent Pragmatic Theories of Scientific Rationality." In N. Rescher, ed., *Scientific Explanation and Understanding: Essays on Reasoning in Science*, pp. 106–142, Pittsburgh: University Press of America.

Doppelt, G. (1988). "The Philosophical Requirements for an Adequate Conception of Scientific Rationality." *Philosophy of Science* **55**: 104–133.

Doppelt, G. (1988). "Relativism and the Reticulational Model of Scientific Rationality." *Synthese* **69**: 225–252.

Doppelt, G. (1990). "The Naturalist Conception of Methodological Standards in Science: A Critique." *Philosophy of Science* **57**: 1-19.

Drago, A. (1990). "The Four Scientific Models of Physical Reality." *Epistemologia* **13**: 303–324.

Drago, A. (1999). "Incommensurability as a Bound of Hermeneutics in Science." In M. Fehér, O. Kiss and L. Ropolyi, eds., *Hermeneutics and Science*, pp. 135–155, Dordrecht: Kluwer.

Edison, J. (1985). "Popperian and Kuhnian Theories of Truth and The Imputation of Relativism." *Indian Philosophical Quarterly* **12**: 9–22.

Enç, B. (1976). "Reference of Theoretical Terms." *Nous* **10**: 261–282.

English, J. (1978). "Partial Interpretation and Meaning Change." *Journal of Philosophy* **75**: 57–76.

Estany, A. and D. Quesada. (1997). *Actas del II Congreso de la Sociedad de Lógica, Metodología y Filosofía de la Ciencia en España*. Barcelona: Servicio de Publicaciones de la Universidad Autónoma Barcelona.

Falkenburg, B. (1997). "Incommensurability and Measurement." *Theoria* **12**: 467–491.

Fernández-Moreno, L. (1995). "La noción de inconmensurabilidad en Kuhn." In ed., L. Lull, pp. 441–456.

Fernández-Moreno, L. (1997). "La inconmensurabilidad de teorías: Kuhn versus Putnam." In A. Estany and D. Quesada, eds., *Actas del II Congreso de la Sociedad de Lógica, Metodología y Filosofía de la Ciencia en España*, pp. 177–180, Barcelona: Servicio de Publicaciones de la Universidad Autónoma Barcelona.

Fernández-Moreno, L. (1997). "Inconmensurabilidad y relativismo ontológico." In A. Estany and D. Quesada, eds., *Actas del II Congreso de la Sociedad de Lógica, Metodología y Filosofía de la Ciencia en España*, pp. 181–184, Barcelona: Servicio de Publicaciones de la Universidad Autónoma Barcelona.

Fernández-Moreno, L. (1997). "Es la referencia del término "agua" inmutable?" *Theoria* **12**: 493–509.

Fernández Moreno, L. (1998). "Incommensurability, Reference and Truth." In C. Martínez, *et. al.*, eds., *Truth in Perspective*, pp. 399–417, Aldershot: Ashgate.

Fernández-Moreno, L. (1998). "Es la tesis de la inconmensurabilidad incoherente?" In C. Solís, ed., *Alta tensión*, pp. 279–291, Barcelona: Paidós.

Feyerabend, P. (1962). "Explanation, Reduction and Empiricism." In H. Feigl and G. Maxwell, eds., *Scientific Explanation, Space, and Time. Minnesota Studies in the Philosophy of Science, Vol. 3*, pp. 28–97, Minneapolis: University of Minnesota Press.

Feyerabend, P. (1965). "On the "Meaning" of Scientific Terms." *Journal of Philosophy* **12**: 266–274.

Feyerabend, P. (1965). "Reply to Criticism: Comments on Smart, Sellars and Putnam." In R. Cohen and M. Wartofsky, eds., *Proceedings of the Boston Colloquium for the Philosophy of Science 1962–64: In Honor of Philipp Frank. Boston Studies in the Philosophy of Science, Vol. 2*, pp. 223–261, New York: Humanities Press.

Feyerabend, P. (1970). "Consolations for the Specialist." In I. Lakatos and A. Musgrave, eds., *Criticism and the Growth of Knowledge. Proceedings of the International Colloquium in the Philosophy of Science, London, 1965*, pp. 197–230, London: Cambridge University Press.

Feyerabend, P. (1975). *Against Method. Outline of an Anarchistic Theory of Knowledge*. London: New Left Books.

Feyerabend, P. (1976). *Wider den Methodenzwang. Skizze einer anarchistischen Erkenntnistheorie*. Frankfurt: Suhrkamp.

Feyerabend, P. (1977). "Changing Patterns of Reconstruction." *British Journal for the Philosophy of Science* **28**: 351–382.

Feyerabend, P. (1978). *Science in a Free Society*. London: New Left Books.

Feyerabend, P. (1979). *Erkenntnis für freie Menschen*. Frankfurt: Suhrkamp.

Feyerabend, P. (1981). *Realism, Rationalism and Scientific Method. Philosophical Papers, Volume 1*. Cambridge: Cambridge University Press.

Feyerabend, P. (1981). *Problems of Empiricism. Philosophical Papers. Volume 2*. Cambridge: Cambridge University Press.

Feyerabend, P. (1981). "More Clothes from the Emperor's Bargain Basement. A review of Laudan's Progress and its Problems." *British Journal for the Philosophy of Science* **32**: 57–71.
Feyerabend, P. (1987). "Putnam on Incommensurability." *British Journal for the Philosophy of Science* **38**: 75–81.
Feyerabend, P. (1988). *Against Method.* Revised edition. London: Verso.
Feyerabend, P. (1993). *Against Method.* 3rd edition. London: Verso.
Feyerabend, P. (1999). *Conquest of Abundance.* Chicago: University of Chicago Press.
Field, H. (1973). "Theory Change and the Indeterminacy of Reference." *Journal of Philosophy* **70**: 462–481.
Fine, A. (1967). "Consistency, Derivability, and Scientific Change." *Journal of Philosophy* **64**: 231–240.
Fine, A. (1975). "How to Compare Theories: Reference and Change." *Nous* **9**: 17–32.
Flonta, M. (1978). ""A "weak" and a "strong" Version of the Incommensurability Thesis." *Philosophie et Logique* **20**: 395–406.
Flores, F. (1995). "Entre La Identidad y La Inconmensurabilidad, La Diferencia: Aristoteles y Freud: El Caso De La Analogia." *Analogia Filosofica* **9**: 3–26.
Franklin, A. (1984). "Are Paradigms Incommensurable?" *British Journal for the Philosophy of Science* **35**: 57–60.
Fu, D. (1995). "Higher Taxonomy and Higher Incommensurability." *Studies in History and Philosophy of Science* **26**: 273–294.
Fuller, S. (1988). *Social Epistemology.* Bloomington: Indiana University Press.
Fuller, S. and M. De May. (1987). "An Interview with Marc De Mey on the Possibility of Cognitive Science as a Focal Point for Science Studies." *Social Epistemology* **1**: 83–95.
Gaeta, R. (1990). "Significado, referencia e inconmensurabilidad." *Revista Latinoamericana de Filosofia* **16**: 80–86.
Garrison, M. (1988). "Relativity, Complementarity, Indeterminacy, and Psychological Theory." *Journal of Mind and Behavior* **9**: 113–135.
Garver, E. (1987). "Paradigms and Princes." *Philosophy of the Social Sciences* **17**: 21–47.
Geraets, T. (1979). *Rationality To-Day.* Ottawa: University of Ottawa Press.
Ghosh, T. (1998). "Berkeley's Idealism, Internal Realism and the Incommensurability Thesis." *Indian Philosophical Quarterly* **25**: 241–251.
Giedymin, J. (1970). "The Paradox of Meaning Variance." *British Journal for the Philosophy of Science* **21**: 257–268.
Giedymin, J. (1971). "Consolations for the Irrationalist?" *British Journal for the Philosophy of Science* **22**: 39–48.
Gonzalez-Castan, O. (1999). "Reflexion historiografica y tradiciones filosoficas: un conflicto sin resolver." *Anales del Seminario de Historia de la Filosofia* **16**: 35–56.
Gopnik, A. (1988). "Conceptual and Semantic Development as Theory Change: The Case of Object Permanence." *Mind & Language* **3**: 197–216.
Grandy, R. (1983). "Incommensurability: Kinds and Causes." *Philosophica* **32**: 7–24.
Greenwood, J. (1990). "Two Dogmas of Neo-Empiricism: The 'Theory-Informity' of Observation and the Quine-Duhem Thesis." *Philosophy of Science* **57**: 553–574.
Grover, S. (1998). "Incommensurability and the Best of All Possible Worlds." *Monist* **81**: 648–668.
Guarino, T. (1990). "Revelation and Foundationalism: Towards Hermeneutical and Ontological Appropriateness." *Modern Theology* **6**: 221–235.
Gutting, G. (1980). *Paradigms and Revolutions.* Notre Dame: University of Notre Dame Press.
Haack, S. (1997). "The Puzzle of 'Scientific Method'." *Revue Internationale de Philosophie* **51**: 495–505.
Hacking, I. (1979). "Michel Foucault's Immature Science." *Nous* **13**: 39–51.
Hacking, I. (1982). "Language, Truth and Reason." In M. Hollis and S. Lukes, eds., *Rationality and Relativism*, pp. 48–66, Cambridge Mass.: M.I.T. Press.
Hacking, I. (1983). *Representing and Intervening. Introductory Topics in the Philosophy of Natural Science.* Cambridge: Cambridge University Press.
Hacking, I. (1988). "On the Stability of the Laboratory Sciences." *Journal of Philosophy* **85**: 507–514.
Hacking, I. (1992). "The Self-Vindication of the Laboratory Sciences." In A. Pickering, ed., *Science as Practice and Culture*, pp. 29–64, Chicago: University of Chicago Press.
Hacking, I. (1993). "Working in a New World: The Taxonomic Solution." In P. Horwich, ed., *World Changes. Thomas Kuhn and the Nature of Science*, pp. 275–310, Cambridge Mass.: MIT Press.
Hales, S. (1997). "A Consistent Relativism." *Mind* **106**: 33–52.

Harper, W. (1978). "Conceptual Change, Incommensurability and Special Relativity Kinematics." *Acta Philosophica Fennica* **30**: 430–461.

Hatab, L. (1990). *Myth and Philosophy: A Contest of Truths*. Peru: Open-Court.

Hattiangadi, J. (1971). "Alternatives and Incommensurables: the Case of Darwin and Kelvin." *Philosophy of Science* **38**: 502–507.

Hattiangadi, J. (1974). "The Importance of Auxiliary Hypotheses." *Ratio* **16**: 115–120.

Hattiangadi, J. (1976). "Rationality and the Problem of Scieitific Traditions." *Dialectica*.

Hattiangadi, J. (1977). "The Crisis in Methodology: Paul Feyerabend." *Philosophy of the Social Sciences*.

Hattiangadi, J. (1983). "A Methodology without Methodological Rules." In R. Cohen and M. Wartowsky, eds., *Language, Logic and Methodology, Boston Studies in the Philosophy of Science*, Dordrecht: Reidel.

Hattiangadi, J. (1987). *How Is Languatge Possible? An Essay in the Evolution of Expressive Capability in Language from a Biological Perspective*. La Salle: Open Court.

Hautamõki, A. (1983). "Scientific Change and Intensional Logic." *Philosophica* **32**: 25–42.

Heidelberger, M. (1976). "Some Intertheoretic Relations Between Ptolemean and Copernican Astronomy." *Erkenntnis* **10**: 323–336.

Hempel, C. (1977). "Die Wissenschaftstheorie des analytischen Empirismus im Lichte zeitgenössischer Kritik." In G. Patzig, E. Scheibe, and W. Wieland, eds., *Logik, Ethik, Theorie der Geisteswissenschaften*, pp. 20–34, Hamburg: Meiner.

Hempel, C. (1979). "Scientific Rationality: Analytic vs Pragmatic Perspectives." In T. Geraets, ed., *Rationality To-Day*, pp. 46–58, Ottawa: University of Ottawa Press.

Hentschel, K. (1990). *Interpretationen und Fehlinterpretationen der speziellen und der allgemeinen Relativitätstheorie durch Zeitgenossen Albert Einsteins*. Basel: Birkhäuser.

Hentschel, K. (1990). "Inkommensurabilitätseffekte in der Debatte zwischen Oskar Kraus, Philipp Frank und Benno Urbach." In K. Hentschel, ed., *Interpretationen und Fehlinterpretationen der speziellen und der allgemeinen Relativitätstheorie durch Zeitgenossen Albert Einsteins*, pp. 541–549, Basel: Birkhäuser.

Hernández-Iglesias, M. (1994). "Incommensurability without Dogmas." *Dialectica* **48**: 29–45.

Hidalgo, C. (1988). "Inconmensurabilidad y criterios de identidad de esquemas conceptuales." *Analisis Filosofico* **8**: 69–78.

Hintikka, J. (1988). "On the Incommensurability of Theories." *Philosophy of Science* **55**: 25–38.

Hintikka, J. (1998). "Ramsey Sentences and the Meaning of Quantifiers." *Philosophy of Science* **65**: 289–305.

Holcomb III, H. (1989). "Interpreting Kuhn: Paradigm Choice as Objective Value Judgment." *Metaphilosophy* **20**: 51–67.

Horsten, L. (1999). "The Layered Structure of Physics According to Peter Galison." *Tijdschrift voor Filosofie* **61**: 747–778.

Hoyningen-Huene, P. (1990). "Kuhn's Conception of Incommensurability." *Studies in History and Philosophy of Science* **21**: 481–492.

Hoyningen-Huene, P. (1990). "Inkommensurabilität bei Kuhn und Theorienvergleich." In E. Agazzi, ed., *Die Vergleichbarkeit wissenschaftlicher Theorien*, pp. 97–108, Freiburg: Universitätsverlag.

Hoyningen-Huene, P. (1993). *Reconstructing Scientific Revolutions. Thomas S. Kuhn's Philosophy of Science*, trans. A. Levin, Chicago: University of Chicago Press.

Hoyningen-Huene, P. (1998). "On Thomas Kuhn's Philosophical Significance." *Configurations* **6**: 1–14.

Hoyningen-Huene, P. (2000). "Paul Feyerabend and Thomas Kuhn." In J. Preston, G. Munevar and D. Lamb, eds., *The Worst Enemy of Science? Essays in Memory of Paul Feyerabend*, pp. 102–114, Oxford: Oxford University Press.

Hoyningen-Huene, P., E. Oberheim and H. Andersen. (1996). "On Incommensurability." *Studies in History and Philosophy of Science* **27**: 131–141.

Hung, H. (1987). "Incommensurability and Inconsistency of Language." *Erkenntnis* **27**: 323–352.

Irzik, G. and T. Grünberg. (1998). "Carnap and Kuhn: Arch enemies or close allies?" *British Journal for the Philosophy of Science* **46**: 285–309.

Irzik, G. and T. Grünberg. (1998). "Whorfian Variations on Kantian Themes: Kuhn's Linguistic Turn." *Studies in History and Philosophy of Science* **29**: 207–221.

Johansson, I. (1986). "Levels of Intension and Theories of Reference." *Theoria* **52**: 1–15.

Jones, G. (1981). "Kuhn, Popper, and Theory Comparison." *Dialectica* **35**: 389–397.

Jones, K. (1986). "Is Kuhn a Sociologist?" *British Journal for the Philosophy of Science* **37**: 443–452.

Khalidi, A. (2000). "Incommensurability." In W. Newton-Smith, ed., *A Companion to the Philosophy of Science*, pp. 172–180, Oxford: Blackwell Publishers.

Katz, J. (1979). "Semantics and Conceptual Change." *Philosophical Review* **88**: 327–365.

Katz, J. (1989). "Rational Common Ground in the Sociology of Knowledge." *Philosophy of the Social Sciences* **19**: 257–271.

Keenan, J. (1995). "Help Must First Come From The Divine: A Response to Fr. George Eber's Claim of the so Called Incommensurability of Orthodox and Non Orthodox Christian Bioethics." *Christian Bioethics* **1**: 153–161.

Keita, L. (1988). "'Theory Incommensurability' and Kuhn's History of Science. A Critical Analysis." *Diogenes* **143**: 41–65.

Kenshur, O. (1984). "The Rhetoric of Incommensurability." *Journal of Aesthetics and Art Criticism* **42**: 375–382.

Khalidi, M. (1998). "Incommensurability in Cognitive Guise." *Philosophical Psychology* **11**: 29–43.

Khalidi, M. (1998). "Natural Kinds and Crosscutting Categories." *Journal of Philosophy* **95**: 33–50.

Kim, J. (1978). "Supervenience and Nomological Incommensurables." *American Philosophical Quarterly* **15**: 149–156.

Kindi, V. (1994). "Incommensurability, Incomparability, Irrationality." *Methodology and Science* **27**: 40–55.

Kindi, V. (1995). "Kuhn's "The Structure of Scientific Revolutions" Revisited." *Journal for General Philosophy of Science* **26**: 75–92.

Kitcher, P. (1978). "Theories, Theorists and Theoretical Change." *The Philosophical Review* **87**: 519–547.

Kitcher, P. (1983). "Implications of Incommensurability." In P. Asquith and T. Nickles, eds., *PSA 1982. Proceedings of the 1982 Biennial Meeting of the Philosophy of Science Association*, pp. 689–703, East Lansing: Philosophy of Science Association.

Kitcher, P. (1993). *The Advancement of Science. Science Without Legend, Objectivity Without Illusions.* Oxford: Oxford University Press.

Klee, R. (1992). "In Defense of The Quine Duhem Thesis: A Reply to Greenwood." *Philosophy of Science* **59**: 487–491.

Knasas, J. (1991). "Incommensurability and Aquinas's Metaphysics." *American Catholic Philosophical Quarterly* **65**: 179–190.

Koertge, N. (1983). "Theoretical Pluralism and Incommensurability: Implications for Science and Education." *Philosophica* **31**: 85–108.

Kordig, C. (1970). "Objectivity, Scientific Change, and Self-Reference." In R. Buck and R. Cohen, eds., *PSA 1970. In Memory of Rudolf Carnap, Studies in the Philosophy of Science*, pp. 519–523, Dordrecht: Reidel.

Kordig, C. (1971). *The Justification of Scientific Change*. Dordrecht: Reidel.

Kordig, C. (1971). "Scientific Transitions, Meaning Invariance, and Derivability." *Southern Journal of Philosophy* **9**: 119–125.

Kordig, C. (1971). "Stipulative Invariance." *Southern Journal of Philosophy* **9**: 129.

Kordig, C. (1971). "The Comparatibility of Scientific Theories." *Philosophy of Science* **38**: 467–485.

Kordig, C. (1980). "Progress Requires Invariance." *Philosophy of Science* **47**: 141.

Koutougos, A. (1989). "Research Programmes and Paradigms as Dialogue Structures." In K. Gavroglu, Y. Goudaroulis and P. Nicolacopoulos, eds., *Imre Lakatos and Theories of Scientific Change*. Norwell: Kluwer.

Krige, J. (1980). *Science, Revolution and Discontinuity*. Sussex: Harvester.

Kripke, S. (1980). *Naming and Necessity*. Cambridge: Harvard University Press.

Kroon, F. (1985). "Theoretical Terms and the Causal View of Reference." *Australasian Journal of Philosophy* **63**: 143–166.

Kroon, F. (1987). "Causal Descriptivism." *Australasian Journal of Philosophy* **65**: 1–17.

Krüger, L. (1979). "Some Remarks on Realism and Scientific Revolutions." In R. Horstmann, ed., *Transcendental Arguments and Science*, Boston: Reidel.

Kuhn, T. (1962). *The Structure of Scientific Revolutions*. Chicago: University of Chicago Press.

Kuhn, T. (1970). "Logic of Discovery or Psychology of Research?" In I. Lakatos and A. Musgrave, eds., *Criticism and the Growth of Knowledge: Proceedings of the International Colloquium the Philosophy of Science, London, 1965*, pp. 1–23, Cambridge: Cambridge University Press.

Kuhn, T. (1970). "Reflections on my Critics." In I. Lakatos and A. Musgrave, eds., *Criticism and the Growth of Knowledge: Proceedings of the International Colloquium the Philosophy of Science, London, 1965*, pp. 231–278, Cambridge: Cambridge University Press.

Kuhn, T. (1970). *The Structure of Scientific Revolutions*. 2nd edition.Chicago: Chicago University Press.

Kuhn, T. (1974). "Second Thoughts on Paradigms." In F. Suppe, ed., *The Structure of Scientific Theories*, pp. 459–482, Urbana: University of Illinois Press.

Kuhn, T. (1976). *Die Struktur wissenschaftlicher Revolutionen*. Frankfurt: Suhrkamp.

Kuhn, T. (1977). *The Essential Tension: Selected Studies in Scientific Tradition and Change*. Chicago: University of Chicago Press.

Kuhn, T. (1981). "What are Scientific Revolutions?" In L. Krüger, L. Daston and M. Heidelberger, eds., *The Probabilistic Revolution: Ideas in History*, pp. 7–22, Cambridge Mass.: M.I.T. Press.

Kuhn, T. (1983). "Commensurability, Comparability, Communicability." In P. Asquith and T. Nickles, eds., *PSA 1982. Proceedings of the 1982 Biennial Meeting of the Philosophy of Science Association*, pp. 669–688, East Lansing: Philosophy of Science Association.

Kuhn, T. (1983). "Response to Commentaries." In P. Asquith and T. Nickles , eds., *PSA 1982. Proceedings of the 1982 Biennial Meeting of the Philosophy of Science Association*, pp. 712–716, East Lansing: Philosophy of Science Association.

Kuhn, T. (1990). "Dubbing and Redubbing: The Vulnerability of Rigid Designation." In W. Savage, ed., *Scientific Theories. Minnesota Studies in the Philosophy of Science*, pp. 298–318, Minneapolis: University of Minnesota Press.

Kuhn, T. (1991). "The Road Since Structure." In A. Fine, M. Forbes and L. Wessels, eds., *PSA 1990, Proceedings of the 1990 Biennial Meeting of the Philosophy of Science Association*, pp. 2–13, East Lansing: Philosophy of Science Association.

Kuhn, T. (1991). "The Natural and the Human Sciences." In D. Hiley, J. Bohman and R. Shusterman, eds., *The Interpretive Turn*, pp. 17–24, Ithaca: Cornell University Press.

Kuhn, T. (1992). *The Trouble with the Historical Philosophy of Science. Robert and Maurine Rothschild Distinguished Lecture, 19 November 1991*. Cambrige Mass.: An Occasional Publication of the Department of the History of Science, Harvard University.

Kuhn, T. (1993). "Afterwords." In P. Horwich, ed., *World Changes: Thomas Kuhn and the Nature of Science*, pp. 311–341, Cambridge Mass.: M.I.T. Press.

Kuokkanen, M. (1986). "On Conceptual Correlation." *Erkenntnis* **25**: 371–401.

Kvasz, L. (1999). "On Classification of Scientific Revolutions." *Journal for General Philosophy of Science* **30**: 201–232.

Kvilhaug, T. (1997). "Naturens realitet i teorier – Kuhns irrealisme, del I." *Agora* **3–4**: 110–159.

Kvilhaug, T. (1998). "Naturens realitet i teorier – Kuhns irrealisme, del 2." *Agora* **1**: 56–107.

Lakatos, I. (1970). "Falsification and the Methodology of Scientific Research Programmes." In I. Lakatos and A. Musgrave, eds., *Criticism and the Growth of Knowledge: Proceedings of the International Colloquium the Philosophy of Science, London, 1965*, pp. 91–196, Cambridge: Cambridge University Press.

Laudan, L. (1976). "Two Dogmas of Methodology." *Philosophy of Science* **43**: 585–597.

Laudan, L. (1977). *Progress and its Problems. Towards a Theory of Scientific Growth*. Berkeley: University of California Press.

Laudan, L. (1990). *Science and Relativism. Some Key Controversies in the Philosophy of Science*. Chicago: University of Chicago Press.

Leeds, S. (1997). "Incommensurability and Vagueness." *Nous* **31**: 385–407.

Leplin, J. (1988). "Is Essentialism Unscientific?" *Philosophy of Science* **55**: 493–510.

Levin, M. (1979). "On Theory-Change and Meaning-Change." *Philosophy of Science* **46**: 407–424.

Lewis, D. (1970). "How to Define Theoretical Terms." *Journal of Philosophy of Science* **67**: 427–446.

Lewis, D. (1972). "Psychophysical and Theoretical Identifications." *Australasian Journal of Philosophy* **50**: 249–258.

Lewis, D. (1983). *Philosophical Papers, Volume 1*. New York: Oxford University Press.

Lewis, D. (1984). "Putnam's Paradox." *Australasian Journal of Philosophy of Science* **62**: 221–237.

Lewis, D. (1994). "Reduction of Mind." In S. Guttenplan, ed., *A Companion to the Philosophy of Mind*, pp. 412–431, Cambridge Mass.: Blackwell.

Lindberg, J. (1994). "Ever Since Kuhn: Conceptual Incommensurability, Social Constructivism, and the Critics." *Dialogue* **36**: 65–73.

Lodynski, A. (1982). "Some Remarks in Defense of the Incommensurability Thesis." In W. Krajewski, ed., *Polish Essays in Philosophy of the Natural Sciences*, pp. 91–102, Dordrecht: Reidel.

Loland, S. (1992). "The Mechanics and Meaning of Alpine Skiing." *Journal of The Philosophy of Sport* **19**: 55–77.

Loland, S. (1993). "Notes on the Epistemology of Sport Technique." In G. Gebauer, ed., *Die Aktualitat der Sportphilosophie*, Sankt Augustin: Academia.

Lueken, G. (1992). *Inkommensurabilität als Problem rationalen Argumentierens*. Stuttgart: Fromman-Holzboog.

Lueken, G. (1997). "Paul K. Feyerabend (13.01.1924–11.02.1994)." *Dialektik* **3**: 101–114.

MacCormac, E. (1971). "Meaning Variance and Metaphor." *British Journal for the Philosophy of Science* **22**: 145–169.

MacIntyre, A. (1991). "Incommensurability, Truth, and the Conversation Between Confucians and Aristotelians about the Virtues." In E. Deutsch, ed., *Culture and Modernity: East-West Philosophic Perspectives*, pp. 104–122, Honolulu: University of Hawaii Press.

Maia Neto, J. (1991). "Feyerabend's Scepticism." *Studies in History and Philosophy of Science* **22**: 543–555.

Malone, M. (1993). "Kuhn Reconstructed: Incommensurability Without Relativism." *Studies in History and Philosophy of Science* **24**: 69–93.

Malpas, J. (1989). "The Intertranslatability of Natural Languages." *Synthese* **78**: 233–264.

Manaktala, G. (1988). "Verifiers and Theory Choice." *Indian Philosophical Quarterly* **15**: 255–265.

Mandelbaum, M. (1979). "Subjective, Objective, and Conceptual Relativisms." *The Monist* **62**: 403–428.

Margolis, J. (1991). *The Truth About Relativism*. Cambridge: Blackwell.

Martin, M. (1971). "Referential Variance and Scientific Objectivity." *British Journal for the Philosophy of Science* **22**: 17–26.

Martin, M. (1972). "Ontological Variance and Scientific Objectivity." *British Journal for the Philosophy of Science* **23**: 252–256.

Martin, M. (1984). "How To Be a Good Philosopher of Science: A Plea for Empiricism in Matters Methodological." In R. Cohen and M. Wartofsky, eds., *Methodology, Metaphysics and the History of Science; in Memory of Benjamin Nelson, Boston Studies in the Philosophy of Science*, pp. 3–42, Dordrecht: Reidel.

Maruyama, M. (1959). "Communicable and Incommunicable Realities." *British Journal for the Philosophy of Science* **10**: 50–54.

Matteo, A. (1989). "Grounding The Human Conversation." *Thomist* **53**: 235–258.

Maudlin, T. (1995). "The Irrelevance of Incommensurability: Reflections of Torretti's Creative Understanding." *Studies in History and Philosophy of Science* **25**: 1005–1012.

Maudlin, T. (1996). "Kuhn edenté: incommensurabilite et choix entre thories." *Revue Philosophique de Louvain* **94**: 428–446.

Mayo, D. (1996). "Ducks, Rabbits, and Normal Science: Recasting the Kuhn's-eye View of Popper's Demarcation of Science." *British Journal for the Philosophy of Science* **47**: 271–290.

Mayo, D. (1996). *Error and the Growth of Experimental Knowledge*. Chicago: University of Chicago Press.

McEvoy, J. (1975). "A "Revolutionary" Philosophy of Science: Feyerabend and the Degeneration of Critical Rationalism into Sceptical Fallibilism." *Philosophy of Science* **42**: 49–66.

Meyer, M. and A. Lempereur. (1990). *Figures at Conflits Rhetoriques*. Bruxelles: Editions de l'Université de Bruxelles.

Miller, A. (1991). "Have incommensurability and causal theory of reference anything to do with actual science? – incommensurability, no; causal theory, yes." *International Studies in the Philosophy of Science* **5**: 97–108.

Mittelstrass, J. (1984). "Forschung, Begründung, Rekonstruktion. Wege aus dem Begründungsstreit." In H. Schnädelbach, eds., *Rationalität. Philosophische Beiträge*, pp. 117–140, Frankfurt: Suhrkamp.

Moberg, D. (1979). "Are There Rival, Incommensurable Theories?" *Philosophy of Science* **46**: 244–262.

Moya, E. (1994). "Incommensuralbilidad Empirica: Un Enfoque Macrologico." *Daimon, Revista De Filosofia* **8**: 119–130.

Mühlhölzer, F. (1991). "Inkommensurabilität und Erkenntnisfortschritt." *Untersuchungen zur Logik und zur Methodologie* **8**: 22–36.

Mühlhölzer, F. (1994). "On the Assumption that Our Concepts 'Structure the Material of Our Experience'."

In E. Rudolph and I. Stamatescu, eds., *Philosophy, Mathematics and Modern Physics*, pp. 170–185, Berlin: Springer.
Mühlhölzer, F. (1995). "Science without Reference?" *Erkenntnis* **42**: 203–222.
Musgrave, A. (1971). "Kuhn's Second Thoughts." *British Journal for the Philosophy of Science* **22**: 287–297.
Musgrave, A. (1978). "How to Avoid Incommensurability." *Acta Philosophica Fennica* **30**: 336–346.
Navarro Delgado, R. (1997). *La Inconmensurabilidad en el Lenguaje*. Venezuela: University of Zulia Press.
Navarro Delgado, R. (1997). "Variaciones sobre un Tema de Thomas S. Kuhn." *Fronesis* **4**: 9–18.
Nersessian, N. (1979). "The Roots of Epistemological 'Anarchy'." *Inquiry* **22**: 423–440.
Nersessian, N. (1982). "Why is 'incommensurability' a problem?" *Acta Biotheoretica* **31**: 205–218.
Nersessian, N. (1984). *Faraday to Einstein: Constructing Meaning in Scientific Theories*. Dordrecht: Kluwer.
Nersessian, N. (1989). "Conceptual Change in Science and in Science Education." *Synthese* **80**: 163–183.
Nersessian, N. (1989). "Scientific Discovery and Commensurability of Meaning." In K. Gavroglu, Y. Goudaroulis and P. Nicolacopoulos, eds., *Imre Lakatos and Theories of Scientific Change*, pp. 323–334, Dordrecht: Kluwer.
Nersessian, N. and H. Andersen. (1997). "Conceptual Change and Incommensurability: A Cognitive-Historical View." *Danish Yearbook of Philosophy* **32**: 111–151.
Newton-Smith, W. (1981). *The Rationality of Science*. London: Routledge.
Niiniluoto, I. (1997). "Reference, Invariance and Truthlikeness." *Philosophy of Science* **64**: 546–554.
Nola, R. (1980). "Fixing the Reference of Theoretical Terms." *Philosophy of Science* **47**: 505–531.
Nola, R. (1980). "'Paradigms Lost, or The World Regained' An Excursion Into Realism and Idealism in Science." *Synthese* **45**: 317–350.
Nola, R. (1989). "Review of Siegel's Relativism Refuted: A Critique of Contemporary Epistemological Relativism." *British Journal for the Philosophy of Science* **40**: 419–427.
Nordmann, A. (1986). "Comparing Incommensurable Theories." *Studies in History and Philosophy of Science* **17**: 231–246.
Norris, C. (1997). "Ontological relativity and meaning-variance. A critical-constructive view." *Inquiry* **40**: 139–173.
Notturno, M. (1996). "The Choice Between Popper and Kuhn." *Filozofska Istrazivanja* **16**: 741–760.
Oberheim, E. and P. Hoyningen-Huene. (1997). "Incommensurability, Realism and Meta-Incommensurability." *Theoria* **12**: 447–465.
Oberheim, E. and P. Hoyningen-Huene. (1999). "On the Early Feyerabend". *Metascience. The Australian Journal for The Philosophy of Science* **8**: 226–23
Oberheim, E. and P. Hoyningen-Huene. (2000). " Feyerabend's Early Philosophy." *Studies in the History and Philosophy of Science* **31**: 363–375.
Oddie, G. (1988). "On a Dogma Concerning Realism and Incommensurability." In R. Nola, ed., *Relativism and Realism in Science*, pp. 169–203, Dordrecht: Kluwer.
Oddie, G. (1989). "Partial Interpretation, Meaning Variance, and Incommensurability." In K. Gavroglu, Y. Goudaroulis and P. Nicolacopoulos, eds., *Imre Lakatos and Theories of Scientific Change*, pp. 305–322, Dordrecht: Kluwer.
O'Leary-Hawthorne, J. (1994). "A Corrective to the Ramsey-Lewis Account of Theoretical Terms." *Analysis* **54**: 105–110.
Papineau, D. (1979). *Theory and Meaning*. Oxford: Clarendon.
Papineau, D. (1996). "Theory-dependent Terms." *Philosophy of Science* **63**: 1–20.
Parrini, P. (1998). *Knowledge and Reality. An Essay in Positive Philosophy. The Western Ontario Series in Philosophy of Science. Vol. 59*. Dordrecht: Kluwer.
Parsons, K. (1971). "On Criteria of Meaning Change." *British Journal for the Philosophy of Science* **22**: 131–144.
Parsons, K. (1971). "A Note on Meaning Invariance." *Southern Journal of Philosophy* **9**: 126–128.
Partington. (1961–2). *A History of Chemistry*. London: Macmillan & Co.
Paulos, J. (1980). "A Model Theoretic Explication of the Theses of Kuhn and Whorf." *Notre Dame Journal of Formal Logic* **21**: 155–165.
Pearce, D. (1982). "Stegmüller on Kuhn and Incommensurability." *British Journal for the Philosophy of Science* **33**: 389–396.
Pearce, D. (1984). "Research Traditions, Incommensurability and Scientific Progress." *Zeitschrift für*

allgemeine Wissenschaftstheorie **15**: 261–271.
Pearce, D. (1986). "Incommensurability and Reduction Reconsidered." *Erkenntnis* **24**: 293–308.
Pearce, D. (1987). *Roads to Commensurability*. Dordrecht: Reidel.
Pearce, D. (1988). "The Problem of Incommensurability: A Critique of Two Instrumentalist Approaches." In I. Hronszky, M. Fehér and B. Dajka, eds., *Scientific Knowledge Socialized. Selected Proceedings of the 5th International Conference on the History and Philosophy of Science*, pp. 385–398, Dordrecht: Kluwer.
Pearce, D. (1989). "Conceptual Change and the Progress of Science." In J. Fenstad, I. Frolov and R. Hilpinen, eds., *Logic, Methodology and Philosophy of Science*, Amsterdam: North-Holland.
Perovich, A. (1981). "Inkommensurabilität – ihre Unterarten und ihre ontologischen Konsequenzen." In H. Dürr, ed., *Versuchungen. Aufsätze zur Philosophie Paul Feyerabends*, pp. 76–94, Frankfurt: Suhrkamp.
Phillips, D. (1975). "Paradigms and Incommensurability." *Theory and Society* **2**: 37–61.
Phillips, D. (1977). *Wittgenstein and Scientific Knowledge: A Sociological Perspective*. London: Macmillan.
Pickering, A. (1994). "After Representation: Science Studies in The Performative Idiom." *Proceedings of the Biennial Meetings of the Philosophy of Science Association* **2**: 413–419.
Polikarov, A. (1988). "On the Incommensurability of superseding Physical Theories." In V. Cauchy, ed., *Philosophy and Culture, Volume 3*, pp. 124–127, Montreal: Ed-Montmorency.
Polikarov, A. (1993). "Is There an Incommensurability Between Superseding Theories? On the Validity of the Incommensurability Thesis." *Journal for General Philosophy of Science* **24**: 127–146.
Polikarov, A. (1993). "On Various Kinds of Scientific Revolution in Physics." *Epistemologia* **16**: 213–233.
Porus, N. (1988). "Incommensurability, Scientific Realism and Rationalism." In I. Hronszky, M. Fehér and B. Dajka, eds., *Scientific Knowledge Socialized. Selected Proceedings of the 5th International Conference on the History and Philosophy of Science*, pp. 375–384, Dordrecht: Kluwer.
Potter, K. (1984). "Does Indian Epistemology Concern Justified True Belief?" *Journal of Indian Philosophy* **12**: 307–328.
Potter, K. (1991). "The Commensurability of Indian Epistemological Theories." In E. Deutsch, ed., *Culture and Modernity: East-West Philosophic Perspectives*, pp. 123–137, Honolulu: University of Hawaii Press.
Preston, J. (1997). *Feyerabend. Philosophy, Science and Society*. Cambridge: Polity Press.
Preston, J. and G. Munévar and D. Lamb. (2000). *The Worst Enemy of Science? Essays in Memory of Paul Feyerabend*. New York: Oxford University Press.
Preyer, G. (1998). "The Received View, Incommensurability and the Comparison of Theories – Beliefs As the Basis of Theorizing." *Protosociology* **12**: 40–58.
Procee, H. (1978). "Thomas S Kuhn Over Incommensurabiliteit." *Kennis en Methode* **2**: 208–227.
Przelecki, M. (1978). "Commensurable Referents of Incommensurable Terms." *Acta Philosophica Fennica* **30**: 347–365.
Putnam, H. (1975). Mind, Language, and Reality. Philosophical Papers, *Volume 2*. Cambridge: Cambridge University Press.
Putnam, H. (1981). *Reason, Truth and History*. Cambridge: Cambridge University Press.
Quante, M. (1998). "Understanding Conceptual Schemes: Rescher's Quarrel with Davidson." In A. Wustehube, ed., *Pragmatic Idealism: Critical Essays on Nicholas Rescher's System of Pragmatic Idealism*, Amsterdam: Rodopi.
Quine, W. (1969). *Ontological Relativity and Other Essays*. New York: Columbia University Press.
Quine, W. (1987). "Indeterminacy of Translation Again." *Philosophy of Science* **84**: 5–10.
Quitterer, J. (1996). *Kant und die These vom Paradigmenwechsel: Eine Gegenuberstellung seiner Transzendentalphilosophie mit der Wissenschaftstheorie Thomas S. Kuhns*. New York: Lang.
Rabb, J. (1975). "Incommensurable Paradigms and Critical Idealism." *Studies in History and Philosophy of Science* **6**: 343–346.
Rabb, J. (1978). "Incommensurable Paradigms and Psycho Metaphysical Explanation." *Inquiry* **21**: 201–212.
Radnitzky, G. (1980). "From Justifying a Theory to Comparing Theories and Selecting Questions: Popper's Alternative to Foundationalism and Scepticism." *Revue Internationale de Philosophie* **34**: 179–228.
Radnitzky, G. (1990). " Is Kuhn's Revolution in the Philosophy of Science a Pseudo-Revolution?" *International Studies in Philosophy* **22**: 77–78.
Radnitzky, G. (1990). "La revolution kuhnienne est-elle une fausse revolution?" *Archives de Philosophie* **53**: 199–212.

Rakover, S. (1989). "Incommensurability: The Scaling of Mind-Body Theories as a Counter Example." *Behaviorism* **17**: 103–118.
Ramberg, B. (1989). *Donald Davidson's Philosophy of Language: An Introduction.* Cambridge: Blackwell.
Rantala, V. (1995). "Translation and Scientific Change." *Pozna Studies in the Philosophy of the Sciences and the Humanities* **44**: 249–268.
Rasmussen, S. (1987). "Sense, Reference, and Meaning Incommensurability." *Analysis* **47**: 170–173.
Rescher, N. (1982). *Empirical Inquiry.* Totowa: Rowman and Littlefield.
Rescher, N. (1983). *Scientific Explanation and Understanding: Essays on Reasoning in Science.* Pittsburgh: University Press of America.
Rivadulla, A. (1986). *Filosofia actual de la ciencia.* Madrid: Tecnos.
Rivadulla, A. (1990). "El enfoque sociologico de Kuhn de las revoluciones cientificas." In W. Gonzalez, ed., *Aspectos metodologicos de la investigacion cientifica*, Madrid: Ediciones de la UAM-Publicaciones de la Universidad de Murcia.
Rivadulla, A. (1990). "Ludwik Fleck's almost unnoticed foundation of sociological epistemology in the thirties." *Explorations in Knowledge* **7**: 19–28.
Robertson, M. (1995). "Science, Technology, and Progress." *Proceedings of the Heraclitean Society* **17**: 102–109.
Rorty, R. (1979). *Philosophy and the Mirror of Nature.* Princeton: Princeton University Press.
Rosch, E. (1977). "Human Categorization." In N. Warren, ed., *Studies in Cross-Cultural Psychology*, pp. 1–49, London: Academic Press.
Rosch, E. and C. Mervis. (1975). "Family Resemblances: Studies in the Internal Structure of Categories." *Cognitive Psychology* **7**: 573–605.
Rouse, J. (1981). "Kuhn, Heidegger and Scientific Realism." *Man and World* **14**: 269–290.
Rudolph, E. and I. Stamatescu. (1994). *Philosophy, Mathematics and Modern Physics.* Berlin: Springer.
Sagal, P. (1972). "Incommensurability, Then and Now." *Zeitschrift für allgemeine Wissenschaftstheorie* **3**: 298–301.
Sanders, J. (1998). "Incommensurability and Demarcation." In R. Dane, ed., *Criticism and Defense of Rationality in Contemporary Philosophy.* Amsterdam: Rodopi.
Sankey, H. (1990). "In Defence of Untranslatability." *Australasian Journal of Philosophy* **68**: 1–21.
Sankey, H. (1991). "Incommensurability and the Indeterminacy of Translation." *Australasian Journal of Philosophy* **69**: 219–223.
Sankey, H. (1991). "Translation Failure Between Theories." *Studies in History and Philosophy of Science* **22**: 223–236.
Sankey, H. (1991). "Incommensurability, Translation and Understanding." *Philosophical Quarterly* **41**: 414–426.
Sankey, H. (1991). "Feyerabend and the Description Theory of Reference." *Journal of Philosophical Research* **16**: 223–232.
Sankey, H. (1992). "Translation and Languagehood." *Philosophia* **21**: 335–337.
Sankey, H. (1993). "Kuhn's Changing Concept of Incommensurability." *British Journal for the Philosophy of Science* **44**: 759–774.
Sankey, H. (1994). *The Incommensurability Thesis.* Aldershot: Avebury.
Sankey, H. (1994). "Judgement and Rational Theory-Choice." *Methodology and Science* **27**: 167–182.
Sankey, H. (1995). "The Problem of Rational Theory-Choice." *Epistemologia* **18**: 299–312.
Sankey, H. (1996). "Normative Naturalism and the Challenge of Relativism: Laudan versus Worrall on the Justification of Methodological Principles." *International Studies in the Philosophy of Science* **10**: 37–51.
Sankey, H. (1996). "Rationality, Relativism and Methodological Pluralism." *Explorations in Knowledge* **134**: 18–36.
Sankey, H. (1997). *Rationality, Relativism and Incommensurability.* Aldershot: Ashgate.
Sankey, H. (1997). "Incommensurability: The Current State of Play." *Theoria* **12**: 425–445.
Sankey, H. (1997). "Il Problema dell' incommensurabilità delle teorie scientifiche." *Nuova Secondaria* **5**: 61–66.
Sankey, H. (1997). "Kuhn's Ontological Relativism." In D. Ginev and R. Cohen, eds., Issues and Images in the Philosophy of Science, pp. 305–329, Dordrecht: Kluwer Academic Publishers. (Reprinted (2000) In *Science & Education* **9**: 59–75

Sankey, H. (1998). "Taxonomic Incommensurability." *International Studies in Philosophy of Science* **12**: 7–16.
Scheffler, I. (1967). *Science and Subjectivity*. Indianapolis: Hackett.
Scheffler, I. (1972). "Vision and Revolution: A Postscript on Kuhn." *Philosophy of Science* **39**: 366–374.
Scheibe, E. (1988). "The Physicists' Conception of Progress." *Studies in History and Philosophy of Science* **19**: 141–159.
Scheibe, E. (1999). *Die Reduktion physikalischer Theorien. Ein Beitrag zur Einheit der Physik. Teil II: Inkommensurabilität und Grenzfallreduktion*. Berlin: Springer.
Schnädelbach, H. (1984). *Rationalität. Philosophische Beiträge*. Frankfurt: Suhrkamp.
Schomberg, R. (1995). *Der rationale Umgang mit Unsicherheit*. Frankfurt: Peter Lang.
Schroeder, H. and F. Schaefer. (1989). "Reduction, Representation and Commensurability of Theories." *Philosophy of Science* **56**: 130–157.
Shand, J. (1986). "Grayling, Feyerabend and the Constancy of Sense." *Analysis* **46**: 211–212.
Shapere, D. (1964). "The Structure of Scientific Revolutions." *The Philosophical Review* **73**: 383–394.
Shapere, D. (1966). "Meaning and Scientific Change." In R. Colodny, ed., *Mind and Cosmos. Essays in Contemporary Science and Philosophy*, pp. 41–85, Pittsburgh: University of Pittsburgh Press.
Shapere, D. (1971). "The Paradigm Concept." *Science* **172**: 706–709.
Shapere, D. (1984). *Reason and the Search for Knowledge. Investigations in the Philosophy of Science*. Dordrecht: Reidel.
Shapere, D. (1989). "Evolution and Continuity in Scientific Change." *Philosophy of Science* **56**: 419–437.
Shea, W. (1988). *Revolutions in Science: Their Meaning and Relevance*. USA: Science History Publications.
Short, T. (1980). "Peirce and the Incommensurability of Theories." *Monist* **63**: 316–328.
Siegel, H. (1980). "Objectivity, Rationality, Incommensurability, and More." *British Journal for the Philosophy of Science* **31**: 359–384.
Siegel, H. (1987). *Relativism Refuted. A Critique of Contemporary Epistemological Relativism*. Dordrecht: Reidel.
Simmons, L. (1994). "Three Kinds of Incommensurability Theses." *American Philosophical Quarterly* **31**: 119–131.
Sisk, J. (1986). "Doctor Johnson Kicks a Stone." *Philosophy and Literature* **10**: 65–75.
Skagestad, P. (1995). "Discussion: Peirce and The History Of Science." In K. Ketner, ed., *Peirce and Contemporary Thought: Philosophical Inquiries*, New York: Fordham University Press.
Small, R. (1991). "Incommensurability and Recurrence: from Oresme to Simmel." *Journal of the History of Ideas* 121–137.
Smirnov, V. (1986). "Logical Relations Between Theories." *Synthese* **66**: 71–87.
Stegmuller, W. (1975). "Structures and Dynamics of Theories: Some Reflections on J. D. Sneed and T. S. Kuhn." *Erkenntnis* **9**: 75–100.
Stegmüller, W. (1977). *Probleme und Resultate der Wissenschaftstheorie, und Analytische Philosophie. Band 2: Theorie und Erfahrung, Zweiter Halbband: Theorienstrukturen und Theoriendynamik*. Berlin: Springer.
Stegmüller, W. (1986). *Probleme und Resultate der Wissenschaftstheorie und Analytischen Philosophie. Band II. Theorie und Erfahrung. 3. Teilband: Die Entwicklung des neuen Strukturalismus seit 1973*. Berlin: Springer.
Stich, S. (1996). *The Deconstruction of the Mind*. New York: Oxford University Press.
Ströker, E. (1974). "Geschichte als Herausforderung. Marginalien zur jüngsten wissenschaftstheoretischen Kontroverse." *Neue Hefte für Philosophie* **7/8**: 27–66.
Suppe, F. (1974). *The Structure of Scientific Theories*. Urbana: University of Illinois Press.
Szumilewicz, I. (1977). "Incommensurability and the Rationality of the Development of Science." *British Journal for the Philosophy of Science* **28**: 345–350.
Szumilewicz-Lachmann, I. (1985). "Poincaré versus Le Roy on Incommensurability." In G. Andersson, ed., *Rationality in Science and Politics. Boston Studies in the Philosophy of Science*, pp. 261–275, Dordrect: Reidel.
Thagard, P. (1990). "The Conceptual Structure of the Chemical Revolution." *Philosophy of Science* **57**: 183–209.
Thagard, P. (1990). "Modelling Conceptual Change." In J. Tiles, G. McKee and G. Dean, eds., *Evolving Knowledge in Natural Science and Artificial Intelligence*, pp. 201–218, Pitman.

Thagard, P. (1991). "Concepts and Conceptual Change." In J. Fetzer, ed., *Epistemology and Cognition*, pp. 101–120, Dordrecht: Kluwer.

Thagard, P. (1992). *Conceptual Revolutions*. Princeton: Princeton University Press.

Tiles, M. (1985). *Bachelard: Science And Objectivity*. New York: Cambridge University Press.

Tuchanska, B. (1988). "The Idea of Incommensurability and the Copernican Revolution." *The Polish Sociological Bulletin* 65–79.

Tuomela, R. (1973). *Theoretical Concepts*. Vienna and New York: Springer-Verlag.

Van Bendegem, J. (1983). "Incommensurability – An Algorithmic Approach." *Philosophica* **32**: 97–115.

Van Brakel, J. and I. Douven. (1995). "Is Scientific Realism an Empirical Hypothesis?" *Dialectica* **49**: 3–14.

Van der Veken, W. (1983). "Incommensurability in the Structuralist View." *Philosophica* **32**: 43–56.

Veatch, R. and W. Stempsey. (1995). "Incommensurability: Its Implications for the Patient/Physician Relation." *The Journal for Medicine and Philosophy* **20**: 253–269.

Verronon, V. (1986). *The Growth of Knowledge: An Inquiry into the Kuhnian Theory*. Jyvaskyla: University of Jyvaskyla Press.

Vision, G. (1988). *Modern Anti-Realism and Manufactured Truth*. London: Routledge.

Watanabe, S. (1975). "Needed: A Historico-Dynamical View of Theory Change." *Synthese* **32**: 113–134.

Watkins, J. (1970). "Against 'Normal Science'." In I. Lakatos and A. Musgrave, eds., *Criticism and the Growth of Knowledge*, pp. 25–37, London: Cambridge University Press.

Weed, D. (1997). "Underdetermination and Incommensurability in Contemporary Epidemiology." *Kennedy Institute of Ethics Journal* **7**: 107–127.

Weinert, F. (1984). "Contra Res Sempiternas." *Monist* **67**: 376–394.

Wisdom, J. (1974). "The Incommensurability Thesis." *Philosophical Studies* **25**: 299–301.

Wohlrapp, H. (1981/82). "Überlegungen zu einer möglichen Lösung des Problems inkommensurabler Welten mithilfe des Konzepts von der Bewegung des Begriffs." *Hegel-Jahrbuch*.

Wrathall, M. (1999). "Practical Incommensurability and the Phenomenological Basis of Robust." *Inquiry* **42**: 79–88.

Yates, S. (1986). "Are There Alternative Conceptual Frameworks?" *Auslegung* **13**: 51–62.

Zemach, E. (1976). "Putnam's Theory on the Reference of Substance Terms." *The Journal of Philosophy* **73**: 116–127.

Zheng, L. (1988). "Incommensurability and Scientific Rationality." *International Studies in the Philosophy of Science* **2**: 227–236.

NOTES ON CONTRIBUTORS

PETER BARKER, although trained as a philosopher, is currently Professor of the History of Science at the University of Oklahoma. His research interests include the history and historiography of the scientific revolution, and the application of methods from cognitive science to issues in traditional philosophy of science, especially conceptual change. Recent publications include: "Continuity Through Revolutions: A frame-based account of conceptual change" *Philosophy of Science (Proceedings)* **67**: 208–223 (2000) with X. Chen; and "The Role of Religion in the Lutheran Response to Copernicus" in M. Osler, ed., *Rethinking the Scientific Revolution*, pp. 59–88, Cambridge: Cambridge University Press (2000).

RICHARD N. BOYD is Professor of Philosophy and of Science and Technology Studies at Cornell University. His undergraduate (mathematics) and graduate (philosophy) degrees are both from the Massachusetts Institute of Technology. He has held teaching positions at the University of Michigan, the University of California (Berkeley), Harvard, and MIT. His research interests lie in philosophy of science, philosophy of biology, philosophy of mind, metaphysics, epistemology, philosophy of language and meta-ethics.

HAROLD I. BROWN is Professor of Philosophy at Northern Illinois University. He is author of *Perception, Theory and Commitment: The New Philosophy of Science*, *Observation and Objectivity*, *Rationality*, and numerous articles on epistemology and philosophy of science.

MARTIN CARRIER is Professor of Philosophy of Science at Bielefeld University and does research on various issues in the history and philosophy of science. His areas of research include theory-ladenness and testability, intertheoretical relations, scientific realism, as well as topics from the philosophy of biology and philosophy of mind. Major recent publications include (1994) *The Completeness of Scientific Theories*, Dordrecht: Kluwer; (1998) "In Defense of Psychological Laws" *International Studies in the Philosophy of Science* **12**: 217-232; (2000) "Multiplicity and Heterogeneity: On the Relations between Functions and their Realizations" *Studies in History and Philosophy of Biology and the Biomedical Sciences* **31C**: 179-191, (2000) "How to Pile up Fundamental Truths Incessantly: On the Prospect of Reconciling Scientific Realism with Unending Progress" in M. Carrier, G. Massey and L. Ruetsche, eds., *Science at Centrury's End: Philosophical Questions on the Progress and Limits of Science*, pp. 92-109, Pittsburgh: University of Pittsburgh Press.

MICHAEL DEVITT is Distinguished Professor and Executive Officer in the Philosophy Program at the Graduate Center of the City University of New York. He was formerly at University of Maryland and the University of Sydney. He is the author of *Designation*, *Realism and Truth*, *Language and Reality: An Introduction to the Philosophy of*

Language (with Kim Sterelny), and *Coming to Our Senses: A Naturalistic Program for Semantic Localism.*

GERALD DOPPELT is Professor of Philosophy and Director of the Graduate Program in Science Studies at the University of California, San Diego. His research in philosophy of science has focused on the topics of scientific rationality, progress, relativism, theory-change, and the ethics of technology. Papers have appeared in *Inquiry*, *Synthese*, *Philosopy of Science*, *The Journal of Ethics*, and other venues. His research in political philosophy has focused on the theory of justice, Rawlsian liberalism, and communitarianism, rival paradigms of freedom, just war theory, and the issue of group rights. Papers have appeared in *NOUS*, *Inquiry*, *Philosophy and Public Affairs*, *Ethics*, *The Journal of Social Philosophy*, and other venues. He is the recipient of the Academic Senate Distinguished Teaching Award at the University of California, San Diego, the Linbach Award for Distinguished Teaching at the University of Pennsylvania, and two National Endowment for the Humanities Research Fellowships.

PAUL HOYNINGEN-HUENE is Director of the Center for Philosophy and Ethics of Science at the University of Hannover, Germany. He is the author of *Reconstructing Scientific Revolutions: Thomas S. Kuhn's Philosophy of Science,* Chicago: Chicago University Press (1993).

FREDERICK KROON teaches at the University of Auckland and has published widely in the areas of philosophy of language, epistemology, metaphysics, and computability theory. His current research focuses on the problem of negative existentials and a pretense-theoretic approach to issues surrounding the reference of empty names.

HUGH LACEY is Professor of Philosophy at Swarthmore College (U.S.A.) and a frequent visiting professor at the University of São Paulo (Brazil). His most recent work is *Is Science Value Free? Values and Scientific Understanding*, London: Routledge (1999).

NANCY J. NERSESSIAN is Professor of Cognitive Science at Georgia Institute of Technology. She is appointed jointly in the School of Public Policy and the College of Computing. She is author of numerous publications including the book *Faraday to Einstein: Constructing Meaning in Scientific Theories*, Kluwer (1984, reprinted 1990); *Model-Based Reasoning in Scientific Discovery*, ed. with L. Magnani and P. Thagard (1999). She is currently at work on a book, *Creating Science: A cognitive-historical approach to conceptual change,* to be published by MIT Press. She has been a Fulbright Research Scholar at the University of Leiden, the Netherlands, a Fellow of the Netherlands Institute for Advanced Study, a Senior Fellow of the Pittsburgh Center for the Philosophy of Science, a Senior Fellow of the Dibner Institute at MIT and has taught at several institutions, including the Technical University of Twente, the Netherlands and Princeton University. She is series editor for Kluwer Academic Publishers of the *Science and Philosophy* book series. She was co-chair of the international conference, MBR '98: Model-based Reasoning in Scientific Discovery in 1998 (http://philos.

unipv.it/courses/progra1.html) and is currently at work on MBR '01: Model-based Reasoning: Science, Technology, and Values, to be held in May 2001.

ROBERT NOLA teaches at the University of Auckland and has published papers in philosophy of science and on nineteenth century philosophers. His most recent books are an edited collection of papers entitled *Foucault* (1998), and a collection edited with Howard Sankey on scientific method entitled *After Popper, Kuhn and Feyerabend* (2000).

HOWARD SANKEY is Associate Professor in Philosophy of Science in the Department of History and Philosophy of Science at the University of Melbourne. A native of southern California, he completed a Ph.D. at Melbourne after undergraduate studies in philosophy at the University of Otago in New Zealand. He has held visiting research positions at the Center for Philosophy of Science at the University of Pittsburgh, the Center for Philosophy and Ethics of Science at the University of Hannover, and the Philosophy Department at Saint David's University College, Lampeter. His main interests lie in the general philosophy of science, with special emphasis on semantic aspects of conceptual change, scientific realism, rationality of scientific theory choice, and naturalized accounts of the epistemic warrant of methodological norms. His major publications include two books, *The Incommensurability Thesis* (1994)and *Rationality, Relativism and Incommensurability* (1997). He is the editor of *Causation and Laws of Nature* (1999) and *After Popper, Kuhn and Feyerabend: Recent Issues in Theories of Scientific Method* (2000) (with Robert Nola).

DUDLEY SHAPERE is Z. Smith Reynolds Professor of the Philosophy and History of Science at Wake Forest University. He is now completing a book, *The Rational Dynamics of Knowledge*, which incorporates and extends, in a unified and systematic way, the ideas which he has been developing since 1982 (that is, since the publication of his essay, "The Concept of Observation in Science and Philosophy" in *Philosophy of Science*. During the first eight months of 2001, he will be an invited guest at the Institute for Advanced Study, Princeton, N.J., with appointment as Otto Neugebauer Fellow in both the School of Natural Sciences and the School of Historical Studies.

HARVEY SIEGEL is Professor of Philosophy at the University of Miami. His areas of specialization include epistemology, philosophy of science, and philosophy of education. He is the author of many papers in these and related areas of philosophy, and of the books *Relativism Refuted: A Critique of Contemporary Epistemological Relativism*, Kluwer (1987); *Educating Reason: Rationality, Critical Thinking, and Education*, Routledge (1988); and *Rationality Redeemed?: Further Dialogues on an Educational Ideal*, Routledge (1997); and is the editor of *Reason and Education: Essays in Honor of Israel Scheffler*, Kluwer (1997).

INDEX OF NAMES

Boston Studies in the Philosophy of Science

Editor: Robert S. Cohen, *Boston University*

1. M.W. Wartofsky (ed.): *Proceedings of the Boston Colloquium for the Philosophy of Science, 1961/1962.* [Synthese Library 6] 1963 ISBN 90-277-0021-4
2. R.S. Cohen and M.W. Wartofsky (eds.): *Proceedings of the Boston Colloquium for the Philosophy of Science, 1962/1964.* In Honor of P. Frank. [Synthese Library 10] 1965 ISBN 90-277-9004-0
3. R.S. Cohen and M.W. Wartofsky (eds.): *Proceedings of the Boston Colloquium for the Philosophy of Science, 1964/1966.* In Memory of Norwood Russell Hanson. [Synthese Library 14] 1967 ISBN 90-277-0013-3
4. R.S. Cohen and M.W. Wartofsky (eds.): *Proceedings of the Boston Colloquium for the Philosophy of Science, 1966/1968.* [Synthese Library 18] 1969 ISBN 90-277-0014-1
5. R.S. Cohen and M.W. Wartofsky (eds.): *Proceedings of the Boston Colloquium for the Philosophy of Science, 1966/1968.* [Synthese Library 19] 1969 ISBN 90-277-0015-X
6. R.S. Cohen and R.J. Seeger (eds.): *Ernst Mach, Physicist and Philosopher.* [Synthese Library 27] 1970 ISBN 90-277-0016-8
7. M. Čapek: *Bergson and Modern Physics.* A Reinterpretation and Re-evaluation. [Synthese Library 37] 1971 ISBN 90-277-0186-5
8. R.C. Buck and R.S. Cohen (eds.): *PSA 1970.* Proceedings of the 2nd Biennial Meeting of the Philosophy and Science Association (Boston, Fall 1970). In Memory of Rudolf Carnap. [Synthese Library 39] 1971 ISBN 90-277-0187-3; Pb 90-277-0309-4
9. A.A. Zinov'ev: *Foundations of the Logical Theory of Scientific Knowledge (Complex Logic).* Translated from Russian. Revised and enlarged English Edition, with an Appendix by G.A. Smirnov, E.A. Sidorenko, A.M. Fedina and L.A. Bobrova. [Synthese Library 46] 1973 ISBN 90-277-0193-8; Pb 90-277-0324-8
10. L. Tondl: *Scientific Procedures.* A Contribution Concerning the Methodological Problems of Scientific Concepts and Scientific Explanation.Translated from Czech. [Synthese Library 47] 1973 ISBN 90-277-0147-4; Pb 90-277-0323-X
11. R.J. Seeger and R.S. Cohen (eds.): *Philosophical Foundations of Science.* Proceedings of Section L, 1969, American Association for the Advancement of Science. [Synthese Library 58] 1974 ISBN 90-277-0390-6; Pb 90-277-0376-0
12. A. Grünbaum: *Philosophical Problems of Space and Times.* 2nd enlarged ed. [Synthese Library 55] 1973 ISBN 90-277-0357-4; Pb 90-277-0358-2
13. R.S. Cohen and M.W. Wartofsky (eds.): *Logical and Epistemological Studies in Contemporary Physics.* Proceedings of the Boston Colloquium for the Philosophy of Science, 1969/72, Part I. [Synthese Library 59] 1974 ISBN 90-277-0391-4; Pb 90-277-0377-9
14. R.S. Cohen and M.W. Wartofsky (eds.): *Methodological and Historical Essays in the Natural and Social Sciences.* Proceedings of the Boston Colloquium for the Philosophy of Science, 1969/72, Part II. [Synthese Library 60] 1974 ISBN 90-277-0392-2; Pb 90-277-0378-7
15. R.S. Cohen, J.J. Stachel and M.W. Wartofsky (eds.): *For Dirk Struik.* Scientific, Historical and Political Essays in Honor of Dirk J. Struik. [Synthese Library 61] 1974 ISBN 90-277-0393-0; Pb 90-277-0379-5
16. N. Geschwind: *Selected Papers on Language and the Brains.* [Synthese Library 68] 1974 ISBN 90-277-0262-4; Pb 90-277-0263-2
17. B.G. Kuznetsov: *Reason and Being.* Translated from Russian. Edited by C.R. Fawcett and R.S. Cohen. 1987 ISBN 90-277-2181-5

Boston Studies in the Philosophy of Science

18. P. Mittelstaedt: *Philosophical Problems of Modern Physics*. Translated from the revised 4th German edition by W. Riemer and edited by R.S. Cohen. [Synthese Library 95] 1976
ISBN 90-277-0285-3; Pb 90-277-0506-2
19. H. Mehlberg: *Time, Causality, and the Quantum Theory*. Studies in the Philosophy of Science. Vol. I: *Essay on the Causal Theory of Time*. Vol. II: *Time in a Quantized Universe*. Translated from French. Edited by R.S. Cohen. 1980 Vol. I: ISBN 90-277-0721-9; Pb 90-277-1074-0
Vol. II: ISBN 90-277-1075-9; Pb 90-277-1076-7
20. K.F. Schaffner and R.S. Cohen (eds.): *PSA 1972*. Proceedings of the 3rd Biennial Meeting of the Philosophy of Science Association (Lansing, Michigan, Fall 1972). [Synthese Library 64] 1974 ISBN 90-277-0408-2; Pb 90-277-0409-0
21. R.S. Cohen and J.J. Stachel (eds.): *Selected Papers of Léon Rosenfeld*. [Synthese Library 100] 1979 ISBN 90-277-0651-4; Pb 90-277-0652-2
22. M. Čapek (ed.): *The Concepts of Space and Time*. Their Structure and Their Development. [Synthese Library 74] 1976 ISBN 90-277-0355-8; Pb 90-277-0375-2
23. M. Grene: *The Understanding of Nature*. Essays in the Philosophy of Biology. [Synthese Library 66] 1974 ISBN 90-277-0462-7; Pb 90-277-0463-5
24. D. Ihde: *Technics and Praxis*. A Philosophy of Technology. [Synthese Library 130] 1979
ISBN 90-277-0953-X; Pb 90-277-0954-8
25. J. Hintikka and U. Remes: *The Method of Analysis*. Its Geometrical Origin and Its General Significance. [Synthese Library 75] 1974 ISBN 90-277-0532-1; Pb 90-277-0543-7
26. J.E. Murdoch and E.D. Sylla (eds.): *The Cultural Context of Medieval Learning*. Proceedings of the First International Colloquium on Philosophy, Science, and Theology in the Middle Ages, 1973. [Synthese Library 76] 1975 ISBN 90-277-0560-7; Pb 90-277-0587-9
27. M. Grene and E. Mendelsohn (eds.): *Topics in the Philosophy of Biology*. [Synthese Library 84] 1976 ISBN 90-277-0595-X; Pb 90-277-0596-8
28. J. Agassi: *Science in Flux*. [Synthese Library 80] 1975
ISBN 90-277-0584-4; Pb 90-277-0612-3
29. J.J. Wiatr (ed.): *Polish Essays in the Methodology of the Social Sciences*. [Synthese Library 131] 1979 ISBN 90-277-0723-5; Pb 90-277-0956-4
30. P. Janich: *Protophysics of Time*. Constructive Foundation and History of Time Measurement. Translated from German. 1985 ISBN 90-277-0724-3
31. R.S. Cohen and M.W. Wartofsky (eds.): *Language, Logic, and Method*. 1983
ISBN 90-277-0725-1
32. R.S. Cohen, C.A. Hooker, A.C. Michalos and J.W. van Evra (eds.): *PSA 1974*. Proceedings of the 4th Biennial Meeting of the Philosophy of Science Association. [Synthese Library 101] 1976 ISBN 90-277-0647-6; Pb 90-277-0648-4
33. G. Holton and W.A. Blanpied (eds.): *Science and Its Public*. The Changing Relationship. [Synthese Library 96] 1976 ISBN 90-277-0657-3; Pb 90-277-0658-1
34. M.D. Grmek, R.S. Cohen and G. Cimino (eds.): *On Scientific Discovery*. The 1977 Erice Lectures. 1981 ISBN 90-277-1122-4; Pb 90-277-1123-2
35. S. Amsterdamski: *Between Experience and Metaphysics*. Philosophical Problems of the Evolution of Science. Translated from Polish. [Synthese Library 77] 1975
ISBN 90-277-0568-2; Pb 90-277-0580-1
36. M. Marković and G. Petrović (eds.): *Praxis*. Yugoslav Essays in the Philosophy and Methodology of the Social Sciences. [Synthese Library 134] 1979
ISBN 90-277-0727-8; Pb 90-277-0968-8

Boston Studies in the Philosophy of Science

37. H. von Helmholtz: *Epistemological Writings.* The Paul Hertz / Moritz Schlick Centenary Edition of 1921. Translated from German by M.F. Lowe. Edited with an Introduction and Bibliography by R.S. Cohen and Y. Elkana. [Synthese Library 79] 1977
ISBN 90-277-0290-X; Pb 90-277-0582-8
38. R.M. Martin: *Pragmatics, Truth and Language.* 1979
ISBN 90-277-0992-0; Pb 90-277-0993-9
39. R.S. Cohen, P.K. Feyerabend and M.W. Wartofsky (eds.): *Essays in Memory of Imre Lakatos.* [Synthese Library 99] 1976 ISBN 90-277-0654-9; Pb 90-277-0655-7
40. Not published.
41. Not published.
42. H.R. Maturana and F.J. Varela: *Autopoiesis and Cognition.* The Realization of the Living. With a Preface to "Autopoiesis' by S. Beer. 1980 ISBN 90-277-1015-5; Pb 90-277-1016-3
43. A. Kasher (ed.): *Language in Focus: Foundations, Methods and Systems.* Essays in Memory of Yehoshua Bar-Hillel. [Synthese Library 89] 1976
ISBN 90-277-0644-1; Pb 90-277-0645-X
44. T.D. Thao: *Investigations into the Origin of Language and Consciousness.* 1984
ISBN 90-277-0827-4
45. F.G.-I. Nagasaka (ed.): *Japanese Studies in the Philosophy of Science.* 1997
ISBN 0-7923-4781-1
46. P.L. Kapitza: *Experiment, Theory, Practice.* Articles and Addresses. Edited by R.S. Cohen. 1980 ISBN 90-277-1061-9; Pb 90-277-1062-7
47. M.L. Dalla Chiara (ed.): *Italian Studies in the Philosophy of Science.* 1981
ISBN 90-277-0735-9; Pb 90-277-1073-2
48. M.W. Wartofsky: *Models.* Representation and the Scientific Understanding. [Synthese Library 129] 1979 ISBN 90-277-0736-7; Pb 90-277-0947-5
49. T.D. Thao: *Phenomenology and Dialectical Materialism.* Edited by R.S. Cohen. 1986
ISBN 90-277-0737-5
50. Y. Fried and J. Agassi: *Paranoia.* A Study in Diagnosis. [Synthese Library 102] 1976
ISBN 90-277-0704-9; Pb 90-277-0705-7
51. K.H. Wolff: *Surrender and Cath.* Experience and Inquiry Today. [Synthese Library 105] 1976
ISBN 90-277-0758-8; Pb 90-277-0765-0
52. K. Kosík: *Dialectics of the Concrete.* A Study on Problems of Man and World. 1976
ISBN 90-277-0761-8; Pb 90-277-0764-2
53. N. Goodman: *The Structure of Appearance.* [Synthese Library 107] 1977
ISBN 90-277-0773-1; Pb 90-277-0774-X
54. H.A. Simon: *Models of Discovery* and Other Topics in the Methods of Science. [Synthese Library 114] 1977 ISBN 90-277-0812-6; Pb 90-277-0858-4
55. M. Lazerowitz: *The Language of Philosophy.* Freud and Wittgenstein. [Synthese Library 117] 1977 ISBN 90-277-0826-6; Pb 90-277-0862-2
56. T. Nickles (ed.): *Scientific Discovery, Logic, and Rationality.* 1980
ISBN 90-277-1069-4; Pb 90-277-1070-8
57. J. Margolis: *Persons and Mind.* The Prospects of Nonreductive Materialism. [Synthese Library 121] 1978 ISBN 90-277-0854-1; Pb 90-277-0863-0
58. G. Radnitzky and G. Andersson (eds.): *Progress and Rationality in Science.* [Synthese Library 125] 1978 ISBN 90-277-0921-1; Pb 90-277-0922-X
59. G. Radnitzky and G. Andersson (eds.): *The Structure and Development of Science.* [Synthese Library 136] 1979 ISBN 90-277-0994-7; Pb 90-277-0995-5

Boston Studies in the Philosophy of Science

60. T. Nickles (ed.): *Scientific Discovery.* Case Studies. 1980 ISBN 90-277-1092-9; Pb 90-277-1093-7
61. M.A. Finocchiaro: *Galileo and the Art of Reasoning.* Rhetorical Foundation of Logic and Scientific Method. 1980 ISBN 90-277-1094-5; Pb 90-277-1095-3
62. W.A. Wallace: *Prelude to Galileo.* Essays on Medieval and 16th-Century Sources of Galileo's Thought. 1981 ISBN 90-277-1215-8; Pb 90-277-1216-6
63. F. Rapp: *Analytical Philosophy of Technology.* Translated from German. 1981 ISBN 90-277-1221-2; Pb 90-277-1222-0
64. R.S. Cohen and M.W. Wartofsky (eds.): *Hegel and the Sciences.* 1984 ISBN 90-277-0726-X
65. J. Agassi: *Science and Society.* Studies in the Sociology of Science. 1981 ISBN 90-277-1244-1; Pb 90-277-1245-X
66. L. Tondl: *Problems of Semantics.* A Contribution to the Analysis of the Language of Science. Translated from Czech. 1981 ISBN 90-277-0148-2; Pb 90-277-0316-7
67. J. Agassi and R.S. Cohen (eds.): *Scientific Philosophy Today.* Essays in Honor of Mario Bunge. 1982 ISBN 90-277-1262-X; Pb 90-277-1263-8
68. W. Krajewski (ed.): *Polish Essays in the Philosophy of the Natural Sciences.* Translated from Polish and edited by R.S. Cohen and C.R. Fawcett. 1982 ISBN 90-277-1286-7; Pb 90-277-1287-5
69. J.H. Fetzer: *Scientific Knowledge.* Causation, Explanation and Corroboration. 1981 ISBN 90-277-1335-9; Pb 90-277-1336-7
70. S. Grossberg: *Studies of Mind and Brain.* Neural Principles of Learning, Perception, Development, Cognition, and Motor Control. 1982 ISBN 90-277-1359-6; Pb 90-277-1360-X
71. R.S. Cohen and M.W. Wartofsky (eds.): *Epistemology, Methodology, and the Social Sciences.* 1983. ISBN 90-277-1454-1
72. K. Berka: *Measurement.* Its Concepts, Theories and Problems. Translated from Czech. 1983 ISBN 90-277-1416-9
73. G.L. Pandit: *The Structure and Growth of Scientific Knowledge.* A Study in the Methodology of Epistemic Appraisal. 1983 ISBN 90-277-1434-7
74. A.A. Zinov'ev: *Logical Physics.* Translated from Russian. Edited by R.S. Cohen. 1983 [*see also* Volume 9] ISBN 90-277-0734-0
75. G-G. Granger: *Formal Thought and the Sciences of Man.* Translated from French. With and Introduction by A. Rosenberg. 1983 ISBN 90-277-1524-6
76. R.S. Cohen and L. Laudan (eds.): *Physics, Philosophy and Psychoanalysis.* Essays in Honor of Adolf Grünbaum. 1983 ISBN 90-277-1533-5
77. G. Böhme, W. van den Daele, R. Hohlfeld, W. Krohn and W. Schäfer: *Finalization in Science.* The Social Orientation of Scientific Progress. Translated from German. Edited by W. Schäfer. 1983 ISBN 90-277-1549-1
78. D. Shapere: *Reason and the Search for Knowledge.* Investigations in the Philosophy of Science. 1984 ISBN 90-277-1551-3; Pb 90-277-1641-2
79. G. Andersson (ed.): *Rationality in Science and Politics.* Translated from German. 1984 ISBN 90-277-1575-0; Pb 90-277-1953-5
80. P.T. Durbin and F. Rapp (eds.): *Philosophy and Technology.* [*Also* Philosophy and Technology Series, Vol. 1] 1983 ISBN 90-277-1576-9
81. M. Marković: *Dialectical Theory of Meaning.* Translated from Serbo-Croat. 1984 ISBN 90-277-1596-3
82. R.S. Cohen and M.W. Wartofsky (eds.): *Physical Sciences and History of Physics.* 1984. ISBN 90-277-1615-3

Boston Studies in the Philosophy of Science

83. É. Meyerson: *The Relativistic Deduction.* Epistemological Implications of the Theory of Relativity. Translated from French. With a Review by Albert Einstein and an Introduction by Milič Čapek. 1985 ISBN 90-277-1699-4
84. R.S. Cohen and M.W. Wartofsky (eds.): *Methodology, Metaphysics and the History of Science.* In Memory of Benjamin Nelson. 1984 ISBN 90-277-1711-7
85. G. Tamás: *The Logic of Categories.* Translated from Hungarian. Edited by R.S. Cohen. 1986 ISBN 90-277-1742-7
86. S.L. de C. Fernandes: *Foundations of Objective Knowledge.* The Relations of Popper's Theory of Knowledge to That of Kant. 1985 ISBN 90-277-1809-1
87. R.S. Cohen and T. Schnelle (eds.): *Cognition and Fact.* Materials on Ludwik Fleck. 1986 ISBN 90-277-1902-0
88. G. Freudenthal: *Atom and Individual in the Age of Newton.* On the Genesis of the Mechanistic World View. Translated from German. 1986 ISBN 90-277-1905-5
89. A. Donagan, A.N. Perovich Jr and M.V. Wedin (eds.): *Human Nature and Natural Knowledge.* Essays presented to Marjorie Grene on the Occasion of Her 75th Birthday. 1986 ISBN 90-277-1974-8
90. C. Mitcham and A. Hunning (eds.): *Philosophy and Technology II.* Information Technology and Computers in Theory and Practice. [*Also* Philosophy and Technology Series, Vol. 2] 1986 ISBN 90-277-1975-6
91. M. Grene and D. Nails (eds.): *Spinoza and the Sciences.* 1986 ISBN 90-277-1976-4
92. S.P. Turner: *The Search for a Methodology of Social Science.* Durkheim, Weber, and the 19th-Century Problem of Cause, Probability, and Action. 1986. ISBN 90-277-2067-3
93. I.C. Jarvie: *Thinking about Society.* Theory and Practice. 1986 ISBN 90-277-2068-1
94. E. Ullmann-Margalit (ed.): *The Kaleidoscope of Science.* The Israel Colloquium: Studies in History, Philosophy, and Sociology of Science, Vol. 1. 1986 ISBN 90-277-2158-0; Pb 90-277-2159-9
95. E. Ullmann-Margalit (ed.): *The Prism of Science.* The Israel Colloquium: Studies in History, Philosophy, and Sociology of Science, Vol. 2. 1986 ISBN 90-277-2160-2; Pb 90-277-2161-0
96. G. Márkus: *Language and Production.* A Critique of the Paradigms. Translated from French. 1986 ISBN 90-277-2169-6
97. F. Amrine, F.J. Zucker and H. Wheeler (eds.): *Goethe and the Sciences: A Reappraisal.* 1987 ISBN 90-277-2265-X; Pb 90-277-2400-8
98. J.C. Pitt and M. Pera (eds.): *Rational Changes in Science.* Essays on Scientific Reasoning. Translated from Italian. 1987 ISBN 90-277-2417-2
99. O. Costa de Beauregard: *Time, the Physical Magnitude.* 1987 ISBN 90-277-2444-X
100. A. Shimony and D. Nails (eds.): *Naturalistic Epistemology.* A Symposium of Two Decades. 1987 ISBN 90-277-2337-0
101. N. Rotenstreich: *Time and Meaning in History.* 1987 ISBN 90-277-2467-9
102. D.B. Zilberman: *The Birth of Meaning in Hindu Thought.* Edited by R.S. Cohen. 1988 ISBN 90-277-2497-0
103. T.F. Glick (ed.): *The Comparative Reception of Relativity.* 1987 ISBN 90-277-2498-9
104. Z. Harris, M. Gottfried, T. Ryckman, P. Mattick Jr, A. Daladier, T.N. Harris and S. Harris: *The Form of Information in Science.* Analysis of an Immunology Sublanguage. With a Preface by Hilary Putnam. 1989 ISBN 90-277-2516-0
105. F. Burwick (ed.): *Approaches to Organic Form.* Permutations in Science and Culture. 1987 ISBN 90-277-2541-1

Boston Studies in the Philosophy of Science

106. M. Almási: *The Philosophy of Appearances.* Translated from Hungarian. 1989
ISBN 90-277-2150-5
107. S. Hook, W.L. O'Neill and R. O'Toole (eds.): *Philosophy, History and Social Action.* Essays in Honor of Lewis Feuer. With an Autobiographical Essay by L. Feuer. 1988
ISBN 90-277-2644-2
108. I. Hronszky, M. Fehér and B. Dajka: *Scientific Knowledge Socialized.* Selected Proceedings of the 5th Joint International Conference on the History and Philosophy of Science organized by the IUHPS (Veszprém, Hungary, 1984). 1988 ISBN 90-277-2284-6
109. P. Tillers and E.D. Green (eds.): *Probability and Inference in the Law of Evidence.* The Uses and Limits of Bayesianism. 1988 ISBN 90-277-2689-2
110. E. Ullmann-Margalit (ed.): *Science in Reflection.* The Israel Colloquium: Studies in History, Philosophy, and Sociology of Science, Vol. 3. 1988
ISBN 90-277-2712-0; Pb 90-277-2713-9
111. K. Gavroglu, Y. Goudaroulis and P. Nicolacopoulos (eds.): *Imre Lakatos and Theories of Scientific Change.* 1989 ISBN 90-277-2766-X
112. B. Glassner and J.D. Moreno (eds.): *The Qualitative-Quantitative Distinction in the Social Sciences.* 1989 ISBN 90-277-2829-1
113. K. Arens: *Structures of Knowing.* Psychologies of the 19th Century. 1989
ISBN 0-7923-0009-2
114. A. Janik: *Style, Politics and the Future of Philosophy.* 1989 ISBN 0-7923-0056-4
115. F. Amrine (ed.): *Literature and Science as Modes of Expression.* With an Introduction by S. Weininger. 1989 ISBN 0-7923-0133-1
116. J.R. Brown and J. Mittelstrass (eds.): *An Intimate Relation.* Studies in the History and Philosophy of Science. Presented to Robert E. Butts on His 60th Birthday. 1989
ISBN 0-7923-0169-2
117. F. D'Agostino and I.C. Jarvie (eds.): *Freedom and Rationality.* Essays in Honor of John Watkins. 1989 ISBN 0-7923-0264-8
118. D. Zolo: *Reflexive Epistemology.* The Philosophical Legacy of Otto Neurath. 1989
ISBN 0-7923-0320-2
119. M. Kearn, B.S. Philips and R.S. Cohen (eds.): *Georg Simmel and Contemporary Sociology.* 1989 ISBN 0-7923-0407-1
120. T.H. Levere and W.R. Shea (eds.): *Nature, Experiment and the Science.* Essays on Galileo and the Nature of Science. In Honour of Stillman Drake. 1989 ISBN 0-7923-0420-9
121. P. Nicolacopoulos (ed.): *Greek Studies in the Philosophy and History of Science.* 1990
ISBN 0-7923-0717-8
122. R. Cooke and D. Costantini (eds.): *Statistics in Science.* The Foundations of Statistical Methods in Biology, Physics and Economics. 1990 ISBN 0-7923-0797-6
123. P. Duhem: *The Origins of Statics.* Translated from French by G.F. Leneaux, V.N. Vagliente and G.H. Wagner. With an Introduction by S.L. Jaki. 1991 ISBN 0-7923-0898-0
124. H. Kamerlingh Onnes: *Through Measurement to Knowledge.* The Selected Papers, 1853-1926. Edited and with an Introduction by K. Gavroglu and Y. Goudaroulis. 1991
ISBN 0-7923-0825-5
125. M. Čapek: *The New Aspects of Time: Its Continuity and Novelties.* Selected Papers in the Philosophy of Science. 1991 ISBN 0-7923-0911-1
126. S. Unguru (ed.): *Physics, Cosmology and Astronomy, 1300–1700.* Tension and Accommodation. 1991 ISBN 0-7923-1022-5

Boston Studies in the Philosophy of Science

127. Z. Bechler: *Newton's Physics on the Conceptual Structure of the Scientific Revolution.* 1991 ISBN 0-7923-1054-3
128. É. Meyerson: *Explanation in the Sciences.* Translated from French by M-A. Siple and D.A. Siple. 1991 ISBN 0-7923-1129-9
129. A.I. Tauber (ed.): *Organism and the Origins of Self.* 1991 ISBN 0-7923-1185-X
130. F.J. Varela and J-P. Dupuy (eds.): *Understanding Origins.* Contemporary Views on the Origin of Life, Mind and Society. 1992 ISBN 0-7923-1251-1
131. G.L. Pandit: *Methodological Variance.* Essays in Epistemological Ontology and the Methodology of Science. 1991 ISBN 0-7923-1263-5
132. G. Munévar (ed.): *Beyond Reason.* Essays on the Philosophy of Paul Feyerabend. 1991 ISBN 0-7923-1272-4
133. T.E. Uebel (ed.): *Rediscovering the Forgotten Vienna Circle.* Austrian Studies on Otto Neurath and the Vienna Circle. Partly translated from German. 1991 ISBN 0-7923-1276-7
134. W.R. Woodward and R.S. Cohen (eds.): *World Views and Scientific Discipline Formation.* Science Studies in the [former] German Democratic Republic. Partly translated from German by W.R. Woodward. 1991 ISBN 0-7923-1286-4
135. P. Zambelli: *The Speculum Astronomiae and Its Enigma.* Astrology, Theology and Science in Albertus Magnus and His Contemporaries. 1992 ISBN 0-7923-1380-1
136. P. Petitjean, C. Jami and A.M. Moulin (eds.): *Science and Empires.* Historical Studies about Scientific Development and European Expansion. ISBN 0-7923-1518-9
137. W.A. Wallace: *Galileo's Logic of Discovery and Proof.* The Background, Content, and Use of His Appropriated Treatises on Aristotle's *Posterior Analytics.* 1992 ISBN 0-7923-1577-4
138. W.A. Wallace: *Galileo's Logical Treatises.* A Translation, with Notes and Commentary, of His Appropriated Latin Questions on Aristotle's *Posterior Analytics.* 1992 ISBN 0-7923-1578-2
Set (137 + 138) ISBN 0-7923-1579-0
139. M.J. Nye, J.L. Richards and R.H. Stuewer (eds.): *The Invention of Physical Science.* Intersections of Mathematics, Theology and Natural Philosophy since the Seventeenth Century. Essays in Honor of Erwin N. Hiebert. 1992 ISBN 0-7923-1753-X
140. G. Corsi, M.L. dalla Chiara and G.C. Ghirardi (eds.): *Bridging the Gap: Philosophy, Mathematics and Physics.* Lectures on the Foundations of Science. 1992 ISBN 0-7923-1761-0
141. C.-H. Lin and D. Fu (eds.): *Philosophy and Conceptual History of Science in Taiwan.* 1992 ISBN 0-7923-1766-1
142. S. Sarkar (ed.): *The Founders of Evolutionary Genetics.* A Centenary Reappraisal. 1992 ISBN 0-7923-1777-7
143. J. Blackmore (ed.): *Ernst Mach – A Deeper Look.* Documents and New Perspectives. 1992 ISBN 0-7923-1853-6
144. P. Kroes and M. Bakker (eds.): *Technological Development and Science in the Industrial Age.* New Perspectives on the Science–Technology Relationship. 1992 ISBN 0-7923-1898-6
145. S. Amsterdamski: *Between History and Method.* Disputes about the Rationality of Science. 1992 ISBN 0-7923-1941-9
146. E. Ullmann-Margalit (ed.): *The Scientific Enterprise.* The Bar-Hillel Colloquium: Studies in History, Philosophy, and Sociology of Science, Volume 4. 1992 ISBN 0-7923-1992-3
147. L. Embree (ed.): *Metaarchaeology.* Reflections by Archaeologists and Philosophers. 1992 ISBN 0-7923-2023-9
148. S. French and H. Kamminga (eds.): *Correspondence, Invariance and Heuristics.* Essays in Honour of Heinz Post. 1993 ISBN 0-7923-2085-9
149. M. Bunzl: *The Context of Explanation.* 1993 ISBN 0-7923-2153-7

Boston Studies in the Philosophy of Science

150. I.B. Cohen (ed.): *The Natural Sciences and the Social Sciences.* Some Critical and Historical Perspectives. 1994 ISBN 0-7923-2223-1
151. K. Gavroglu, Y. Christianidis and E. Nicolaidis (eds.): *Trends in the Historiography of Science.* 1994 ISBN 0-7923-2255-X
152. S. Poggi and M. Bossi (eds.): *Romanticism in Science.* Science in Europe, 1790–1840. 1994 ISBN 0-7923-2336-X
153. J. Faye and H.J. Folse (eds.): *Niels Bohr and Contemporary Philosophy.* 1994 ISBN 0-7923-2378-5
154. C.C. Gould and R.S. Cohen (eds.): *Artifacts, Representations, and Social Practice.* Essays for Marx W. Wartofsky. 1994 ISBN 0-7923-2481-1
155. R.E. Butts: *Historical Pragmatics.* Philosophical Essays. 1993 ISBN 0-7923-2498-6
156. R. Rashed: *The Development of Arabic Mathematics: Between Arithmetic and Algebra.* Translated from French by A.F.W. Armstrong. 1994 ISBN 0-7923-2565-6
157. I. Szumilewicz-Lachman (ed.): *Zygmunt Zawirski: His Life and Work.* With Selected Writings on Time, Logic and the Methodology of Science. Translations by Feliks Lachman. Ed. by R.S. Cohen, with the assistance of B. Bergo. 1994 ISBN 0-7923-2566-4
158. S.N. Haq: *Names, Natures and Things.* The Alchemist Jābir ibn Ḥayyān and His *Kitāb al-Aḥjār* (Book of Stones). 1994 ISBN 0-7923-2587-7
159. P. Plaass: *Kant's Theory of Natural Science.* Translation, Analytic Introduction and Commentary by Alfred E. and Maria G. Miller. 1994 ISBN 0-7923-2750-0
160. J. Misiek (ed.): *The Problem of Rationality in Science and its Philosophy.* On Popper vs. Polanyi. The Polish Conferences 1988–89. 1995 ISBN 0-7923-2925-2
161. I.C. Jarvie and N. Laor (eds.): *Critical Rationalism, Metaphysics and Science.* Essays for Joseph Agassi, Volume I. 1995 ISBN 0-7923-2960-0
162. I.C. Jarvie and N. Laor (eds.): *Critical Rationalism, the Social Sciences and the Humanities.* Essays for Joseph Agassi, Volume II. 1995 ISBN 0-7923-2961-9
Set (161–162) ISBN 0-7923-2962-7
163. K. Gavroglu, J. Stachel and M.W. Wartofsky (eds.): *Physics, Philosophy, and the Scientific Community.* Essays in the Philosophy and History of the Natural Sciences and Mathematics. In Honor of Robert S. Cohen. 1995 ISBN 0-7923-2988-0
164. K. Gavroglu, J. Stachel and M.W. Wartofsky (eds.): *Science, Politics and Social Practice.* Essays on Marxism and Science, Philosophy of Culture and the Social Sciences. In Honor of Robert S. Cohen. 1995 ISBN 0-7923-2989-9
165. K. Gavroglu, J. Stachel and M.W. Wartofsky (eds.): *Science, Mind and Art.* Essays on Science and the Humanistic Understanding in Art, Epistemology, Religion and Ethics. Essays in Honor of Robert S. Cohen. 1995 ISBN 0-7923-2990-2
Set (163–165) ISBN 0-7923-2991-0
166. K.H. Wolff: *Transformation in the Writing.* A Case of Surrender-and-Catch. 1995 ISBN 0-7923-3178-8
167. A.J. Kox and D.M. Siegel (eds.): *No Truth Except in the Details.* Essays in Honor of Martin J. Klein. 1995 ISBN 0-7923-3195-8
168. J. Blackmore: *Ludwig Boltzmann, His Later Life and Philosophy, 1900–1906.* Book One: A Documentary History. 1995 ISBN 0-7923-3231-8
169. R.S. Cohen, R. Hilpinen and R. Qiu (eds.): *Realism and Anti-Realism in the Philosophy of Science.* Beijing International Conference, 1992. 1996 ISBN 0-7923-3233-4
170. I. Kuçuradi and R.S. Cohen (eds.): *The Concept of Knowledge.* The Ankara Seminar. 1995 ISBN 0-7923-3241-5

Boston Studies in the Philosophy of Science

171. M.A. Grodin (ed.): *Meta Medical Ethics*: The Philosophical Foundations of Bioethics. 1995 ISBN 0-7923-3344-6
172. S. Ramirez and R.S. Cohen (eds.): *Mexican Studies in the History and Philosophy of Science*. 1995 ISBN 0-7923-3462-0
173. C. Dilworth: *The Metaphysics of Science*. An Account of Modern Science in Terms of Principles, Laws and Theories. 1995 ISBN 0-7923-3693-3
174. J. Blackmore: *Ludwig Boltzmann, His Later Life and Philosophy, 1900–1906* Book Two: The Philosopher. 1995 ISBN 0-7923-3464-7
175. P. Damerow: *Abstraction and Representation*. Essays on the Cultural Evolution of Thinking. 1996 ISBN 0-7923-3816-2
176. M.S. Macrakis: *Scarcity's Ways: The Origins of Capital*. A Critical Essay on Thermodynamics, Statistical Mechanics and Economics. 1997 ISBN 0-7923-4760-9
177. M. Marion and R.S. Cohen (eds.): *Québec Studies in the Philosophy of Science*. Part I: Logic, Mathematics, Physics and History of Science. Essays in Honor of Hugues Leblanc. 1995 ISBN 0-7923-3559-7
178. M. Marion and R.S. Cohen (eds.): *Québec Studies in the Philosophy of Science*. Part II: Biology, Psychology, Cognitive Science and Economics. Essays in Honor of Hugues Leblanc. 1996 ISBN 0-7923-3560-0
Set (177–178) ISBN 0-7923-3561-9
179. Fan Dainian and R.S. Cohen (eds.): *Chinese Studies in the History and Philosophy of Science and Technology*. 1996 ISBN 0-7923-3463-9
180. P. Forman and J.M. Sánchez-Ron (eds.): *National Military Establishments and the Advancement of Science and Technology*. Studies in 20th Century History. 1996 ISBN 0-7923-3541-4
181. E.J. Post: *Quantum Reprogramming*. Ensembles and Single Systems: A Two-Tier Approach to Quantum Mechanics. 1995 ISBN 0-7923-3565-1
182. A.I. Tauber (ed.): *The Elusive Synthesis: Aesthetics and Science*. 1996 ISBN 0-7923-3904-5
183. S. Sarkar (ed.): *The Philosophy and History of Molecular Biology: New Perspectives*. 1996 ISBN 0-7923-3947-9
184. J.T. Cushing, A. Fine and S. Goldstein (eds.): *Bohmian Mechanics and Quantum Theory: An Appraisal*. 1996 ISBN 0-7923-4028-0
185. K. Michalski: *Logic and Time*. An Essay on Husserl's Theory of Meaning. 1996 ISBN 0-7923-4082-5
186. G. Munévar (ed.): *Spanish Studies in the Philosophy of Science*. 1996 ISBN 0-7923-4147-3
187. G. Schubring (ed.): *Hermann Günther Graßmann (1809–1877): Visionary Mathematician, Scientist and Neohumanist Scholar*. Papers from a Sesquicentennial Conference. 1996 ISBN 0-7923-4261-5
188. M. Bitbol: *Schrödinger's Philosophy of Quantum Mechanics*. 1996 ISBN 0-7923-4266-6
189. J. Faye, U. Scheffler and M. Urchs (eds.): *Perspectives on Time*. 1997 ISBN 0-7923-4330-1
190. K. Lehrer and J.C. Marek (eds.): *Austrian Philosophy Past and Present*. Essays in Honor of Rudolf Haller. 1996 ISBN 0-7923-4347-6
191. J.L. Lagrange: *Analytical Mechanics*. Translated and edited by Auguste Boissonade and Victor N. Vagliente. Translated from the *Mécanique Analytique, novelle édition* of 1811. 1997 ISBN 0-7923-4349-2
192. D. Ginev and R.S. Cohen (eds.): *Issues and Images in the Philosophy of Science*. Scientific and Philosophical Essays in Honour of Azarya Polikarov. 1997 ISBN 0-7923-4444-8

Boston Studies in the Philosophy of Science

193. R.S. Cohen, M. Horne and J. Stachel (eds.): *Experimental Metaphysics*. Quantum Mechanical Studies for Abner Shimony, Volume One. 1997 ISBN 0-7923-4452-9
194. R.S. Cohen, M. Horne and J. Stachel (eds.): *Potentiality, Entanglement and Passion-at-a-Distance*. Quantum Mechanical Studies for Abner Shimony, Volume Two. 1997 ISBN 0-7923-4453-7; Set 0-7923-4454-5
195. R.S. Cohen and A.I. Tauber (eds.): *Philosophies of Nature: The Human Dimension*. 1997 ISBN 0-7923-4579-7
196. M. Otte and M. Panza (eds.): *Analysis and Synthesis in Mathematics*. History and Philosophy. 1997 ISBN 0-7923-4570-3
197. A. Denkel: *The Natural Background of Meaning*. 1999 ISBN 0-7923-5331-5
198. D. Baird, R.I.G. Hughes and A. Nordmann (eds.): *Heinrich Hertz: Classical Physicist, Modern Philosopher*. 1999 ISBN 0-7923-4653-X
199. A. Franklin: *Can That be Right?* Essays on Experiment, Evidence, and Science. 1999 ISBN 0-7923-5464-8
200. D. Raven, W. Krohn and R.S. Cohen (eds.): *The Social Origins of Modern Science*. 2000 ISBN 0-7923-6457-0
201. Reserved
202. Reserved
203. B. Babich and R.S. Cohen (eds.): *Nietzsche, Theories of Knowledge, and Critical Theory*. Nietzsche and the Sciences I. 1999 ISBN 0-7923-5742-6
204. B. Babich and R.S. Cohen (eds.): *Nietzsche, Epistemology, and Philosophy of Science*. Nietzsche and the Science II. 1999 ISBN 0-7923-5743-4
205. R. Hooykaas: *Fact, Faith and Fiction in the Development of Science*. The Gifford Lectures given in the University of St Andrews 1976. 1999 ISBN 0-7923-5774-4
206. M. Fehér, O. Kiss and L. Ropolyi (eds.): *Hermeneutics and Science*. 1999 ISBN 0-7923-5798-1
207. R.M. MacLeod (ed.): *Science and the Pacific War*. Science and Survival in the Pacific, 1939-1945. 1999 ISBN 0-7923-5851-1
208. I. Hanzel: *The Concept of Scientific Law in the Philosophy of Science and Epistemology*. A Study of Theoretical Reason. 1999 ISBN 0-7923-5852-X
209. G. Helm; R.J. Deltete (ed./transl.): *The Historical Development of Energetics*. 1999 ISBN 0-7923-5874-0
210. A. Orenstein and P. Kotatko (eds.): *Knowledge, Language and Logic*. Questions for Quine. 1999 ISBN 0-7923-5986-0
211. R.S. Cohen and H. Levine (eds.): *Maimonides and the Sciences*. 2000 ISBN 0-7923-6053-2
212. H. Gourko, D.I. Williamson and A.I. Tauber (eds.): *The Evolutionary Biology Papers of Elie Metchnikoff*. 2000 ISBN 0-7923-6067-2
213. S. D'Agostino: *A History of the Ideas of Theoretical Physics*. Essays on the Nineteenth and Twentieth Century Physics. 2000 ISBN 0-7923-6094-X
214. S. Lelas: *Science and Modernity*. Toward An Integral Theory of Science. 2000 ISBN 0-7923-6303-5

Boston Studies in the Philosophy of Science

215. E. Agazzi and M. Pauri (eds.): *The Reality of the Unobservable*. Observability, Unobservability and Their Impact on the Issue of Scientific Realism. 2000 ISBN 0-7923-6311-6
216. P. Hoyningen-Huene and H. Sankey (eds.): *Incommensurability and Related Matters*. 2001 ISBN 0-7923-6989-0

Also of interest:

R.S. Cohen and M.W. Wartofsky (eds.): *A Portrait of Twenty-Five Years Boston Colloquia for the Philosophy of Science, 1960-1985*. 1985 ISBN Pb 90-277-1971-3

Previous volumes are still available.

KLUWER ACADEMIC PUBLISHERS – DORDRECHT / BOSTON / LONDON